ENERGY FUTURES AND URBAN AIR POLLUTION

Challenges for China and the United States

Committee on Energy Futures and Air Pollution in Urban China and the United States

Development, Security and Cooperation

Policy and Global Affairs

NATIONAL ACADEMY OF ENGINEERING
NATIONAL RESEARCH COUNCIL
OF THE NATIONAL ACADEMIES

In collaboration with

CHINESE ACADEMY OF ENGINEERING
CHINESE ACADEMY OF SCIENCES

THE NATIONAL ACADEMIES PRESS
Washington, D.C.
www.nap.edu

THE NATIONAL ACADEMIES PRESS 500 Fifth Street, N.W. Washington, D.C. 20001

NOTICE: The project that is the subject of this report was approved by the Governing Board of the National Research Council, whose members are drawn from the councils of the National Academy of Sciences, the National Academy of Engineering, and the Institute of Medicine. The members of the committee responsible for the report were chosen for their special competences and with regard for appropriate balance.

This study was supported by funding from the National Academies. Any opinions, findings, conclusions, or recommendations expressed in this publication are those of the author(s) and do not necessarily reflect the views of the organizations or agencies that provided support for the project.

Suggested citation: National Academy of Engineering and National Research Council. 2008. *Energy Futures and Urban Air Pollution Challenges for China and the United States*. Washington, D.C.: The National Academies Press.

International Standard Book Number-13: 978-0-309-11140-9
International Standard Book Number-10: 0-309-11140-4

Additional copies of this report are available from the National Academies Press, 500 Fifth Street, N.W., Lockbox 285, Washington, D.C. 20055; (800) 624-6242 or (202) 334-3313 (in the Washington metropolitan area); Internet, http://www.nap.edu

Printed in the United States of America

THE NATIONAL ACADEMIES

Advisers to the Nation on Science, Engineering, and Medicine

The **National Academy of Sciences** is a private, nonprofit, self-perpetuating society of distinguished scholars engaged in scientific and engineering research, dedicated to the furtherance of science and technology and to their use for the general welfare. Upon the authority of the charter granted to it by the Congress in 1863, the Academy has a mandate that requires it to advise the federal government on scientific and technical matters. Dr. Ralph J. Cicerone is president of the National Academy of Sciences.

The **National Academy of Engineering** was established in 1964, under the charter of the National Academy of Sciences, as a parallel organization of outstanding engineers. It is autonomous in its administration and in the selection of its members, sharing with the National Academy of Sciences the responsibility for advising the federal government. The National Academy of Engineering also sponsors engineering programs aimed at meeting national needs, encourages education and research, and recognizes the superior achievements of engineers. Dr. Charles M. Vest is president of the National Academy of Engineering.

The **Institute of Medicine** was established in 1970 by the National Academy of Sciences to secure the services of eminent members of appropriate professions in the examination of policy matters pertaining to the health of the public. The Institute acts under the responsibility given to the National Academy of Sciences by its congressional charter to be an adviser to the federal government and, upon its own initiative, to identify issues of medical care, research, and education. Dr. Harvey V. Fineberg is president of the Institute of Medicine.

The **National Research Council** was organized by the National Academy of Sciences in 1916 to associate the broad community of science and technology with the Academy's purposes of furthering knowledge and advising the federal government. Functioning in accordance with general policies determined by the Academy, the Council has become the principal operating agency of both the National Academy of Sciences and the National Academy of Engineering in providing services to the government, the public, and the scientific and engineering communities. The Council is administered jointly by both Academies and the Institute of Medicine. Dr. Ralph J. Cicerone and Dr. Charles M. Vest are chair and vice chair, respectively, of the National Research Council.

www.national-academies.org

COMMITTEE ON ENERGY FUTURES AND AIR POLLUTION IN URBAN CHINA AND THE UNITED STATES

U.S. Committee

John WATSON, *Chair*, Desert Research Institute, Nevada
Dave ALLEN, University of Texas at Austin, Texas
Roger BEZDEK, Management Information Services, Inc., Washington, D.C.
Judith CHOW, Desert Research Institute, Nevada
Bart CROES, California Air Resources Board, California
Glen DAIGGER, CH2M Hill, Inc., Colorado
David HAWKINS, Natural Resources Defense Council, Washington, D.C.
Philip HOPKE, Clarkson University, New York
Jana MILFORD, University of Colorado at Boulder, Colorado
Armistead RUSSELL, Georgia Institute of Technology, Georgia
Jitendra J. SHAH, The World Bank, Washington, D.C.
Michael WALSH, Consultant, Virginia

Staff

Jack FRITZ, Senior Program Officer, National Academy of Engineering (through April 2006)
Lance DAVIS, Executive Officer, National Academy of Engineering
Proctor REID, Director, Program Office, National Academy of Engineering
John BORIGHT, Executive Director, International Affairs, National Research Council
Derek VOLLMER, Program Associate, Policy and Global Affairs, National Research Council

Chinese Committee

ZHAO Zhongxian, C*hair*, Institute of Physics, Chinese Academy of Sciences, Beijing
AN Zhisheng, Institute of Earth Environment, Chinese Academy of Sciences, Xi'an
CAI Ruixian, Institute of Engineering Thermophysics, Chinese Academy of Sciences, Beijing
CAO Junji, Institute of Earth Environment, Chinese Academy of Sciences, Xi'an
FAN Weitang, China National Coal Association, Beijing
HE Fei, Peking University, Beijing
JIN Hongguang, Institute of Engineering Thermophysics, Chinese Academy of Sciences, Beijing
TANG Xiaoyan, Peking University, Beijing
WANG Fosong, Academic Divisions, Chinese Academy of Sciences

WANG Yingshi, Institute of Engineering Thermophysics, Chinese Academy of Sciences, Beijing
XU Xuchang, Tsinghua University, Beijing
YAN Luguang, Institute of Electrical Engineering, Chinese Academy of Sciences
YOU Changfu, Tsinghua University, Beijing
YU Zhufeng, China Coal Research Institute, Beijing

Preface

In relation to studies and understanding of broad energy and pollution management issues, the U.S. National Academies have had an on-going program of cooperation with the Chinese Academies (Chinese Academy of Sciences and Chinese Academy of Engineering) for a number of years. Joint study activities date to the late 1990s and led to the publication in 2000 of *Cooperation in the Energy Futures of China and the United States.* This volume was the first examination of the broad energy questions facing both nations at the turn of the new millennium.

The Energy Futures study was followed in 2003 with a study publication titled *Personal Cars and China*, which sought to provide insight to the Chinese government in the inevitable development of a private car fleet. And, in the fall of 2003, the Chinese and U.S. Academies organized an informal workshop in Beijing to review progress made to date in China in managing urban airsheds. This resulted in a proceedings publication titled *Urbanization, Energy, and Air Pollution in China; The Challenges Ahead*, published in 2004.

As time has evolved it has become abundantly clear that the United States and China are inextricably intertwined through global competition for scarce energy resources and their disproportionate impact on the globe's environmental health. These realities reinforce the need for the United States and Chinese Academies to continue to work closely together on a frequent and more intensive basis. An underlying assumption is that China can benefit from assimilating U.S. lessons learned from a longer history of dealing with the interplay between air pollution and energy production and usage. Moreover, as both countries focus on energy independence, there are significant opportunities to learn from one another and to cooperate on issues of mutual interest.

It is against this backdrop that the current study was developed. Following the 2003 workshop which first explored the role of urbanization in China's energy use and air pollution, it was concluded that a full-scale consensus study should be carried out to compare the United States and Chinese experiences. Both countries' respective Academies established committees comprised of leading experts in the fields of energy and air quality to jointly carry out this task. Specifically, this study was to compare strategies for the management of airsheds in similar locales, namely ones located in highly industrial, coal-rich areas, as exemplified by Pittsburgh and Huainan, and others located in more modern, coastal/port and car-oriented areas, as exemplified by Los Angeles and Dalian. It was anticipated that a comparative analysis focusing at the local level should reveal how national and regional (state/provincial) policies affect local economies and their populations.

Visits to all four cities by the U.S. and Chinese committee members were organized to learn as much as possible about the experiences of each city. The teams met with city government officials, local university and research personnel, and with key private-sector actors. The teams toured local industrial plants, power plants, research laboratories, transportation control centers, and air quality monitoring facilities. In order to understand local policy and compliance aspects, the teams also met with local, regional, and national regulatory officials. This report has been prepared on the basis of those visits, as well as on the basis of the professional expertise of the U.S. and Chinese committee members and the trove of data available on worldwide energy resources and consumption and environmental regimes and challenges in the United States and China.

This study could not examine in detail the related and increasingly significant issue of greenhouse gas (GHG) emissions and global climate change. We do, however, attempt to highlight the fact that this will be a central issue, perhaps *the* issue, in discussions of energy and air pollution in the future. We also give attention to opportunities to mitigate GHG emissions and some of the strategies that cities are able to and are already employing. This is an area where continued cooperation between the U.S. and Chinese Academies will be particularly useful. Similarly, we did not focus on the impacts of long-range pollution transport, but we acknowledge that this is an important global issue, and one that links our two countries.

As the goals and priorities of both countries evolve with respect to energy and air pollution, it is clear that there will be a number of different strategies available, though certainly no magic bullets. This large and diverse bilateral effort was designed to represent the different (and sometimes competing) viewpoints that might support these various strategies; throughout the process, each side learned valuable lessons from the other and came away with a better understanding of the circumstances unique to each country. We hope that the resultant report is of value to policy and decision makers not only in China but also in the United States, and that the lessons learned may be instructive to other countries currently experiencing rapid urbanization. We were honored to serve as chairs of these distinguished committees, and we compliment the U.S. and Chinese committee members for their efforts throughout this study process.

John G. Watson
National Academy of Engineering
National Research Council

Zhao Zhongxian
Chinese Academy of Sciences

Acknowledgments

We wish to thank the late Alan Voorhees, member of the National Academy of Engineering, the U.S. National Academies, the Chinese Academy of Sciences, the Chinese Academy of Engineering, the Energy Foundation, and the Ford Motor Company for their financial support of this project. The committee also wishes to thank officials of the cities of Huainan and Dalian for agreeing to participate in this study and for welcoming the committee during its October 2005 study tour. In particular, we wish to thank Mayor Zhu Jili, Vice Mayor Dong Zhongbing, and the rest of the Huainan Municipal government; the CPC Huainan Committee; Huainan Mining Group; Huainan Chemical Industrial Group; the Pingwei Power Plant; Zhao Baoqing and others at the Huainan Environmental Protection Bureau; Mayor Xia Deren and the rest of the Dalian Municipal government; Hua Xiujing and others at the Dalian Environmental Protection Bureau; the Dalian Traffic Direction and Control Center; the Dalian Environmental Monitoring Center; the CAS Institute of Chemical Physics; Dalian Steel Factory; Huaneng Power Factory; and the Xianghai Thermal Power Factory.

On the U.S. side, we wish to thank Lee Schipper and Wei-Shiuen Ng of EMBARQ; Dale Evarts of the U.S. EPA; Todd Johnson and Sarath Guttikunda of the World Bank; Allegheny County Chief Executive Dan Onorato; Stephen Hepler of the Pennsylvania Department of Environmental Protection; Mark Freeman and others at DOE's National Energy Technology Laboratory; Cliff Davidson and others at Carnegie Mellon University; Jayme Graham, Roger Westman, and others at the Allegheny County Health Department; Rachel Filippini of the Group Against Smog and Pollution; FirstEnergy Bruce Mansfield Power Plant; U.S. Steel Clairton Works; ALCOSAN; Bellefield Boiler Plant; Dave Nolle of DQE Energy Services; Michael Kleinman, Scott Samuelson, and Barbara Finlayson-Pitts of the University of California-Irvine; ARB El Monte; Elaine Chang and others at the South Coast Air Quality Management District; Art Wong and others at the Port of Long Beach; Walter Neal of the BP Refinery; Alan Foley and others at the Southeast Resource Recovery Facility; and Art Rosenfeld of the California Energy Commission.

We would like to recognize the contributions made by Jack Fritz, former Staff Officer at the NAE and the original director of this study, Lance Davis and Derek Vollmer for carrying on this work, as well as Kathleen McAllister and Mike Whitaker, who assisted with research, compilation, and report review process. Cui Ping and Li Bingyu of the CAS Institute of Engineering Thermophysics also deserve recognition for their work in coordinating the efforts of this bilateral group.

This report has been reviewed in draft form by individuals chosen for their diverse perspectives and technical expertise, in accordance with procedures approved by the National Academies' Report Review Committee. The purpose of this independent review is to provide candid and critical comments that will assist the institution in making its published report as sound as possible and to ensure that the report meets institutional standards for objectivity, evidence, and responsiveness to the study charge. The review comments and draft manuscript remain confidential to protect the integrity of the process.

We wish to thank the following individuals for their review of this report: Xuemei Bai, Commonwealth Scientific and Industrial Research Organisation, Australia; Hal Harvey, Hewlett Foundation; Jiming Hao, Tsinghua University; Peter Louie, Hong Kong Environmental Protection Department; Wei-Ping Pan, Western Kentucky University; Mansour Rahimi, University of Southern California; Kirk Smith, University of California, Berkeley; David Streets, Argonne National Laboratory; Richard Suttmeier, University of Oregon; Wenxing Wang, Global Environmental Institute; Yi-Ming Wei, Chinese Academy of Sciences; and Xiliang Zhang, Tsinghua University.

Although the reviewers listed above have provided many constructive comments and suggestions, they were not asked to endorse the conclusions or recommendations, nor did they see the final draft of the report before its release. The review of this report was overseen by Maxine Savitz (Retired), Honeywell, Inc., and Lawrence Papay, PQR, Inc. Appointed by the National Academies, they were responsible for making certain that an independent examination of this report was carried out in accordance with institutional procedures and that all review comments were carefully considered. Responsibility for the final content of this report rests entirely with the authoring committee and the institution.

Contents

Acronyms and Abbreviations

$(NH_4)_2SO_4$	Ammonium Sulfate
NH_4HSO_4	Ammonium Bisulfate
°C	Degrees Celsius
μm	Micrometers
ACCD	Allegheny Conference on Community Development, Pittsburgh, U.S.
ACHD	Allegheny County Health Department, Pittsburgh, U.S.
ACI	Activated Carbon Injection for Hg removal
ANL	Argonne National Laboratory, U.S.
APA	Administrative Procedure Act, U.S.
API	Air Pollution Index
AQM	Air Quality Management
AWMA	Air & Waste Management Association
CAA	Clean Air Act, U.S.
CAAQS	California Ambient Air Quality Standards, U.S.
CAIR	Clean Air Interstate Rule, U.S.
CAMD	Clean Air Markets Database, U.S.
CAMR	Clean Air Mercury Rule, U.S.
CARB	California Air Resources Board, U.S.
CAVR	Clean Air Visibility Rule, also called Regional Haze Rule, U.S.
CAS	Chinese Academy of Sciences, China
CBM	Coal Bed Methane
CCP	Chinese Communist Party, China
CEM	Continuous Emission Monitor

CEC	California Energy Commission, U.S.
CEQ	Council on Environmental Quality, U.S.
CHP	Combined Heat and Power
CCHP	Combined Cooling, Heating and Power
CFB	Circulating Fluidized Bed coal combustion
CI	Compression Ignition
CMAQ	Community Multiscale Air Quality Model
CMB	Chemical Mass Balance receptor model
CNEMC	China National Environmental Monitoring Center
CNG	Compressed Natural Gas
CO	Carbon Monoxide
CO_2	Carbon Dioxide
COG	Coke-Oven Gas
CSC	China Standard Certification Center
CTL	Coal to Liquids
CTM	Chemical Transport Model
CUEC	Comprehensive Urban Environmental Control, China
DE	Distributed Energy production
DOE	Department of Energy, U.S.
DOI	Department of Interior, U.S.
DOT	Department of Transportation, U.S.
DRB	Demonstrated Reserve Base, U.S.
EC	Elemental Carbon
ECL	Energy Conservation Law, China
EIA	Environmental Impact Assessment
EIA	Energy Information Administration, U.S.
EIS	Environmental Impact Statement
ELI	Efficient Lighting Institute, China
EPA	Environmental Protection Agency, U.S.
EPACT	Energy Policy Act of 2005, U.S.
EPB	Environmental Protection Bureau, China
ERS	Environmental Responsibility System, China
ESP	Electrostatic Precipitator
FBC	Fluidized Bed Combustion
FERC	Federal Energy Regulatory Commission, U.S.
FGD	Flue Gas Desulfurization
FON	Friends of Nature, China
FYP	Five-Year Plan, China
g/km	Grams per Kilometer

GASP	Group Against Smog and Pollution, Pittsburgh, U.S.
GDP	Gross Domestic Product
GEF	Global Environment Facility, China
GHG	Greenhouse Gases
H_2O	Water/Water Vapor
HAPs	Hazardous Air Pollutants
Hg	Mercury
HC	Hydrocarbon
HEW	Department of Health, Education, and Welfare, U.S.
HTS	High-Temperature Superconductivity transmission lines
ICR	Information Collection Request
IEA	International Energy Agency
IFC	International Finance Corporation
IGCC	Integrated Gasification Combined Cycle coal power plant
IMPROVE	Interagency Monitoring of PROtected Visual Environments, U.S.
kHz	Kilohertz
kW	Kilowatt
LADWP	Los Angeles Department of Water and Power, U.S.
LAPCD	Los Angeles Air Pollution Control District, U.S.
LEVII	Low Emission Vehicle Phase II, U.S.
LFSO	Limestone with Forced Oxidation SO_2 removal
LNG	Liquefied Natural Gas
MANE-VU	Mid Atlantic, Northeast Visibility Union, U.S.
MLR	Ministry of Land and Resources, China
MOST	Ministry of Science and Technology, China
NAAQS	National Ambient Air Quality Standard, U.S.
NAE	National Academy of Engineering, U.S.
NAMS	National Air Monitoring Stations, U.S.
NAS	National Academy of Science, U.S.
NBB	National Biodiesel Board, U.S.
NCC	National Coal Council, U.S.
NDRC	National Development and Reform Commission, China
NEET	New and Emerging Environmental Technologies Data Base, U.S.
NEPA	National Environmental Policy Act, U.S.
NETL	National Energy Technology Laboratory, U.S.
NGO	Non-Governmental Organization
NREL	National Renewable Energy Laboratory, U.S.

NH_3	Ammonia
NH_4NO_3	Ammonium Nitrate
NMCEP	National Model City of Environmental Protection, China
NO	Nitrogen Oxide
NO_2	Nitrogen Dioxide
NO_3^-	Nitrate
NO_x	Oxides of Nitrogen (Nitrogen Oxides)
NPC	National Peoples' Congress, China
NPC	National Petroleum Council, U.S.
NRC	National Research Council, U.S.
NSF	National Science Foundation, U.S.
NSPS	New Source Performance Standards, U.S.
NSR	New Source Review, U.S.
ns	Nanosecond
O_3	Ozone
OBD	On-Board Diagnostics for motor vehicle monitoring
ORNL	Oak Ridge National Laboratory, U.S.
OTAG	O_3 Transport Assessment Group, U.S.
OTR	O_3 Transport Region, U.S.
PAC	Powdered Activated Carbon for Hg removal
PAMS	Photochemical Assessment Monitoring Stations, U.S.
PaDNR	Pennsylvania Department of Natural Resources, U.S.
Pb	Lead
PC	Pulverized Coal power plant
PM	Particulate Matter, includes TSP, PM_{10}, $PM_{2.5}$, and UP
PM_{10}	Particles with aerodynamic diameters < 10 μm
$PM_{2.5}$	Particles with aerodynamic diameters < 2.5 μm (also fine PM)
PMF	Positive Matrix Factorization receptor model
POLA	Port of Los Angeles, U.S.
PRC	Peoples Republic of China
QESCCUE	Quantitative Examination System on Comprehensive Control of Urban Environment
RH	Relative Humidity
RMB	Renminbi, Chinese currency unit ≈0.13 dollar. Also termed the yuan.
RPO	Regional Planning Organization, U.S.
RVP	Reid Vapor Pressure gasoline fuel specification
SBQTS	State Bureau of Quality and Technical Standards, China
SCAG	Southern California Association of Governments, U.S.

SCAQMD	South Coast Air Quality Management District, Los Angeles, U.S.
SCE	Southern California Edison, U.S.
SCIO	State Council Information Office, China
SCR	Selective Catalytic Reduction NO_x removal
SCRAM	Support Center for Regulatory Monitoring, U.S.
SEPA	State Environmental Protection Agency, China
SERC	State Electricity Regulatory Commission, China
SERRF	Southeast Resource Recovery Facility, California, U.S.
SETC	State Economic and Trade Commission, China
SIP	State Implementation Plan, U.S.
SLAMS	State and Local Air Monitoring Stations, U.S.
SNCR	Selective Non-Catalytic Reduction
SO_2	Sulfur Dioxide
SO_4^{2-}	Sulfate
SoCAB	South Coast Air Basin, Los Angeles and surrounding cities, U.S.
STN	Speciation Trends Network, U.S.
SUV	Sports Utility Vehicle
TOD	Transit-Oriented Development
TSP	Total Suspended Particulate, particles with aerodynamic diameters ~<30 μm
UCS	Union of Concerned Scientists
UN	United Nations
UNCHE	United Nations Conference on the Human Environment
UNDP	United Nations Development Programme
UNEP	United Nations Environment Programme
UP	Ultrafine Particles with aerodynamic diameters < 0.1 μm
U.S.	United States
USC	Ultra SuperCritical coal combustion
USC	United Smoke Council, U.S.
USDA	Department of Agriculture, U.S.
USFS	Forest Service, U.S.
USGS	Geological Survey, U.S.
VMT	Vehicle Miles Traveled
VOC	Volatile Organic Compound
WHO	World Health Organization
WRAP	Western Regional Air Partnership, U.S.

Summary

The United States and China are the number one and two energy consumers in the world. China is the largest emitter of sulfur dioxide (SO_2) worldwide, and the two countries lead the world in carbon dioxide (CO_2) emissions. Energy consumption on a grand scale and the concomitant air pollution it can cause have myriad effects, from local to global, and there are a number of underlying issues which have a profound impact on their interplay. Both countries possess massive coal reserves and intend to continue utilizing these resources, which have been a major source of pollution. In spite of energy security concerns, the United States is still the world's largest consumer of petroleum, though China's skyrocketing demand has made it the second largest consumer and a major source of demand growth. This is, of course, being driven by rapid urbanization and, in particular, by the rise of personal vehicle use.

The United States has made great strides in improving air quality since the early part of the 20th century, by reducing domestic and transportation coal use and by refining combustion conditions in large centralized facilities. Further improvements were achieved during the last half of the 20th century by better understanding the relationships between emissions and air quality, developing and applying pollution controls, increasing energy efficiency, and instituting a management framework to monitor airsheds and to enforce regulations. U.S. ambient levels of SO_2, nitrogen dioxide (NO_2), carbon monoxide (CO), and lead (Pb) have largely been reduced to levels that comply with air quality standards. However, ozone (O_3), suspended particulate matter (PM), mercury (Hg), and a large list of Hazardous Air Pollutants are still at levels of concern. O_3 and a large portion of PM are not directly emitted, but form in the atmosphere from other emissions, including SO_2, oxides of nitrogen (NO_x), volatile organic compounds (VOCs), and ammonia (NH_3). The relationships between direct emissions and ambient concentrations are not linear and involve large transport distances, thereby complicating air quality management.

China has focused on directly emitted PM and SO_2 emissions and concentrations, with less regulatory attention being given to secondary pollutants such as O_3 or the sulfate, nitrate, and ammonium components of PM. China has made great progress over the last 25 to 30 years in reducing emissions per unit of fuel use or production. However, rapid growth in all energy sectors means more fuel use and product, which counteracts reductions for individual units. Shuttering obsolete facilities, which are often the most offensive polluters, has been an effective strategy, as well as adopting modern engine designs and requiring cleaner fuels (e.g., low sulfur coal). While necessary measures, these represent the "low-hanging fruit," and greater reductions for a larger number of emitters and economic sectors will be needed to attain healthful air quality. The responsibility for developing and instituting many air quality and energy strategies rests with local and regional governments. The importance of national policies and actions should not be overlooked, but the most appropriate solutions in China will require local knowledge, willpower, and implementation.

To examine the challenges faced today by China and the United States in terms of energy use and urban air pollution, the U.S. National Academies, in cooperation with the Chinese Academy of Engineering and the Chinese Academy of Sciences, developed this comparative study. In addition to informing national policies in both countries, the study is intended to assist Chinese cities in assessing their challenges, which include meeting increased energy demands, managing the growth in motor vehicle use, and improving air quality, all while maintaining high rates of economic growth. This report is geared towards policy and towards decision makers involved in urban energy and air quality issues. It identifies lessons learned from the case studies of four cities (Pittsburgh and Los Angeles in the United States, Huainan and Dalian in China), addresses key technological and institutional challenges and opportunities, and highlights areas for continued cooperation between the United States and China. Owing to the small number of case studies, the committee decided against making many recommendations specifically tailored to the case study cities, or to cities in general, based solely on the experience of the four case studies. Instead, the case studies provide insight into how energy use and air quality are managed at a local level, and how our cities might learn from one another's experience. This study does not examine in detail the related and increasingly significant issue of global climate change. It does acknowledge that this will be a central issue in future discussions of energy and air pollution, and an area where continued cooperation between the U.S. and Chinese Academies will be critical. The study committee, composed of leading experts on energy and air quality from both countries, began its work in 2005.

ENERGY RESOURCES, CONSUMPTION AND PROJECTIONS

In both countries, fossil fuels continue to dominate energy production. Renewable energy offers the potential to decrease this dependence, but, except for hydropower and wood, has not yet been heavily exploited in either country.[1] Due in large part to its abundance in both countries, coal has played an important role in electricity production and industrial processes, and its combustion has been a major source of air pollution. Coal has been and will continue to be primarily used for power production in the United States and China, but it can also be used to create gaseous and liquid fuels, as well as other feed stocks, and may play a larger role, depending on prices, as an alternative to natural gas and petroleum. Therefore, a primary challenge for both countries is to seek ways to utilize their coal resources in an environmentally acceptable manner. Petroleum accounts for nearly 40 percent of the U.S. primary energy consumption, mostly for liquid fuels in the transportation sector. China's energy consumption is still dominated by industry (70 percent) and is supplied by coal (69 percent), but petroleum demand has increased rapidly in recent years in tandem with the burgeoning transportation sector (Figure S-1).

Neither country has sufficient domestic petroleum reserves to satisfy current demand; in a business as usual scenario, both countries will be increasingly dependent upon imports. Natural gas has played an important role in the United States, primarily due to environmental concerns; but limited supplies and higher prices have led to renewed interest in coal-fired power plant development. In China, natural gas is not used widely, though China does possess large reserves of natural gas and of coalbed methane (CBM) and is taking steps to develop these energy sources. For both countries, future natural gas consumption will likely rely on advances in liquefied natural gas technologies and trade. Finally, nuclear power, which is the second largest source of electricity in the United States, has been receiving renewed interest, owing to higher energy prices and concerns over CO_2 emissions. However, it is still unclear whether or not this sector will expand in the United States, and it still constitutes a small portion of total power production in China.

Energy forecasting has proved challenging in both countries, owing to limited data and inaccurate projections of available resources and consumption. Energy consumption and projection data are also used as the basis for creating emission inventories used in air quality management. Energy security is a primary concern for both countries, and projected increases in fuel imports (notably petroleum) are a primary driver for the United States and China to pursue energy efficiency improvements and fuel substitution strategies. Energy prices have an important impact on decisions regarding fuel consumption. Rising natural gas prices in the United States have led to renewed interest in coal-fired capacity; and, in China,

[1]There are notable exceptions, including western states in the United States which have reduced their fossil fuel dependence relative to the rest of the country.

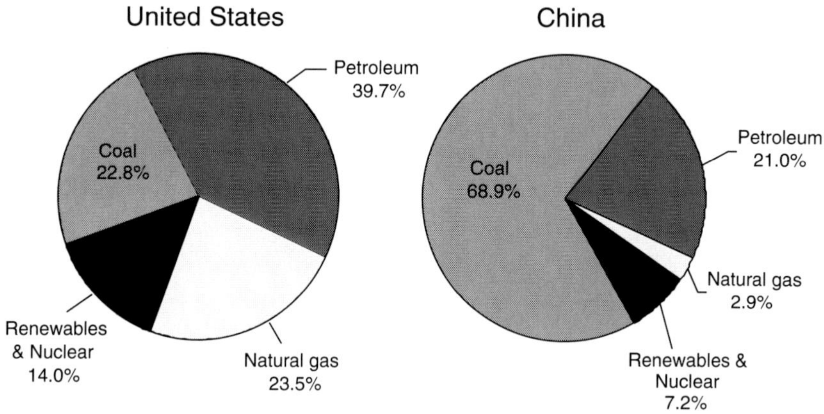

FIGURE S-1 Primary commercial energy consumption by fuel type, 2005.
NOTE: China's nuclear power production represents less than 1 percent of total consumption.

the rising cost of delivered coal, due to escalating costs of transportation by train, has led some coastal cities to import cheaper coal from other countries. Rising fossil energy prices will also affect the development and use of alternative energy resources, such as biofuels.

In terms of energy consumption, industrial uses continue to dominate in China, although buildings (residential and commercial) and transportation will increase their share in the coming years. Buildings are a large consumer of energy in the United States, in terms of electricity consumption for lighting and appliances and energy for heating and cooling (40 percent of total energy consumed). Transportation is also an important energy consumer in the United States (nearly 30 percent), almost exclusively in petroleum-based fuels. China's transportation sector currently consumes 8 percent of total energy, but this proportion is certain to increase along with the increase in personal vehicle use, air travel, and goods shipment (Figure S-2 and Figure S-3). As such, fuel quality will be an important issue, in addition to its availability. In many parts of China, fuel quality remains poor, especially diesel fuel, and consequently transportation fuels have a disproportionate impact on air quality.

AIR POLLUTION TRENDS AND EFFECTS

The United States and China both regulate air pollution because of its effects on human health, visibility, and the environment. Both countries have adopted air quality standards for individual pollutants, although China's air pollution index contains five separate classes, allowing for "compliance" at levels less stringent

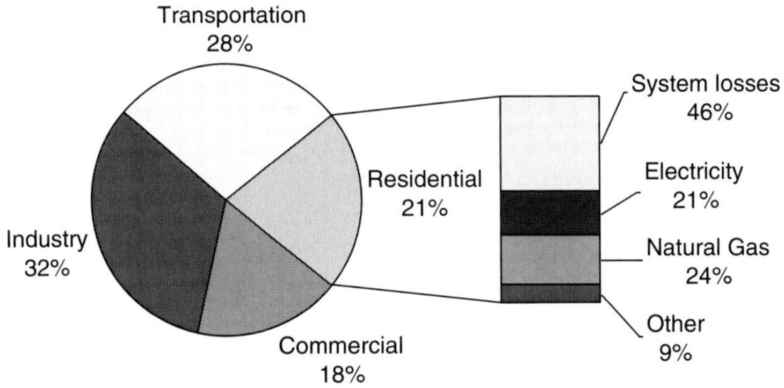

FIGURE S-2 United States Energy consumption by sector, 2006.

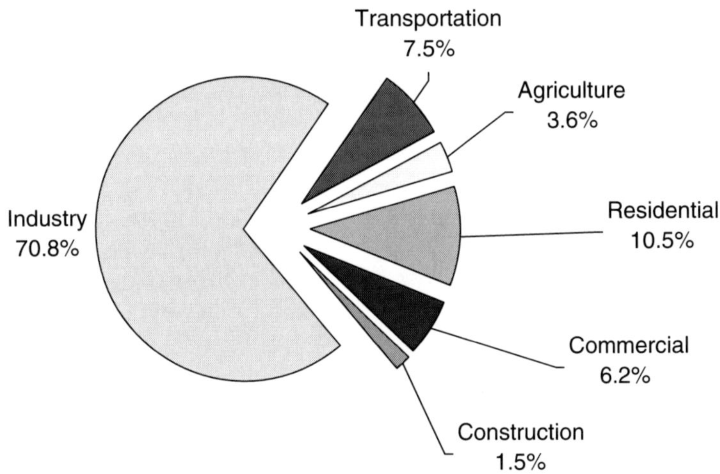

FIGURE S-3 China Energy consumption by sector, 2005.

than international standards. In the United States, National Ambient Air Quality Standards (NAAQS) have been established for O_3, CO, SO_2, NO_2, Pb, $PM_{2.5}$ (< 2.5 μm aerodynamic diameter), and PM_{10} (<10 μm aerodynamic diameter), based on their adverse health effects. Indoor air pollution, largely associated with the use of coal for heating and cooking in China and with smoking, building materials, wood burning, and natural gas cooking in both countries—is an important health concern that is not regulated. Respiratory and cardiovascular sickness and death rates are significantly higher in polluted compared to non-polluted areas in both countries. It is estimated that nearly 50 percent of respiratory ailments are related to excessive air pollution and that, by 2020, China may be devoting 13 percent of its projected GDP to healthcare costs associated with coal burning. Like the United States, China is ultimately bearing some of the external costs of air pollution through healthcare costs. In the United States, acid deposition and visibility impairment are being reduced, but it will still take decades and larger emission reductions to attain desired levels. Plant life is more sensitive than humans are to O_3 and this has important implications for forest ecosystems and agricultural crop production. China is currently studying the agricultural impacts of O_3 exposure; by some projections, O_3 could cause 20-30 percent crop losses for soybeans and winter wheat by 2020.

Largely as a result of air pollution regulation, the United States has witnessed substantial reductions in emissions and ambient concentrations of PM_{10}, CO, SO_2, NO_x, and Pb. However, $PM_{2.5}$ and O_3 exceed healthful levels in many parts of the United States and China. These require both local and regional emission reductions of directly emitted $PM_{2.5}$, SO_2, NO_x, and VOCs, which lead to secondary ozone formation. In addition to controlling industrial sources (including power plants), the United States has instituted pollution controls for mobile sources and specifications for motor vehicle fuels. This has led to marked decreases in Pb emissions (China is currently experiencing similar decreases) and CO levels. China's emissions are predominantly industrial; SO_2 emissions have been increasing, although soot and dust (the other two currently regulated emissions) have remained slightly more stable since the mid-1990s. Although some Chinese cities measure and report O_3 and other pollutants, local governments are only required to report on CO, NO_2, SO_2, and PM_{10}. Of these, PM_{10} has most often been associated with unhealthy air quality. However, regional and local studies in urbanized regions have observed excessive O_3 and $PM_{2.5}$. $PM_{2.5}$ constitutes a large part of PM_{10} (50-70 percent) and therefore is an important urban and regional air pollutant, which is currently unregulated in China.

An important lesson learned is that air pollution damage imposes major economic costs, through premature mortality, increased sickness and lost productivity, as well as decreased crop yields and ecosystem impacts. Cost-benefit analyses in the United States show that emission reduction programs have provided much greater benefits than their costs, by a ratio of up to 40 to 1, according to some estimates.

INSTITUTIONAL AND REGULATORY FRAMEWORKS

The United States has strong federal leadership and enforcement (U.S. Environmental Protection Agency [EPA]) for NAAQS attainment. This resulted from the realization that air pollution crossed political boundaries, and that some states and localities were not sufficiently controlling their emissions. There is a partnership between federal, state, and local agencies that addresses different types of emissions, with partial federal financing available to state and local pollution control agencies. Federal highway funds can be withheld from areas that do not make good faith efforts to attain pollution standards. In China, the central authority (State Environmental Protection Agency, SEPA) plays a minor role in air quality management in cities, with most activities carried out by local Environmental Protection Bureaus (EPBs). Cities and provinces have little motivation to reduce emissions that might affect neighboring regions. Pollution reduction laws have been ineffective in the absence of enforcement, emissions monitoring, and ambient air monitoring. Thus, monitoring and enforcement are key challenges for China. The central government recognizes the importance of air quality and has enacted a series of regulations aimed at reducing pollutant emissions. However, the local EPBs charged with the responsibility for enforcement often lack the necessary funding, technical capacity for monitoring, and/or the will to perform appropriately. Moreover, local and provincial leaders are evaluated primarily on economic performance that does not consider the costs of pollution, often leading to short-sighted decisions favoring economics over pollution control. As a result, air quality management has been inconsistent.

Emission controls are often less costly to implement than first envisioned. Control costs are also not purely costs, as they create opportunities (e.g., manufacturing and sales of pollution control and energy efficient equipment) that result in economic growth. Appropriate programs can lead to economically efficient approaches for improving the environment, thereby further reducing costs. In particular, both countries are experiencing a trend towards market-based approaches to air quality management (in contrast to the earlier command and control approach). The U.S. successful SO_2 "cap and trade" program is being adopted elsewhere, including in China. Other tools, such as emission taxes and fees, can also be utilized to achieve air quality goals, but these likewise require judicious monitoring and enforcement. China has made important strides in closing down inefficient and heavily polluting industries, and SEPA has recently become influential in reviewing environmental impact assessments and even in halting major construction projects. Still, challenges remain in terms of managing remaining infrastructure and in planning for future growth.

Aside from the EPA and SEPA, other agencies in both countries play roles in air quality management. Energy policies also impact air quality. In the United States, the Department of Energy (DOE) plays a dominant role in setting policy, as well as in conducting key research; but in China energy responsibilities are more diffuse. Both countries might benefit from increased coordination between energy and air quality research and policy making. While much data and information about emissions, ambient concentrations, and energy use are publicly available in the United States (many of them over the Internet), such data are often sequestered in China. The EPA has converted older data management methods to modern web-based systems. The U.S. Energy Information Administration has a similar compilation of energy data. Public and scientific scrutiny of these data has led to improved quality and utility over time. Many of these modern concepts can be applied in China. Although China has made progress in reporting air quality indices to the public, the data needed for successful energy and air quality management are still difficult to obtain and analyze. Non-governmental organizations (NGOs) have also played important roles in setting air quality and energy priorities in the United States; environmental NGOs are on the rise in China, but their active involvement is predicated on access to information.

KEY INTERVENTIONS

Energy Efficiency

Improved energy efficiency provides benefits for air quality and energy security, while reducing costs. Energy efficiency can provide gains similar to, or greater than those provided by specific pollution controls and can reduce the need for new power generators. Cost-effective technology is currently available to greatly improve energy efficiency across all energy use sectors. Overall, energy intensity (a measure of energy consumption divided by GDP) has been declining in the United States over the past 20 years; China's intensity also declined from 1985 to 2000, but, since 2000, it has been increasing. However, this broad measure does not always accurately reflect changes in energy efficiency. The U.S. economy has experienced a reduction in energy-intensive industries, as part of a transition to a more service-based economy, and in many cases these energy-intensive industries have relocated in China. Still, both countries have made important sectoral improvements, which could be implemented more broadly. Energy efficiency has been an underutilized resource in both the United States and China.

China can make substantial and immediate gains through improvements in supply-side energy efficiency. Its power generation and industrial sectors have lagged behind international standards for energy efficiency, although there is increasing interest in utilizing more efficient coal technologies (ultra-supercritical pulverized coal combustion or integrated gasification combined cycle (IGCC) coal combustion). China has made strong efforts to integrate energy systems, such as in combined heat and power (CHP) and in combined cooling, heat, and power (CCHP) plants, both of which efficiently capture waste heat from power generation and utilize it to provide heating and cooling for residential and commercial buildings. CHP plants represent roughly 12 percent of total installed electrical capacity in China, and there are plans to double this share by 2020.

Efficiency in the transportation sector is another area in which both the United States and China can improve. In the United States, fuel economy standards imposed in the 1970s led to rapid improvements in vehicle fuel efficiency, but owing to the popularity of less stringently regulated light duty trucks coupled with low fuel prices, overall fleet fuel efficiency has declined since the early 1990s. China has developed fuel economy standards which surpass those of the United

States, though it is not yet clear how effectively these are being or will be enforced. Hybrids, which combine electric batteries with conventional fuel tanks, are available in both countries and offer substantial fuel savings. However, higher initial costs and battery replacement costs make these vehicles prohibitively expensive for some consumers. One additional means of improving efficiency in the urban transportation sector is by decreasing congestion and increasing the use of more efficient modes (e.g., public transportation).

Combustion and Pollution Control Technologies

It is less costly to plan for and implement pollution controls up front than to install them later. Due to a lack of knowledge of pollution effects and controls, the United States did not act early enough to provide for emission controls on stationary and mobile sources. Thus, retrofitting is an important but expensive part of the U.S. strategy to meet current air quality goals. Fortunately, in the U.S. experience, pollution control costs have declined and equipment costs are now anywhere from one-half to one-tenth the cost of older systems, and are more effective at pollutant removal. China is mandating SO_2 scrubbers on new power plants, and this is an important first step. But monitoring and enforcement will be needed to ensure that controls are properly installed, maintained, and continually operated. Future solutions to air quality goals may necessitate additional retrofits in China, such as adding scrubbers to existing plants and reducing NO_x emissions with low-NO_x burners, or through selective catalytic reduction systems. Coal-fired boilers have long lifespans ($\geq$ 50 years) and decisions made at the time of construction persist for many decades; this is particularly important, given the rate at which China is currently constructing new coal-fired power sets. Lack of available technical expertise, supply bottlenecks, financing, short-sighted economic decisions, and/or political opposition may continue to limit the up-front implementation of the best available control technology; but leaving room for it in the future will make it easier to install when the necessary resources are available.

Future pollution controls for stationary sources in the United States will focus on further reducing SO_2 and NO_x emissions from older facilities, reducing Hg emissions from coal-fired power stations, and decreasing the introduction of CO_2 into the atmosphere. Mercury capture is, in some cases, a co-benefit of other installed pollution controls, but ongoing research is focused on improving technologies specifically designed for mercury control (e.g., activated carbon). Carbon capture and sequestration, though not currently mandated, are being studied and could be regulated in the United States in the future. It is for this reason that IGCC technology is of great interest, as it permits the most efficient capture of CO_2 and other pollutants from coal gas, before it is used to drive a turbine. China has been a world leader in developing coal gasification technologies, though it is currently used almost exclusively for chemical production. One notable project involving both countries is FutureGen, a DOE-led venture which seeks to utilize IGCC with carbon capture and sequestration, to produce electricity, hydrogen from coal, and to realize co-benefits such as the use of the captured CO_2 as a medium to drive enhanced oil recovery.

Renewable Energy

Renewable energy sources, including solar, wind, geothermal, waste-to-energy, and biofuels, constitute important, but not large, fractions of energy portfolios in both countries. But the current rate of growth in renewables is insufficient to meet the projected needs for fossil fuel energy. Hydropower and wood to produce electricity are the dominant renewable resources currently being utilized, and are projected to remain so—although other technologies, notably wind turbines, have been improving and their use is rapidly expanding. Several applications, such as solar water heating and wind turbines to generate electricity, are economical in the long term, but can require larger up-front investments and backup power versus more conventional sources. Therefore, energy prices influence the market penetration of renewable technologies. Government mandates also play a role, as both countries (including state and local governments) have set targets for renewable energy consumption. For the time being, except for hydroelectric, renewable electricity generation sources mostly fulfill niche applications, but they are showing promise as distributed or off-grid energy supplies, as they are cleaner and can be more cost-effective than extending existing power lines. China has been expanding its capacity of small hydropower units, in order to electrify remote areas. China has also made great strides in developing its domestic capacity to produce wind turbines, and it is already the world leader in production and use of solar water heaters. Renewable technologies will also be critical to the eventual pursuit of a hydrogen economy. Hydrogen can currently be produced economically from natural gas for industrial purposes, but large-scale production will almost certainly rely on renewable energy for production, if hydrogen is to be considered a clean alternative energy carrier.

It is unclear whether some biofuels, including ethanol from non-cellulosic sources, provide more renewable energy than they consume in non-renewable energy for their production. Biodiesel production has been increasing, but it still constitutes a minor fraction of total biofuel production. In the United States, ethanol is predominantly derived from corn, while in China its sources are slightly more diversified, but still grain-based. In both cases, this production is viewed as competitive with food markets and, ultimately, the future of ethanol as a viable petroleum alternative will depend on advances in cellulose-based production technologies and their successful commercialization. However, ethanol has been effectively used as an additive for reformulated gasoline (RFG) for a number of years, in order to reduce certain harmful emissions; experiences in U.S. metropolitan areas have shown that use of ethanol in RFG can help reduce total CO emissions, as well as toxics such as benzene.

LESSONS FROM CITIES

In the United States, many cities, including Pittsburgh and Los Angeles, have successfully implemented policies and technologies to reduce various emissions and to improve air quality. Local pollution prevention measures showed benefits as early as the 1940s in Pittsburgh, when smoke controls in place likely saved the city from a severe air pollution episode that caused loss of life in nearby Donora. Civil society played an important role in Pittsburgh's approach to air quality management. Early activist groups raised awareness of air pollution issues and paved the way for an open stakeholder process which allowed NGOs, such as the Group Against Smog and Pollution, to take part in policy formulation.

Pittsburgh has diversified its economy since its industrial prime. As local pollution sources have been cleaned up or closed down, the city has focused more on regional pollution issues such as O_3 and $PM_{2.5}$. Indeed, as many U.S. cities remediated local air pollution problems, it became apparent that some issues required regional solutions, as current pollution levels derive from a variety of energy uses and sectors on local and regional scales. All of these sectors must participate in solutions to pollution. As demonstrated in Los Angeles, emission controls can be applied to many small and medium-size sources that collectively have a large effect on pollution levels. Federal intervention often leads to local regulations to solve what are ultimately regional challenges. Air pollution does not obey boundaries, and while many Chinese cities are pointing out the impact that regional pollution has on local conditions, to date there have been few examples of regional cooperation. In the United States, the Los Angeles situation is more common, where regional and statewide organizations such as the South Coast Air Quality Management District and the California Air Resources Board both play critical and complementary roles in air quality management.

Both U.S. and Chinese cities have benefited from research, development, and technology transfer efforts in their universities, research institutes, and professional associations. These efforts also provide local expertise for states and provinces and train professionals needed for regulatory, industrial, and educational enterprises. Pittsburgh and Los Angeles both continue to rely on their local universities and research institutes to address emerging challenges in energy and air pollution. An ongoing challenge for many U.S. cities is that U.S. transportation and economic development policies have created the need to drive long distances, resulting in high personal vehicle use and automobile emissions. A similar pattern is now occurring in many Chinese cities, and their response has been to build more roads to alleviate congestion. The rapid growth of traffic in Dalian and in similar Chinese cities will repeat the air quality and energy consumption mistakes of Los Angeles and other U.S. cities, if not better managed. Chinese cities can benefit from their greater densities (relative to most U.S. cities) and take steps to limit the need for personal vehicle use, as the cities continue to grow. Some U.S. cities are attempting to undo the effects of their sprawling development, but these efforts are slow and costly.

Huainan and Dalian can also set examples for other Chinese and U.S. cities. Both cities benefited from efforts to relocate key industries away from urban centers and from closing down inefficient, highly polluting industries. However, the net impacts of industrial relocation are not yet fully understood; moving polluting industries away from densely populated city centers has lowered the risk and exposure for numerous city dwellers, but the relocated industries may transfer the risk to rural or suburban residents; moreover, depending upon the location of the industries, the air quality impacts may not be fully reflected by data generated at urban air quality monitoring stations. Huainan has improved its air quality, though future plans to develop the city into a regional base for energy and chemical production will necessitate further strengthening of the air quality management system. As a coal-rich city, Huainan has benefited from local research and development, which has allowed it to begin harnessing coalbed methane and to utilize coal gasification technologies. These may be usefully applied to the energy needs of numerous other cities. Dalian has enjoyed a reputation as one of the cleanest cities in China, and it has often established environmental quality standards which exceed national standards. Other Chinese cities could benefit from adopting and pursuing similar aggressive standards.

KEY RECOMMENDATIONS

To meet the challenges of increasing energy consumption while achieving air quality goals, the U.S. and Chinese governments (national and local) should consider the committee's specific recommendations in 15 areas. Some of the key recommendations are presented below. A discussion of all of the committee's recommendations and study findings may be found in Chapter 12.

1. Learn from experience. China should learn from the successes and failures of the United States and other developed countries in reducing the influence of energy use on air quality. Mistakes already made in the United States and elsewhere should be identified (as this report has attempted to do) and avoided in China (Recommendation 1-a). Continued dialogue and information exchange among U.S. and Chinese scientists and policy makers should be promoted through professional organizations, government support programs, and the National Academies in both countries, to promote joint development of energy and pollution control strategies (Recommendation 1-b).

2. Recognize and respond to external costs of energy production and use. Both countries need to improve permitting policies and economic mechanisms that reflect the external costs of pollution that are being paid by others (e.g., through adverse health effects and degraded quality of life). These might include high enough taxes on emissions to make the addition of controls economically attractive, and rebates or subsidies to encourage use of higher efficiency and renewable technologies (Recommendation 2-a).

3. *Establish and implement standards that protect human health.* Both the United States and China should adopt minimum standards based on healthful air quality, which may require revising currently accepted standards. Local governments should be able to enact more stringent local standards, but there should *not* be a sliding scale based on the level of economic development (Recommendation 3-a). $PM_{2.5}$ control should be emphasized over, but not at the expense of, PM_{10} and O_3 reductions (Recommendation 3-f).

4. *Address pollution sources comprehensively.* There has to be participation in emissions reductions by all sectors, not just by the major industries. Enforcement and monitoring, as well as incentives, are needed to assure that emission reductions are implemented and maintained (Recommendation 4-e). Governments must improve policy incentives to adopt specific control technologies. Policies requiring the implementation of pollution controls are a positive first step, but these policies must be developed in tandem with appropriate incentives to overcome financial or other barriers (Recommendation 4-g).

5. *Strengthen SEPA's role in overseeing air quality planning and enforcement.* The Chinese government needs to expand SEPA's staff and influence over local air quality surveillance, management, and enforcement. Better coordination is needed between national and provincial authorities (Recommendation 5-a). As in the United States, China needs formal emission reduction plans specific to cities and regions that are independently evaluated and enforced at the national level. These plans should specify the activities that will bring areas into compliance with standards and that will keep areas already in compliance from becoming more polluted (Recommendation 5-b).

6. *Realize the potential of energy efficiency improvements.* The United States and China should consider evaluating the best energy efficiency standards for all energy sectors that have been formulated by each country, by their states/provinces, or by other countries. Efficiency standards, like air quality standards, will need to be properly enforced in order to be effective (Recommendation 6-a).

7. *Promote efficient transportation systems and sustainable urban design.* Transit-oriented design and smart growth policies should be implemented to develop new urban areas or to redevelop existing areas, particularly in rapidly developing cities with high projected growth. Bus rapid transit (BRT) should be considered in a number of U.S. and Chinese cities, as it represents a low-cost (relative to subways and light-rail) transit system easily adapted to existing infra-

structure, with proven success in other parts of the world (Recommendation 7-a). Traffic management systems, such as the system in place in Dalian, should be implemented in other Chinese cities, in order to manage the rapidly expanding vehicle fleets and to limit congestion (Recommendation 7-c).

8. Accelerate improvements in fuel economy and reductions in mobile source emissions. The United States should examine the present Corporate Average Fuel Economy (CAFE) standards or alternative incentives to improving fuel economy, to develop standards tailored to the U.S. market and vehicle stock (Recommendation 8-a). China should enforce their fuel economy standards and consider other, possibly more effective alternatives as well (Recommendation 8-b). China should continue to increase its vehicle emission standards and to enforce those standards; China should also improve the quality of its refined fuels (Recommendation 8-d).

9. Improve energy efficiency in buildings. Building codes in both countries should be updated to require energy-saving technologies (e.g., CCHP; Recommendation 9-a). Subsidies, incentives, and low cost financing should be enhanced in both countries to encourage up-front investments in energy efficient technologies that will be paid back in future cost savings (Recommendation 9-b). Both countries should allow or encourage utilities to decouple profits from energy sales. This is occurring to some degree in the United States, but needs to be accelerated, and must be implemented in China (Recommendation 9-c).

10. Promote cleaner technologies for heat and power generation. Incentives are needed in the United States and China to implement cleaner coal conversion technologies (e.g., IGCC), more efficient generation methods, and productive use of waste heat (Recommendation 10-a). Coal washing and sieving rules should be implemented and enforced in all sectors of the coal industry in China, to reduce SO_2 and to increase combustion efficiency (Recommendation 10-b). Following the example of cities such as Huainan, coal-rich areas should implement systems to recover and make effective use of CBM and coke oven gas (Recommendation 10-d).

11. Plan in advance for pollution control. Better evaluation tools need to be promulgated, specific to the United States and China, which assist project designers in evaluating the costs and benefits of different energy conservation/pollution control alternatives (Recommendation 11-a). Projects need to be planned with the expectation that pollution controls and retrofits may be required, or deemed economical, in the future, even if benefits do not exceed costs by today's standards (Recommendation 11-b).

12. Accelerate development and use of renewable energy sources. Both countries should continue to encourage the development, production, and use of renewable energy wherever possible, through various policy instruments (e.g., renewable portfolio standards, tax rebates, preferential purchasing) (Recommendation 12-a).

13. Expand public participation in Chinese air quality management efforts. SEPA needs to convince public officials that the advantages of disseminating energy use, emissions, and air quality data outweigh the disadvantages. Such transparency will result in better data quality, by providing feedback on deficiencies to data generators (Recommendation 13-a). SEPA and provincial agencies in China should continue to increase their efforts in outreach and education to engage the public in helping address air pollution problems, and to encourage public participation in environmental impact studies and decisions affecting the environment (Recommendation 13-d). Local governments in China should encourage more volunteer groups focused on improving the environment (Recommendation 13-f).

14. Improve capacity to address current and future issues through research and education. Both countries need to strengthen research and development in clean energy, energy efficiency, and air quality research. There is also a need for improved research across disciplines, in order to better understand the linkages between energy and air quality (Recommendation 14-a). Chinese cities need to develop local and regional technical training centers and professional education centers, in order to build the capacity to operate and maintain pollution controls and advanced technologies (Recommendation 14-e).

15. Expand cooperation on energy and air quality issues, including efforts to reduce greenhouse gas emissions. Given the existing interest in climate change, it is imperative that the United States and China begin substantial cooperation on issues to reduce greenhouse gas emissions. In addition to energy efficiency, there is great potential for collaborative research on improving CO_2 capture and sequestration technologies (Recommendation 15-a). China will benefit from further cooperation on developing regional air quality management. Future activities should complement the ongoing work between Guangdong and Hong Kong, and efforts to develop SEPA's regional offices. Research universities and governments should also increase collaboration on measuring and monitoring $PM_{2.5}$ and O_3, as well as air quality forecasting (Recommendation 15-c).

1

Introduction

The United States and China are the two largest consumers of energy in the world, and projections for both countries indicate that their consumption will continue to rise in the foreseeable future. Both countries are mostly dependent upon fossil fuels for their energy supplies (over 85 percent in the United States and over 90 percent in China); thus, in addition to meeting increasing energy demands, the United States and China must confront the air quality challenges that result from fossil fuel consumption on such a large scale. While the United States has made progress in remediating much of the air pollution experienced in its heavy industrial period in the late 19th and early 20th centuries, it still faces air pollution challenges resulting from electricity production, vehicle use, and numerous other sources. China is presently in the midst of a period of rapid industrialization accompanied by meteoric economic growth on a very short time scale and, thus, it is experiencing similar if not more severe pollution than has plagued the United States.

Air pollution has historically been viewed as a nuisance, but also, in some cases, as a sign of economic progress; it was and sometimes still is thought to be a requisite to development. However, research on air pollution effects has led to an increased understanding of the linkages between fossil fuel combustion and air quality and, more importantly, the links between health and air quality. Health studies in the United States and elsewhere in the mid-20th century raised awareness that air pollution, whether from industrial sources or from a then emerging new source, the automobile, had major impacts on morbidity and mortality. This improved understanding paved the way for regulation and other efforts to combat air pollution in its various forms, and further research has exposed its effects on ecosystems, agriculture, and general quality of life. This research, in turn, has been translated into cost-benefit analyses, which now influence the decisions made on balancing the interplay between energy consumption and air quality. On the whole, the U.S. experience provides some rich lessons which China, with the benefit of this hindsight, may incorporate into its quest for environmentally sustainable development.

China presents a particularly interesting case, because, in addition to its well-known economic growth and industrial transition, it is also undergoing a demographic transition of rapid urbanization, which will play a central role in its ability to manage its energy use and air quality. China's urban population in 1980 was less than 20 percent of its total population; today approximately 40 percent of residents live in cities (compared to over 80 percent in the United States), and this share will increase to 60 percent of the population by 2030 (UN, 2005). China is home to over 100 cities with 1 million or more residents in each city—fewer than half of which achieve China's own minimum standards for air quality (SEPA, 2007). Further complicating this trend is the fact that urbanization in other countries has brought with it increased rates of energy consumption and vehicle use.

Although the United States continues to face air quality challenges, the lessons it has learned (successes and failures) in managing air quality should be relevant to the Chinese experience. Additionally, there are lessons to be learned from developments within China, which might be instructive to any number of developing cities facing similar challenges. Finally, in consideration of the globalized economy, increasing competition for finite resources, and a shared global environment, it is important to keep in mind that the decisions that one country or city makes today can certainly have a lasting impact on the opposite side of the world.

In order to examine the challenges faced today by China and the United States in terms of energy use and urban air pollution, the U.S. National Academies, in cooperation with the Chinese Academy of Engineering and the Chinese Academy of Sciences, developed this comparative study, building on nearly a decade of inter-Academy collaboration. In addition to informing national policies in both countries, the study is intended to assist Chinese cities in assessing their challenges, including the dual challenges of continued use of coal as the dominant source of energy and the rapidly increasing use of private vehicles, in the context of rapid economic growth, preservation of the environment, and ensuring the quality of life for their citizens. This report is geared towards policy and decision makers at all levels of government, as they seek to balance urban energy consumption with air quality management. It identifies lessons learned from the case studies of four cities (Pittsburgh and Los Angeles in the United States, Huainan and Dalian in China); the study addresses key technological and institutional challenges and opportunities, and highlights areas for continued cooperation between the United States and China on energy and air quality issues. Specifically, the study was designed to:

- Describe current and planned energy uses for different economic sectors in China and the United States and their effects on air quality;
- Compare and contrast the evolution of energy use and air quality management between two pairs of cities in China and the United States;
- Specify energy strategies that have been successful and unsuccessful in improving urban air quality, and identify leapfrogging opportunities; and
- Compare successful energy and air quality approaches with current policies in China and the United States and recommend potential modifications to current trends.

These issues are discussed in detail in the chapters that follow. One ancillary benefit of the study process was that it provided useful cross-sector and cross-country exchanges, particularly among local stakeholders. Having local environmental managers interact with scientific and technical experts, and being able to "kick the tires" on a technology, is valuable to any city seeking to meet its energy needs without compromising air quality. However, owing to the small number of case studies, the committee decided against making many recommendations specifically tailored to the case study cities, or to cities in general, based solely on the experience of the four case studies. Instead, the case studies provide insight into how energy use and air quality are managed at a local level, and how our cities might learn from one another's experience.

MAJOR ISSUES

Coal supplies are abundant in both countries, but conventional coal combustion is a major source of criteria air pollutants (e.g., SO_2), as well as of CO_2 emissions. In recent decades, the United States has favored natural gas as a cleaner-burning substitute, but supply constraints and increased prices have led to renewed interest in coal and other alternatives. China's rapid economic growth has largely been supported by coal (~ 2 billion metric tons[1] consumed annually); but China is also seeking cleaner sources such as natural gas and renewable or alternative fuels for its transport sector and for use in selected urban areas. Petroleum remains the dominant transportation fuel in both countries, and rising imports have cautioned both countries to focus on energy security as part of their overall energy policy strategies. Chapter 2 examines these and other issues related to major energy resources, specifically coal, petroleum, natural gas, and nuclear. In addition to a discussion of current resources and consumption, the chapter looks at future projections for the continued use of energy resources.

The impacts of urban air pollution as a result of conventional coal combustion are substantial and well known. The traditional mix of particulate matter (PM)

[1]Due to the broad range of units used in energy calculations, between countries and across sectors, energy figures will be reported in both the commonly accepted unit as well as a standardized unit (exajoules or EJ) for comparison. See Appendix D for information on energy conversions.

and SO_2 from large stationary sources is augmented by emissions from mobile sources, which are now a major source of numerous pollutants in the United States and an increasingly important source in Chinese cities. Chapter 3 reviews air pollution effects, providing the context for how energy use and air pollution are interrelated and why policy makers and the general public are increasingly concerned about air quality's impacts on health, the economy, and the environment. Chapter 3 also highlights sources of emissions, ambient concentration levels, and the differences between the countries in terms of what is measured and how it is measured. Appendix C provides a related discussion of a series of source-apportionment studies carried out in China. For the United States, current air quality challenges include meeting the 1997 standards for ozone and particulate, the 1999 standard for regional haze, and the revised 24-hour $PM_{2.5}$ standard. Presently, U.S. environmental management is increasingly focusing on a broad range of "emerging" issues such as toxic air pollutants (mercury and other hazardous air pollutants), the health effects of chronic exposure to low pollutant concentrations, issues related to environmental justice, fragile ecosystems, multistate and cross-border pollutants, and climate change.

Translating this information into appropriate policies and actions is generally left to government agencies guided by overarching statutory frameworks. Thus, Chapter 4 looks at the institutional and regulatory frameworks in each country, exploring the relationship between local, regional, and national monitoring and regulation and the impact that regulation has on energy use and air quality. It highlights the differences between the two countries' capacity and approaches to air quality management (AQM), and offers lessons learned from nearly four decades of AQM experience in the United States. In addition to the regulatory aspects, Chapter 4 addresses the capacity for research and development (R&D) in air pollution abatement. Slowly, R&D capacity in this and in related areas is being developed at select universities and research institutes in China—but again there are some useful examples from the U.S. experience. As energy use and air quality challenges change, R&D capacity will be increasingly important. A central theme throughout this chapter is the importance of data availability and scrutiny and, thus, Appendix A of this report provides a series of useful web-based resources focusing on energy and air pollution.

Following Chapters 2-4, which cover the broad issues of energy and air quality management, the next three chapters focus on interventions which offer promise to better meet energy needs without compromising air quality. Chapter 5 details both countries' experiences with improving energy efficiency; it examines supply-side and demand-side options for improved efficiency, which benefits air quality and energy security at low cost. Reducing energy intensity, or the amount of energy required to produce a unit of measure (e.g., GDP), has been one of the goals of Chinese policymakers for some time and, although the United States has improved in a number of sectors, U.S. policymakers have not paid sufficient attention to energy efficiency.

Owing to the importance of coal in both countries, in terms of consumption and resultant emissions, Chapter 6 is devoted to a closer look at coal combustion technologies and pollution controls. China has made progress in controlling its power sector air pollution by using modern, state-of-the-art plants to meet new capacity requirements. As an example, World Bank lending in this sector was accompanied by strict environmental oversight of local efforts by environmental authorities to control emissions and to procure pollution-control equipment. This has resulted in substantial reductions in both SO_2 and particulate emissions from modern power plants (Jia et al., 2000). The remaining challenges are for power plant operators to continue to operate downstream pollution-control equipment as mandated, and to retrofit or decommission old installed coal combustion capacity. After decades of mostly gas-fired power plant construction, the United States has shown renewed interest in developing coal-fired capacity and, thus, there are important opportunities for the countries to collaborate on clean coal technologies.

Renewable energy technologies and hydrogen are considered separately in Chapter 7, as they hold potential to dramatically influence the energy/air pollution scenario. Currently, renewable technologies serve mostly niche applications, but these applications have served, for example, to electrify remote or off-grid areas, and can substantially reduce emissions near populated areas vis-à-vis conventional power sources. Recent attention has also focused on renewable energy technologies for liquid fuel production (biofuels), which offer the opportunity to increase energy security by reducing dependence on foreign imports, while simultaneously decreasing air emissions relative to conventional fuel sources. Other potential alternative energy sources, such as oil shale, are not closely examined within the body of the report, but are discussed in Appendix B.

THE ROLE AND IMPACT OF URBANIZATION

There are many reasons that this report focuses on the role of cities. First, the U.S. population is overwhelmingly urban; by the 1950s the rate had surpassed 65 percent and has been increasing steadily ever since. By contrast, China's population is still less than half urban, but at the same time, its rapid urbanization, encouraged by the central government, has been described as the largest mass migration in human history. This migration is underpinning the industrialization and economic growth taking place in Chinese cities, and though it presents new challenges in terms of energy use and associated air pollution, it also provides new opportunities for sustainable consumption and improved air quality management.

It is important to have a clear understanding of how a "city" is defined in each country. In the United States, "city" is primarily a legal term meaning an urban area with a degree of autonomy (i.e., a township), rather than meaning an entire large settlement (metropolitan area). China has a more precise definition and classification system for its cities. Most of China's 670 cities are either prefecture-level or county-level cities, the distinction being that the former have more administrative power. However, in both cases, these cities are comprised of an urban district as well as of surrounding rural or less urbanized districts. In this respect, they are more similar to counties in the United States. These urban areas are the loci of energy-intensive industries, automobiles, and high concentrations of residents, and they, therefore, render urban AQM uniquely challenging.

As a consequence, this report does not pay special attention to some of the broader energy and air quality challenges that China is currently facing. In particular, traditional biomass combustion continues to be the primary source of energy for hundreds of millions of rural residents, and the associated health impacts of this combustion (especially indoors) are well known. Agricultural burning is a practice that is widespread in China, and the resulting pollution is not confined to rural areas and can impact nearby urban districts. Dust storms are another important source of air pollution and are a result of increased desertification. These dust storms have affected air quality regionally within Asia as well as globally; regions in the western United States are limited in their ability to achieve visibility goals, due to dust transported from Asia (NRC, 2001). Mitigation measures will require concerted national and global action (NRC, 2004).

Long-range pollution transport is an important regional and global challenge and a relatively new focus in atmospheric science (Akimoto, 2003). Both dust and pollutants related to energy use can be traced, through the use of transport modeling, to sources thousands of kilometers away (e.g., Seinfeld et al., 2004). Similarly, airborne measurements and satellite imagery are increasingly being used to observe intercontinental pollutant transport (Jaffe et al., 1999; Wilkening et al., 2000; Huntreiser et al., 2005). Understanding the contribution of pollutant transport will be another key element in each city's strategy for managing urban air quality.

CITIES AS EXAMPLES

The United States and China represent two vastly different countries with different levels of economic development, institutional priorities, and regulatory frameworks. Recognizing these differences, there are important similarities as well. In terms of energy, both countries possess abundant coal reserves, but are dependent upon petroleum and other imports to meet many of their energy needs. Moreover, though electrification rates and levels of automobile use may differ between the two countries, the technologies employed are mostly similar and thus directly comparable. Regarding AQM, China's institutional and regulatory capacity is still years behind the United States, but successful U.S. approaches such as SO_2 emissions trading are already being copied in China.

Alongside these differing levels of development and institutional capacity are differing levels of risk tolerance. In other words, the United States is now focusing more attention on the lower-risk air toxics, while China is still grappling with controlling major pollutants. The U.S. Environmental Protection Agency's (EPA's) first administrator, William D. Ruckelshaus, credits a National Academy of Sciences report on risk assessment with influencing the agency's transition to risk-based decision making in the 1980s—which aided it in setting priorities (NRC, 1983; EPA, 1993). Conventional wisdom suggests that as China's economy continues to grow, and in particular, as the middle class expands, China too will increasingly focus on additional pollutants. Still, China is facing increasing international pressure over its emissions, including those it does not currently regulate, such as mercury and CO_2. In some of China's more developed coastal cities, risk tolerance is already changing, as the informed middle class becomes more aware of the research and activities taking place in the developed world. Thus, China finds itself in a position where it is being challenged to do it all at once; and while this may not always be feasible, there are lessons from certain cities which may bear repeating, as well as certain win-win opportunties (e.g., energy efficiency) which effectively satisfy numerous objectives.

Some of the challenges posed by these energy consumption and air quality issues necessarily require management at the national level, which is reflected in a number of the chapters herein. However, this study is intended to assist Chinese cities, many of which are decentralized to a degree not common in the United States; therefore, it was important to also focus on lessons learned at the local level, in order to inform the hundreds of developing cities in China. Admittedly, one cannot encompass the multitude of variables characterizing the nearly 700 cities in China by selecting only two, nor can one convey the breadth of experience with managing energy and air quality in the United States by selecting two cities. Still, the cities of Pittsburgh (Chapter 8) and Los Angeles (Chapter 10) were chosen to be illustrative, based on their well-known experiences: Pittsburgh, on the one hand, is a previously heavily polluted industrial city seeking to continue to modernize its economy without irreparably degrading the environment, while Los Angeles, on the other hand, is a widely sprawled port city fraught with a continuously increasing vehicle fleet, and a particularly challenging local topography for shedding pollution and a heightened need to manage pollution on a regional basis.

Pittsburgh represents the historically industrial U.S. cities. Its air pollution problems were well known by the early 20th century, and as such, its efforts to ameliorate this pollution and address more recent challenges (e.g., ozone) are also well documented. Moreover, Pittsburgh's reliance on coal, combined with its strong industrial roots, make it a city to which hundreds of Chinese cities might relate. Los Angeles, on the other hand, represents the more "modern" U.S. city, though its industrial roots have also had an impact on its air pollution history. Los Angeles' air quality is similarly well studied and documented, and its profile is in many ways similar to that emerging in dozens of coastal and major cities in China, which are already experiencing the effects of a large fleet of personal vehicles.

Huainan (Chapter 9) and Dalian (Chapter 11) were selected as case study cities in China, as they somewhat mirror the circumstances of Pittsburgh and Los Angeles, respectively, and they are also broadly representative of a number of other cities in China and elsewhere in the developing world. Huainan is a major coal and industrial base in east-central China. Like Pittsburgh, it is benefiting economically from proximity to abundant coal resources; and as industries and power generation have increased in recent years, so too have pollution levels. It has, however, made important strides in improving air quality and in reducing coarse particulate matter, while still growing its economy. Dalian, a coastal city in northern China, has a more diversified economy and is now beginning to face the challenges of increased motorization (with nearly 500,000 cars on the road). It has long been considered a model city in China for its environmental management, and thus its successes may be replicable in other cities.

The United States and China have benefited from increased cooperation in the last three decades on numerous issues of mutual interest and while, due to their many differences, lessons learned on one side may not be directly applicable to the other, we believe comparative studies such as this one are still of considerable benefit in creating a broader understanding and in informing future decisions.

REFERENCES

Akimoto, H. 2003. Global air quality and pollution. Science 302(5651):1716-1719.

EPA (Environmental Protection Agency). 1993. William D. Ruckelshaus: Oral History Interview, January.

Huntreiser, H., J. Heland, H. Schlager, C. Forster, A. Stohl, H. Aufmhoff, F. Arnold, H.E. Scheel, M. Campana, S. Gilge, R. Eixmann, and O. Cooper. 2005. Intercontinental air pollution transport from North America to Europe: Experimental evidence from airborne measurements and surface observations. Journal of Geophysical Research 110:D01305.

Jaffe, D., T. Anderson, and I. Uno. 1999. Transport of Asian air pollution to North America. Geophysical Research Letters. 26(6):711-714.

Jia, L., B. Baratz, and J.J. Fritz. 2000. Environmental Performance of Bank-Financed Coal-Fired Power Plants in China. East Asia Environment and Social Development Unit, The World Bank, February.

NRC (National Research Council). 1983. Risk Assessment in the Federal Government: Managing the Process. Washington, D.C.: National Academy Press.

NRC. 2001. Global Air Quality: An Imperative for Long-Term Observational Strategies. Washington, D.C.: National Academy Press.

NRC. 2004. Air Quality Management in the United States. Washington, D.C.: The National Academies Press.

Seinfeld, J.H., G.R. Carmichael, R. Arimoto, W.C. Conant, F.J. Brechtel, T.S. Bates, T.A. Cahill, A.D. Clarke, S.J. Doherty, P.J. Flatau, B.J. Huebert, J. Kim, K.M. Markowicz, P.K. Quinn, L.M. Russell, P.B. Russell, A. Shimizu, Y. Shinozuka, C.H. Song, Y. Tang, I. Uno, A.M. Vogelman, R.J. Weber, J.H. Woo, and X.Y. Zhang. 2004. ACE-ASIA: Regional Climatic and Atmospheric Chemical Effects of Asian Dust and Pollution. Bulletin of the American Meteorological Society, March.

SEPA (State Environmental Protection Administration). 2007. Report on State of the Environment in China 2006.

UN (United Nations). 2005. World Urbanization Prospects 2005 Update, UN Population Division.

Wilkening, K., L.A. Barrie, and M. Engle. 2000. Atmospheric science: Trans-Pacific air pollution. Science 290(5489):65-66.

2

Energy Resources

This chapter summarizes the major sources and consumption of energy for the United States and China, as well as corresponding energy forecasts. Both countries' energy profiles are presently dominated by hydrocarbon resources, and large-scale changes in the system are difficult to implement quickly. Traditional biomass also constitutes an important source of energy and of emissions throughout much of China, but not in the United States—this is not well represented in national inventories and is discussed separately in Chapter 7. This chapter focuses on the current major energy resources for each country. It is not intended to be an authoritative energy review, but the context is useful for comparing the resources that each country possesses, some of the factors at play which influence energy prices and consumption, and the dynamic tensions between a desire for energy security and clean air.

MAJOR ENERGY RESOURCES

The United States and China are no longer energy independent, and in a globalized economy, one country's energy consumption can have a dramatic impact on the other's policy, as well as on world prices. As will be explored, both countries possess domestic reserves (most notably coal) but changing demands, dwindling supplies, and concerns over emissions all factor into each country's distinct energy scenario. Fossil fuels constitute a large majority in both countries and will continue to do so, though renewable sources and cleaner alternatives are poised to contribute a slightly higher percentage of total energy in the next 25 years (EIA, 2006b). Figure 2-1 shows the relative energy consumption in China

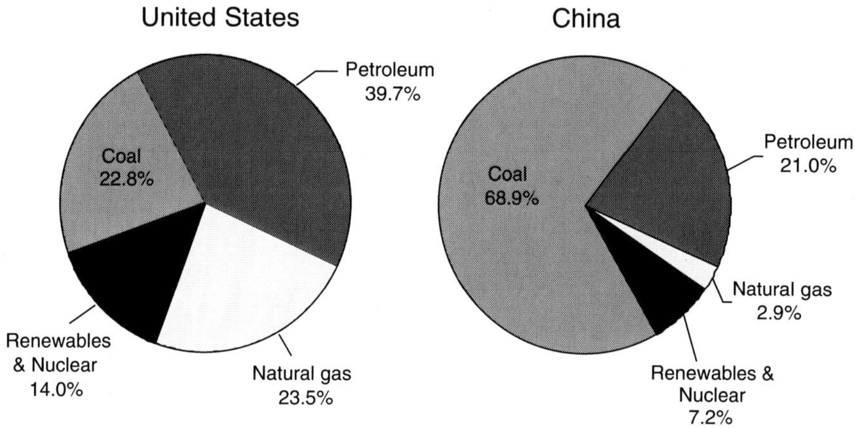

FIGURE 2-1 Primary commercial energy consumption by fuel type.
NOTE: China's nuclear power production represents less than 1 percent of total consumption.
SOURCES: EIA, 2006a; NBS, 2005a.

and the United States by fuel type. This figure illustrates some discrepancies in the fuel consumption of the two countries:

- More than two-thirds of China's energy consumption is derived from coal, whereas the United States derives less than a quarter of its energy from coal.
- The United States relies on natural gas for 24 percent of its energy, whereas China relies on it for only 3 percent of its energy.
- Petroleum supplies 39 percent of U.S. energy needs, but only 21 percent of China's.

Energy data on supplies, consumption, and future projections are largely dependent upon official statistics. In addition to their utility in forecasting trends and in making energy policy adjustments, energy data are also critical in developing air pollution mitigation strategies. A projected increase in coal consumption signals a need for action to address potential increases in SO_2 and CO_2 emissions. Emissions are generally estimated using statistics from the International Energy Agency (IEA), or official statistics from a country (Akimoto et al., 2006). Much of the data presented in this chapter come from official national sources (the Energy Information Administration (EIA) for the United States and the *China Energy Annual Review*). It should be noted, however, that China's National Bureau of Statistics has recently adjusted energy consumption statistics for 2001-2004, and that statistics from 1996-2002 have been called into question and were likely underreported (Sinton and Fridley, 2003; Tu, 2006; Akimoto et al., 2006). Furthermore, traditional biomass, which does not typically reach commercial markets, is not accurately captured in national energy statistics. Though it is associated with rural use, it does play a significant role in China in urban use (e.g., wood-burning stoves) and perhaps more significantly, activities such as agricultural burning may often take place at or near the urban periphery and subsequently affect urban air quality.

United States

Coal

U.S. coal resources are immense. They account for over one-quarter of the world's recoverable coal, more than for Russia and over twice that of China. This compares to the U.S. oil reserves at 2 percent of the world's total, and natural gas at 3 percent. Coal estimates have not been updated since the 1970s, and a reassessment could reveal an even greater coal resource base. In any case, the Department of Energy's (DOE's) estimate of 497.7 billion short tons of coal (over 13,000 EJ) as a demonstrated reserve base (DRB) is a good preliminary estimate for available U.S. coal reserves that will ultimately be recovered.

Coal can be delivered by rail, barge, or truck to almost any location in the United States. Coal's high energy density, ease of transport and storage, widespread abundance, and low cost per energy unit make it a potentially important feedstock for producing liquid fuels, in addition to its use as a solid fuel. While there are substantial coal reserves in numerous states, production comes primarily from existing production regions such as the Powder River Basin, the Rocky Mountains, the Illinois Basin, Central Appalachia, Northern Appalachia, the Great Plains, and Texas.

In 2005 the United States produced 1.13 billion short tons (31 EJ) of coal, second only to China. U.S. coal fields are vast, diverse, and well distributed across the country. DOE reports coal deposits of one or more types or ranks (bituminous, sub-bituminous, lignite, and anthracite) in 33 states. Approximately 21 percent of U.S. coal deposits lie in the Appalachian region, 32 percent in the Interior region,and 47 percent in the Western region. They are found in four major types, also known as "rank." Anthracite comprises approximately 1.5 percent of the DRB, bituminous 53 percent, sub-bituminous 37 percent, and lignite 8.5 percent. Most of the reserve base (68 percent) is recoverable by underground methods, and the rest with surface mining.

Petroleum

The United States was endowed with huge reserves of petroleum, which underpinned U.S. economic growth during the 20th century. However, growing U.S. demand resulted in the peaking of U.S. oil production in the lower 48 states

BOX 2-1
Petroleum Refining Capacity in the United States

U.S. refineries are currently operating near capacity (93 percent) and this is projected to rise to 95 percent by 2030 (EIA, 2006b). No new refineries have been built in the United States since 1976, although substantial expansions and capacity additions have occurred at existing sites. The reasons for this are difficulties in obtaining regulatory permits for expansions and new construction, as well as a lack of investment. This is a critical limitation, since, as demand for refined products grows, the United States will only be able to import as much crude oil as it can refine. Refineries are typically located near crude oil production sites, or alternatively, where demand for refined petroleum products is located (e.g., near major metropolitan areas), which is another challenge to building new refineries. Demand for refined products is expected to outpace domestic capacity increases, leading to a rise in refined petroleum product imports, as Eastern Europe and Asia in particular develop their capacity to meet stringent U.S. standards and demand.

in the early 1970s and in Alaska during the 1980s. With relatively minor exceptions, U.S. oil production has been in continuing decline ever since. Because U.S. demand for petroleum products continued to increase, the United States became an oil importer. The United States currently depends on foreign sources for more than 60 percent of its needs, and future U.S. imports are projected to continue to increase (EIA, 2006c).

U.S. oil production is currently at a record low and has been steadily declining since 1986 (EIA, 2006a). Petroleum production in the lower 48 states decreased from 9.0 million barrels per day (MM bpd) in 1973 to 7.5 MM bpd in 1978. Only the development of oil in Alaska prevented a steep decline in overall production. Production from Prudhoe Bay came on line in significant volumes causing Alaskan production to increase from 464,000 bpd in 1978 to 1.6 MM bpd in 1980 and to peak at 2 MM bpd in 1988. Thus, there were modest gains in overall production that extended from 1980 through 1985.

The U.S. dependence on foreign oil reached a record high in 2005, following a slight decline after September 11, 2001. The U.S. dependency on foreign oil has increased steadily since 1986. Major changes in imports are usually related to changes in the U.S. economy and the U.S. oil production. At the time of the October 1973 Oil Embargo, the United States received a little less than 35 percent of its petroleum supply from imports. In response to higher prices, total petroleum demand declined from 15.8 MM bpd in 1973 to 14.9 MM bpd in 1975. Thereafter, the increase in oil consumption resumed, and by 1978 total petroleum consumption averaged 17.1 MM bpd, 8 percent higher than in 1973. Imports as a percentage of petroleum supply increased at a more or less steady rate, from 35 percent in 1973 to approximately 42 percent in 1978, and exceeded 50 percent in several months.

Dependence on imports grew faster than consumption. The rapid increase in international oil prices starting in late 1978 led to the only sustained period of declining U.S. dependence on imports in the post-1973 Embargo period. Two factors contributed to the decline: higher U.S. oil production and lower consumption caused by substitution, conservation, increased efficiency, and fuel switching. Dependence on imports declined to 27 percent in 1985—the lowest percent in the past four decades. With the oil price collapse in 1986, import dependence once again resumed its upward path to 55 percent in 2001, and to more than 60 percent in 2005, due to falling domestic production and to ever-increasing demand for transportation fuels. Gasoline, diesel, and jet fuel account for most of the increase in petroleum consumption. By 2030, oil imports are forecast to increase to more than 17 MM bpd (EIA, 2006c).

Natural Gas

Natural gas is a critical source of energy and of raw material, permeating virtually all sectors of the U.S. economy. It supplies nearly 25 percent of U.S. energy, generating about 19 percent of electric power, supplying heat to over 60 million households, and providing over 40 percent of all primary energy for industries.

North America is moving to a period in its history in which it will no longer be self-reliant in meeting its growing natural gas needs; production from traditional U.S. and Canadian basins has plateaued. Traditional North American producing areas are expected to provide about 75 percent of long-term U.S. gas needs, but will be unable to meet projected demand. New, large-scale resources such as liquefied natural gas (LNG) and Arctic gas are available and could meet 20-25 percent of demand but are higher-cost, have longer lead times, and face major barriers to development.

Given depletion rates in North American fields, the sources of natural gas supply must change significantly to meet demand growth of more than 4.6 Tcf[1] (5.2 EJ) in only two decades. EIA projections indicate that more than 75 percent of all new incremental demand must be met by a 580 percent increase in LNG imports—increasing such imports to 4.1 Tcf (4.6 EJ). To put such a large amount of LNG in perspective:

- 4.1 Tcf is greater than the entire 2004 natural gas production of the Gulf of Mexico (4.0 Tcf).
- 4.1 Tcf is the Btu equivalent of importing more than 700 million barrels of oil.

[1]One trillion cubic feet (Tcf) is equivalent to 0.0283 trillion cubic meters (Tcm) natural gas.

BOX 2-2
Natural Gas and Electricity

EIA projects that by 2015 more than 22 percent of U.S. power generation will be natural gas (NG) (EIA, 2007). This focus on NG-fueled power plants to meet incremental demand for electricity has had serious implications for the U.S. economy:

- Natural gas plants are increasingly part of baseload generation, especially in states such as Texas, California, and Florida, where NG now supplies more than 40 percent of the electricity.
- Dramatic NG price increases can be directly attributed to meeting baseload demand with NG.
- The domestic competition between various sectors of the economy is especially serious, due to declining production of NG.
- The high cost of NG has resulted in tens of thousands of megawatts of capacity sitting idle because it is too expensive to operate.
- Reserve capacity is increasingly based on NG plants, which greatly increases the vulnerability of the electric supply system to outages and supply shortfalls.

To utilize idle natural gas combined cycle (NGCC) plants, it may be necessary to convert some of them to coal. However, conversion involves many financial, environmental, performance, and technical issues, and the conversion itself involves the alteration of the combined cycle power equipment to utilize the lower-Btu fuel from coal. Conversion also requires capital investment for the turbine modifications and the gasification plant.[a] Fuel switching may also require the renegotiation of environmental permits and the reopening of public discussion—and local public and infrastructure impacts from coal transport also may be an issue. For those locations where the NGCC plant was established primarily for environmental reasons, the difficulty of obtaining a permit to repower may increase. However, for coal gasification plants, emissions are normally well within the ranges of NG and are within Best Available Control Technology limits—a benchmark in the permitting process.[b]

---

[a]If the gasification facility is financed and constructed as a separate fuel-gas supply entity, the overall cost of the produced fuel gas can be as low as 35-40 percent of current NG prices.

[b]A prime consideration in the conversion decision is the accessibility and availability of coal supply. The site must accommodate the logistics of coal delivery, off-loading, coal preparation, and storage of coal, reagents, by-products, and sulfur, and in some cases new environmental permits will be required.

- 4.1 Tcf at the 2005 (January-June) average LNG cost of $6.46 per Mcf would cost the United States at least $27 billion per year, in addition to the current cost of more than $200 billion for oil imports.

The United States has four operating LNG terminals, and a number of proposals for new terminals have been advanced. However, the construction of new terminals demands state and local approvals, and because of environmental concerns and fear of terrorism at LNG facilities, a number of the proposed terminals have been rejected. There are also objections from Mexico, which has been proposed as a host for LNG terminals, to support west coast natural gas demands (Flalka and Gold, 2004). Alternatively, some are considering locating LNG terminals offshore with gas pipelined underwater to land; related costs will be higher, but safety would be enhanced.

While hopes of meeting future demand have turned to LNG imports,[2] LNG presents the same economic cost and national security problems as imported oil. Efforts to import massive amounts of LNG will take time, cost money, and could result in unforeseen consequences. Thus, while LNG is a promising source of new supply, prudent planning suggests the parallel pursuit of other alternatives, given the large number of unanswered questions that surround LNG.

The experience with North American natural gas represents a dramatic example of the risks of overreliance on geological resource projections. Natural gas supplies had been plentiful at real prices of roughly $2/Mcf for almost two decades, and became the fuel of choice for new electric power generation plants. Part of its attractiveness was resource estimates for the United States and Canada that promised growing supply at reasonable prices for the foreseeable future. However, the United States is now experiencing supply constraints and high natural gas prices. Supply difficulties are almost certain for at least the next decade.

Nuclear

Nuclear energy is the second-largest source of electricity in the United States after coal and is the largest emission-free source of electricity. The United States has more than 100 licensed nuclear plants that have a capacity of more than 97,000 megawatts (MW), and they provide more than 700 billion kilowatt-hours (kWh). At present, almost every U.S. home, business, and industry receives part of its electricity from nuclear power plants through a nationwide, interconnected transmission system.

No nuclear power plant has been ordered in the United States since 1978, and the last nuclear power plant to be completed came on line in 1996. In recent years, however, electricity supplies have become increasingly tight and the nuclear power option is currently being re-examined. There is government support for

---

[2]The Alaska natural gas pipeline is at least 10 years from operation, maybe longer.

new construction, so that, within a few years, the United States could begin constructing new plants. A number of utilities have already initiated site planning for new plants.

Average nuclear production costs are declining and have been for more than 10 years. Furthermore, the deregulated, competitive electric generating business creates a powerful business incentive to keep a nuclear plant operating beyond its initial 40-year licensing period since, with deregulation, a fully depreciated nuclear plant is a valuable asset that can sell energy at marginal cost.

The average capacity factor (a measure of utilization) of U.S. nuclear plants has improved steadily. In 1999, it reached a record high of 86.8 percent, increasing from 67.5 percent as recently as 1990, and has continued to gradually increase since then—it is currently about 90 percent. Nationally, each percentage point increase in capacity factor is roughly equivalent to bringing another 1,000 MW of generating capacity on line.

Nevertheless, nuclear energy's future in the United States is uncertain. An especially difficult problem is the long-term storage of high-level nuclear waste; efforts to site a centralized facility at Yucca Mountain, Nevada, have been stalled for more than a decade. Nuclear power also suffers from problems relating to health and safety issues, potential accidents, and other concerns. Further, many environmentalists and special interest groups are strongly opposed to any expansion of nuclear power. Finally, the cost-competitiveness of the proposed new nuclear power plants is not clear.

China

Coal

Coal is much more abundant than other fossil fuels in China (see Figure 2-2). Because of coal's relative abundance, China's reliance on fossil fuels, and an emphasis on sustained economic development, coal will continue to be the dominant source of energy in China. China has not only recognized the strategic significance of its coal resources, but is acting aggressively to realize the full potential of this multi-use fuel and feedstock. Coal is a primary fuel source for the production of electricity and steel, and China has also taken the lead with regard to coal-to-liquids and coal gasification initiatives.

Based on the reports of the Ministry of Land and Resources (MLR), coal reserves at depths $\leq$ 2000 m are estimated at over 5.5 trillion tons (168,000 EJ). This includes predicted recoverable reserves of over 4.5 trillion tons (138,000 EJ), with proven recoverable reserves of 204 billion tons (6200 EJ). But because there is not enough recent exploration of the coal reserves, at present the proven reserves total just 18 percent nationwide, and only 4 percent in western China. The existing proven reserves cannot meet the demand of large-scale coal development. Coal resources are mainly concentrated in the north and northwest of China. The

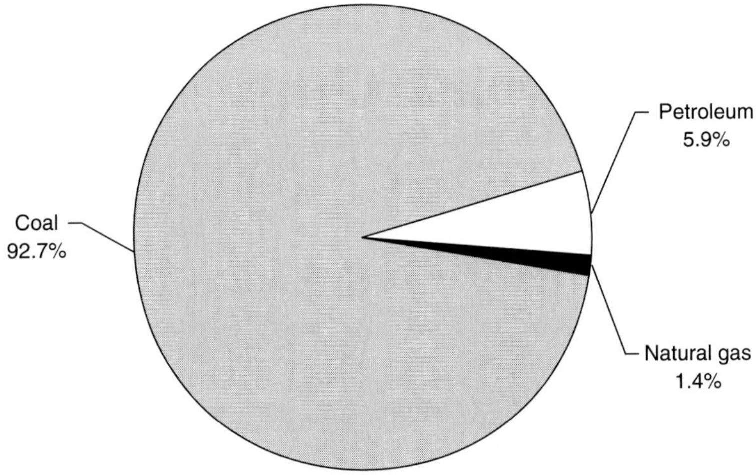

FIGURE 2-2 Recoverable fossil fuel resources by fuel type (in terms of EJ).
SOURCE: Liu, 2002.

proportion of coal resources in northwest China is 47 percent, but the verified rate is 30 percent, and the exploitable rate is less than 15 percent. The coal resources in northern China rank second with 39 percent and a verified rate of 58 percent. Transporting this coal presents additional challenges (Box 2-3).

Coal mining operations will likely shift west in the future. Coal demand is secure, as it provides 75 percent of electric power, 60 percent of chemical industrial fuel, and 80 percent of industrial fuel overall. After a slight decrease of the production and consumption of coal in the 1990s, the production and consumption of coal began to increase in recent years, because of the development of China's economy. The production of coal was 2.19 billion tons (66 EJ) in China in 2005 (see Figure 2-3).

The Chinese coal market has undergone major changes in recent years. Government-led reform and reorganization of the coal industry has promoted the establishment of large coal mining companies. Large coal mining enterprises have taken over and upgraded small and medium-sized coal mines. In other cases small mines (mostly operated by a township as opposed to the state) were closed. As a result of this restructuring, the number of small coal mines decreased from 85,000 in 1996 to 24,000 as of 2006. The general trend has been one of consolidation, in order to expand the scale and scope of coal mining operations. In the process, mechanization rates and safety levels have been gradually enhanced. The mine rates of state-owned key coal mines, state-owned local coal mines, and town coal mines were, proportionally, 39:16:45 in 1996. By 2005 this proportion had shifted to 48:15:37.

BOX 2-3
Coal Transport

Transportation and coal quality are significant issues which impact cost, thermal efficiency, and emissions. Lignite, for example, is not suitable for transporting, as its high moisture content adds excess weight, making transporting it economically impractical. (Moisture content also decreases its energy intensity and thus its heating value.) In China, 90 percent of the coal resources are in the sparsely populated northern and western regions, practically opposite of the heavily populated and economically active regions of the south and east. This, of course, necessitates a great deal of coal transportation and, at present, about 45 percent of railway capacity is used to transport coal. In 2003, the coal transportation load was 1 billion tons; primary railway lines are basically saturated or super-saturated. Transportation by waterway has been insufficient as well. Coal transportation difficulties have created a bottleneck in China's economic development. Furthermore, the increasing cost of transporting energy resources has driven up their prices. In 2004, the price of coal in the Shanxi Coal Mine was 140 RMB/ton, with railway freight charges of 0.15 RMB/t-km and highway freight charges of 0.45 RMB/t-km. So, when transported to the East, 1,000 km away, the price of the coal would be 320 RMB/ton—more than two times the original price. Accounting for other transport-related expenditures, the price of coal could reach 400 RMB/ton for Guangzhou and Fujian provinces; in other words, higher than the international market price.

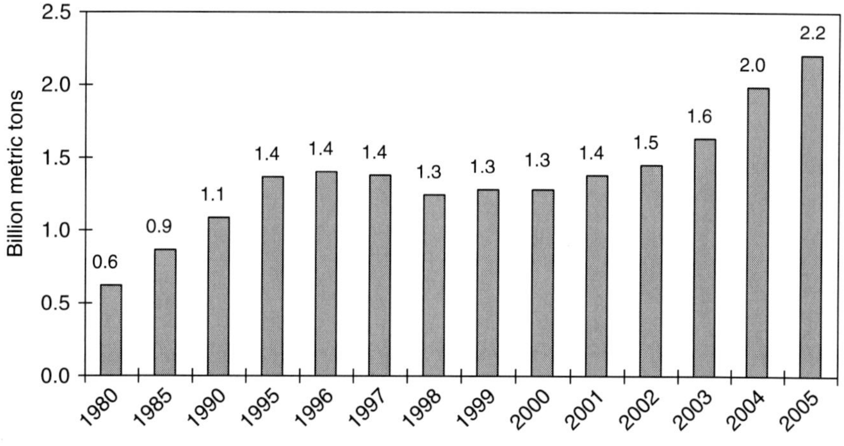

FIGURE 2-3 Coal production in China.
SOURCE: China Statistical Yearbooks (1981-2006).

Still, the prevalence of small and sporadic coal mines presents a serious challenge for China. In 2004 the top five coal companies (Shenhua, Shanxi Jiaomei, Datong, Zhongmei, and Yunkuang companies) had a market share of only 16 percent. Production from small town and village coal mines continues to increase rapidly, which has several disadvantages for the coal industry. In general, the small mines suffer from

- Production inefficiencies,
- Market price fluctuations,
- Lack of regulations,
- Increased rate of mining accidents, and
- Increased environmental degradation.

From 2001 to 2004 China's coal exports totaled approximately 80 million tons annually (see Figure 2-4), though this number decreased in 2005 to 71 million tons. The main market for Chinese coal exports is Asia, which imports about 94 percent of the total. At the same time, as a result of rising prices in the mining industry, coupled with railroad restrictions on coal transportation, China's coal imports totaled 26 million tons (0.79 EJ) in 2005. China is expected to continue exporting some coal in the future, even as it increases imports to meet its own demand. Coking coal demand in particular is estimated to require more than

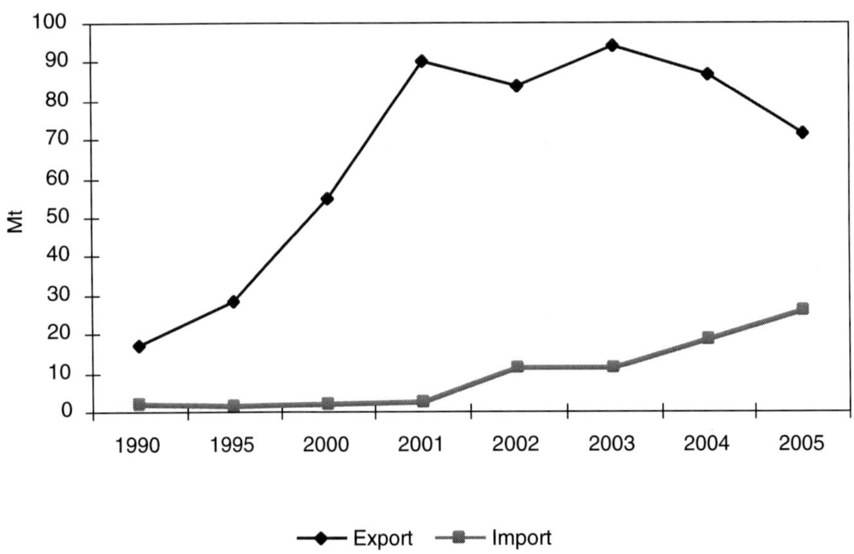

FIGURE 2-4 Coal imports and exports.
SOURCE: NBS, 2006b.

BOX 2-4
Coal to Liquids (CTL)

China considers coal liquefaction an important part of its petroleum substitution strategy. Coal to liquids (CTL) has received less attention in the United States for a variety of reasons, but could play a role in increasing national energy security. It is important to note that the commercial viability of this technology is very dependent on high oil prices (NRC, 2001). There are two basic technologies for producing liquid fuels from coal: direct and indirect liquefaction. Direct liquefaction produces a synthetic crude that must then be refined to produce gasoline and diesel fuel, whereas indirect liquefaction involves gasification of coal to produce a syngas that is then converted into liquid fuels via Fischer-Tropsch (FT) synthesis. Indirect liquefaction is a well-developed technology and has been used by the South African company Sasol for more than five decades. Indirect coal liquefaction is a three-step CTL technology: (1) coal gasification, (2) FT synthesis, and (3) FT product upgrading.

CTL plant analyses assume an output of 70 percent ultraclean diesel fuel and 30 percent naphtha, though some technologies may be able to decrease the proportion of naphtha and thus increase the yield of the higher-value diesel product. FT fuels are biodegradable, essentially zero sulfur, and have low particulate and NO_x emissions profiles (EPA, 2002). These fuels are interchangeable with conventional diesel, requiring no engine modifications, nor do they require a completely separate distribution system (as do some other alternative fuels such as ethanol).

Although coal liquefaction provides a technically feasible alternative to petroleum-based liquid fuels, its environmental impacts may preclude it from becoming a large-scale strategy, at least in the United States. In addition to the traditional concerns over coal mining and transport, CTL operations could significantly increase CO_2 emissions per gallon of fuel produced and consumed. However, measures such as co-generation (providing electricity to the local community), co-processing with locally derived waste biomass, and installing carbon capture and sequestration technologies can help reduce life-cycle CO_2 emissions to levels comparable to gasoline and diesel fuels currently in use (SSEB, 2006; Bartis, 2007; Freerks, 2007).

a seven-fold increase in imports by 2030, leaving China as a net importer of coal (EIA, 2006b). In fact, coal imports outpaced exports in the first quarter of 2007, leading some to speculate that China could become a net importer much sooner.

Petroleum

Because of progress in drilling technology and increasing petroleum demand, China's oil production is increasing. The Chinese oil and gas region is divided into six areas: eastern, central, western, southern, Tibetan, and offshore. In 2004, there were 124 basins whose oil resources totaled 102.1 billion tons (4,580 EJ). The recoverable resource of oil is 6.1 billion tons (274 EJ) (Figure 2-5). The proportion of oil onshore is 61.2 percent.

Since 1993, China has been a net importer of oil. China's oil imports totaled 117 million tons (5.3 EJ) in 2004, making up 40 percent of its supply, and this proportion has been increasing in recent years (Figures 2-6 and 2-7). Chinese oil production rose slightly in 2005, to 181 million tons (8.1 EJ), but present production increases are not able to keep pace with increasing demand, resulting in progressively more imports. The main oil wells in eastern China have entered the latter period of stable production and, therefore, increasing their production is not feasible. Oil wells in Tarim basin are not yet producing. Overall, China's rapid economic growth and rising demand for petroleum, particularly in the burgeoning transportation sector, have resulted in a correspondingly steep increase in imports. Although there are currently no official projections available of the annual increase in China's petroleum imports, by 2020 consumption may reach 500 million tons, with no projected increase in domestic production. Some projections estimate that by 2030, China's imports will have increased four-fold (EIA, 2006b).

China's petroleum refining capacity reached 270 million tons in 2004, which ranks third worldwide. From 1998 to 2004, capacity increased about 120 million tons, and over this same period, refined petroleum production increased 8.3 million tons annually. In 2004, gasoline, kerosene, and diesel made up 168 million

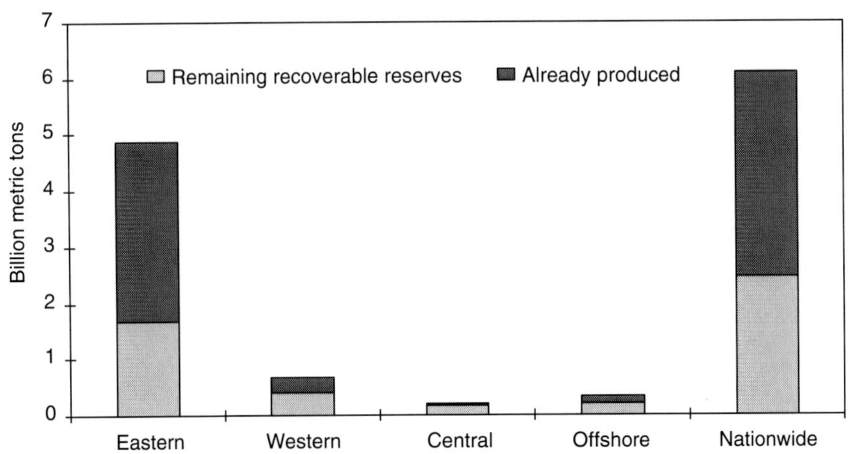

FIGURE 2-5 Recoverable petroleum resources, by location.
NOTE: Due to the scale of the figure, the much smaller identified resources in the southern and Tibetan regions have been omitted.
SOURCE: Liu, 2002.

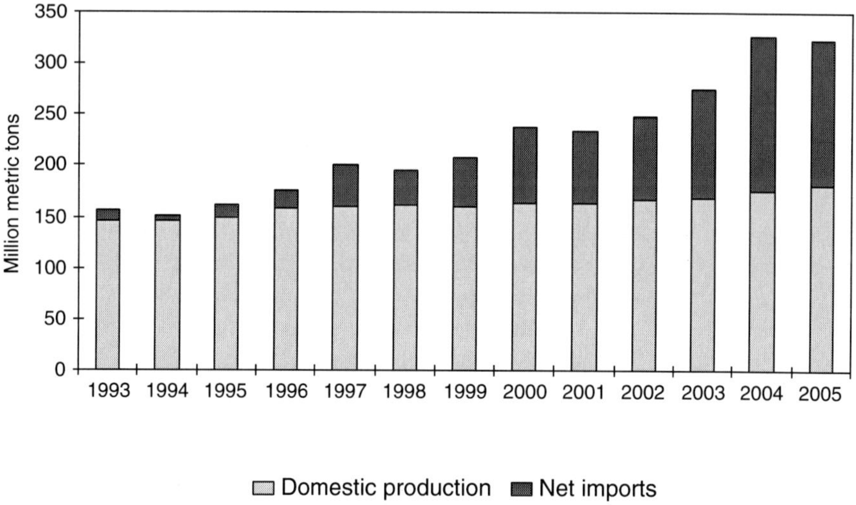

FIGURE 2-6 Petroleum supplies, 1993-2005.
SOURCE: China Statistical Yearbooks.

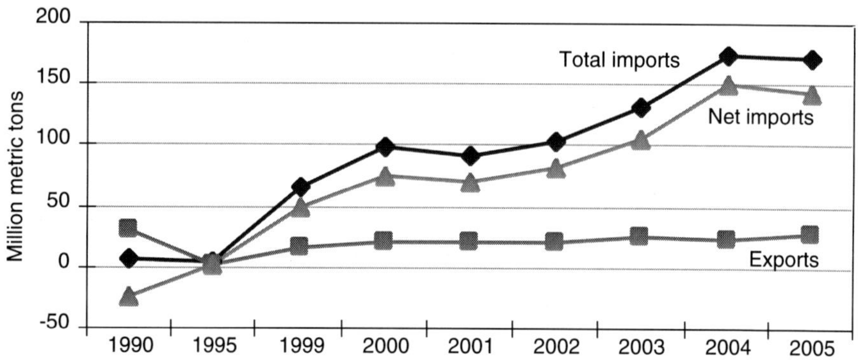

FIGURE 2-7 Petroleum imports and exports.
SOURCE: General Administration of Customs, China.

tons of this total. China plans to continue expanding its refining capacity, in order to meet the rising demand for refined products—and several projects are already under way to expand existing facilities (DOE, 2006).

Most refineries use a basic method of atmospheric distillation of crude oil; the oil is heated and fed into a tower to separate the oil into its many compounds. Increasingly, refineries must also use coking and hydrotreating, in order to comply with tighter restrictions on fuel quality. Coking breaks down heavier crude into elemental carbon, while hydrotreating removes sulfur. Refineries themselves are sources of pollution, not necessarily from their smokestacks but more often from equipment leaks otherwise known as fugitive emissions. As China greatly increases its refining capacity, this raises concern over increased toxic emissions of benzene and other chemicals, in addition to existing concerns over the potential for fires, explosions, or spills.

Natural Gas

As of 2006, China's estimated natural gas resource base was 47 trillion m^3, (1,870 EJ) with recoverable reserves of 2.45 trillion m^3 (97.5 EJ) (MLR, 2006). The average recovery ratio is slightly more than 64 percent, and only 16-23 percent of recoverable natural gas resources have been proven. China also possesses substantial coalbed methane (CBM) resources (36 trillion m^3 /1,430 EJ), though its recoverable reserves (47 billion m^3 /1.87 EJ) still constitute only a small fraction (MLR, 2006). Still, research is under way to improve methods for developing this potential resource, as well as to harness CBM, which is currently vented and released during typical coal mining operations.

China's natural gas resources are mainly located in the midwestern part of the country, as well as offshore. The main production areas of natural gas, representing more than 83 percent of recoverable resources, are Sichuan province, Eerduosi (Ordos), Talimu (Tarim), Chaidamu, Yingge Sea, and the East China Sea. Natural gas resources in the populous eastern and southern coastal parts of China are severely deficient; as is the case worldwide, the supply of natural gas deposits does not align well with the location of demand. Accordingly, in order to develop the natural gas industry, China has implemented a strategy to transport natural gas via the West-to-East pipeline, linking deposits in Eerduosi and Talimu to population centers along the coast. There are also plans to transport gas from offshore to the coastal areas via pipelines. In recent years, owing to these efforts to develop the industry, natural gas production has grown quickly, from 27.2 billion m^3 in 2000 to 50 billion m^3 in 2005 (Figure 2-8).

At present the supply and demand of Chinese natural gas are in balance. However, in order to solve the energy shortages in eastern China, the Zhujiang Delta, the Yangtze River delta, and the Fujian coast are planning to introduce LNG in quantities between 17 and 27 million tons annually.[3] Construction is under way on LNG receiving terminals in Guangdong and Fujian provinces. Negotiations have also been taking place with Russia to construct a natural gas pipeline through China, which could ultimately link Russia to South Korea as well. As demand increases into the future, China could be relying on imports to meet up to 40 percent of its demand by 2030 (EIA, 2006b).

[3]One metric ton of LNG is equivalent to 1,379 m^3 of natural gas (55 GJ).

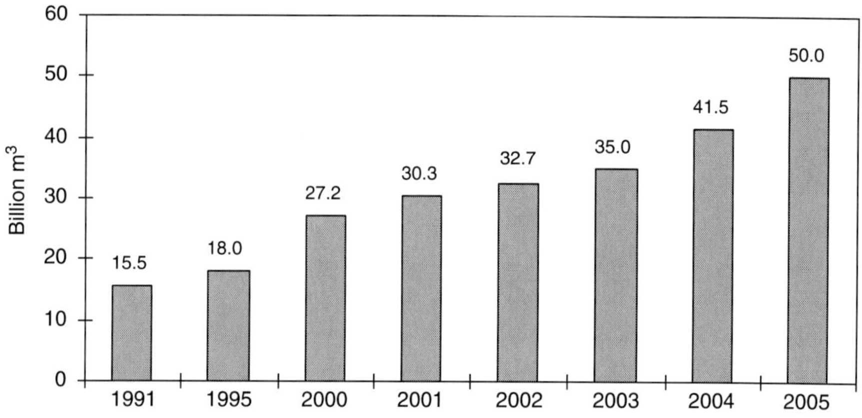

FIGURE 2-8 Natural gas production.
SOURCE: China Statistical Yearbooks.

Nuclear

Development of commercial nuclear power plants began in the late 1980s; currently there are nine nuclear power plants throughout China. In 2004, installed net capacity was 6,940 MW, producing power of 50,100 GWh, or 2.3 percent of China's total power production. In 2005, nuclear power production rose to 52,300 GWh. The domestically built Qinshan nuclear power plant has operated safely for 14 years, and research is under way to develop new generation nuclear power plant technologies. In December 2006, the DOE and China's National Development and Reform Commission (NDRC) signed an agreement which will allow the U.S.-based Westinghouse Electric Company to build four civilian nuclear power plants in China (DOE, 2006).

China has the capability to design, manufacture, and construct pressurized water reactors, although there is a still a quality gap between domestic-built reactors and internationally built reactors. In the near future, China will focus on third-generation pressurized water reactors in an attempt to achieve advanced international standards for 1,000-MW reactors. China's first high-temperature air-cooled reactor with a capacity of 10 MW became fully operational in January 2003. It was also the first block-type high-temperature air-cooled reactor experimental power plant. Huaneng Shidao nuclear power plant plans to build a high-temperature gas-cooled reactor demonstration project, slated to begin operation in 2010.

CONSUMPTION AND ENERGY FORECASTS

The following sections provide a brief overview of the major sources of energy consumption in each country, as well as corresponding forecasts. The U.S.

section also provides historical data, in order to help illustrate the changes it has undergone during the past half-century. Energy consumption in China is classified differently than in the United States; most notably, industrial consumption, which is widely acknowledged as the largest energy consumer, includes electricity consumption. As is discussed, projections are imprecise and liable to change, but they nonetheless provide additional context for the challenges that each country will face in balancing energy security with improving air quality.

United States

Coal

U.S. coal consumption and production increased in tandem over the past half-century: from 500 million short tons (13.7 EJ) per year (TPY) in 1949 to 1.1 billion TPY (30.2 EJ) in 2004 (EIA, 2006a). Consumption is projected to exceed production in about 2016, and after this date the United States will become a net coal importer. By 2030, it is predicted that the United States will be importing about 100 million TPY (2.7 EJ).

Figure 2-9 shows the history of U.S. coal use and its forecast for the next two decades. It suggests that:

- Most future U.S. coal use will be for electric power production.
- By 2020, coal use for coke plants and CTL products will begin to increase significantly.

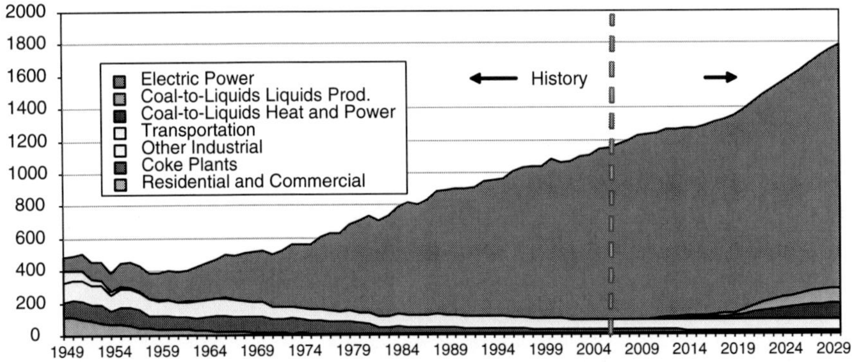

FIGURE 2-9 U.S. coal use by sector: history and forecast.
SOURCE: EIA, 2006a.

Petroleum

U.S. petroleum consumption increased from 5 MM bpd in 1949 to 20 MM bpd in 2004 and is forecast to increase to 28 MM bpd by 2030. U.S. production increased from 5 MM bpd in 1949 to 12 MM bpd in 1970, and then gradually declined to 7 MM bpd in 2004. Oil imports exceeded U.S. production after 1997, and by 2030, oil imports are forecast to total more than 17 MM bpd.

Figure 2-10 shows the history and forecast of refined petroleum products in the United States, 1949-2030. It illustrates the following:

- Motor gasoline has been the dominant product, accounting for as many barrels per day as all of the other products combined, and is forecast to continue to do so.
- Jet fuel and distillate account for about an equal share of total product, and are forecast to continue to do so.
- The other products are of relatively minor importance.

Figure 2-11 shows the history and forecast of refined petroleum products in the United States by end use, 1949-2030. It suggests the following:

- Transportation is, by far, the major end use of petroleum products, and this dominance is forecast to increase through 2030.
- The industrial sector is the second largest petroleum consumer, followed by the residential and commercial sector.
- Little petroleum is used to generate electrical power.

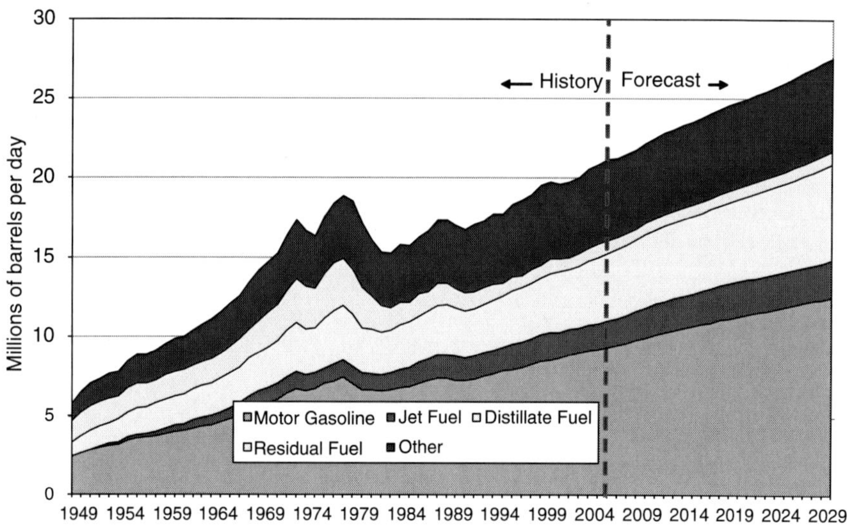

FIGURE 2-10 Refined petroleum products supplied by fuel: history and forecast. SOURCE: EIA, 2006a.

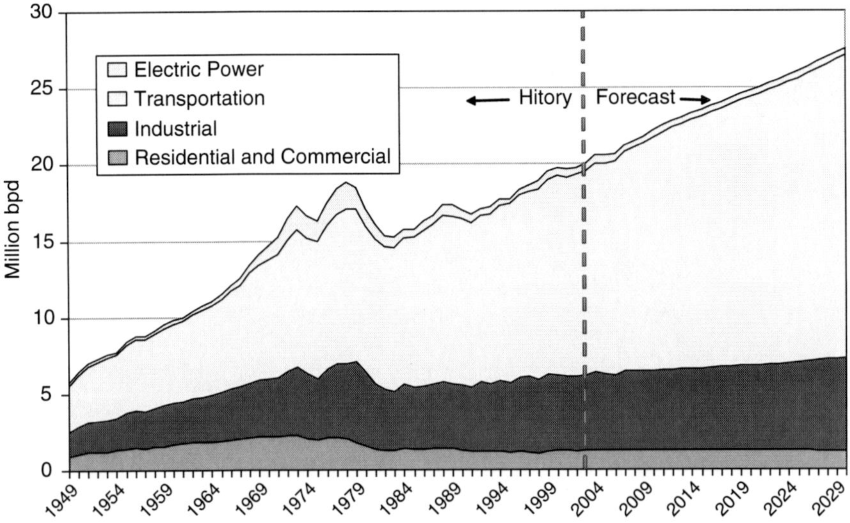

FIGURE 2-11 Refined petroleum products by end use: history and forecast.
SOURCE: EIA, 2006a.

Natural Gas

The demand for natural gas has been affected by three underlying trends (NCC, 2006). First, forecasts of natural gas supply and price were much too optimistic, and government agencies, industry associations, and energy analysts projected that natural gas would be plentiful, stable, and cheap far into the future. Second, demand increased based upon these optimistic forecasts as power plant construction and space heating steadily turned to natural gas as the preferred fuel, and demand for natural gas has steadily increased since 2000. Third, the supply of natural gas from traditional major sources is showing signs of increasing strain. It is increasingly apparent that even recent projections of natural gas production have generally underestimated these difficulties and overestimated future supply.

Natural gas consumption increased from 5 Tcf (5.6 EJ) in 1949 to 23 Tcf (25.9 EJ) in 2004, and is forecast to increase to 27 Tcf (30.4 EJ) by 2030. U.S. natural gas production satisfied consumption requirements until 1988, and imports were negligible. After 1988, imports increased rapidly to 3.5 Tcf (3.9 EJ) in 2004 and are forecast to increase to nearly 6 Tcf (6.7 EJ) by 2030.

Figure 2-12 shows the history and forecast of U.S. natural gas consumption by sector, 1949-2030. It illustrates the following:

- The industrial sector is the major natural gas consumer and is forecast to remain so through 2030.

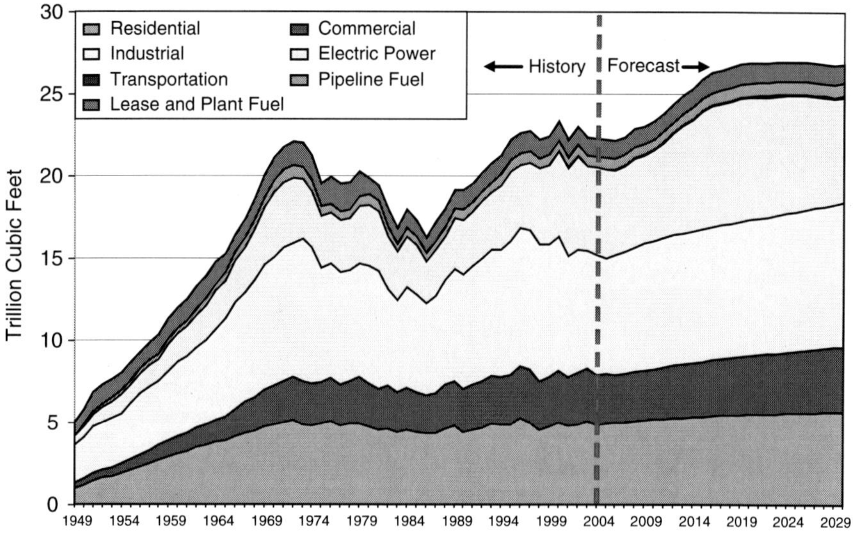

FIGURE 2-12 U.S. natural gas consumption by sector: history and forecast.
SOURCE: EIA, 2006a.

- The consumption of natural gas for electricity generation has increased continually, and this increase is forecast to continue.
- The percent of natural gas used in the residential, commercial, and transportation sectors has remained relatively constant and is forecast to remain so.

Nuclear

Figure 2-13 shows the history and forecast of nuclear power in the United States. It illustrates the following:

- Nuclear power increased very rapidly, from negligible in the 1960s, to 650 billion kWh in 1980, and to 750 billion kWh in 2004.
- Nuclear power is forecast to increase to 875 billion kWh by 2030.
- The share of nuclear power in electrical generation is forecast to increase to nearly 22 percent by 2010, and then to gradually decline to about 17 percent by 2030.

Electricity and Consumption by Sector

Figure 2-14 provides a closer look at current electrical generation by fuel type. Figure 2-15 shows the history and forecast for net generation of electrical power by fuel type. These figures illustrate the following:

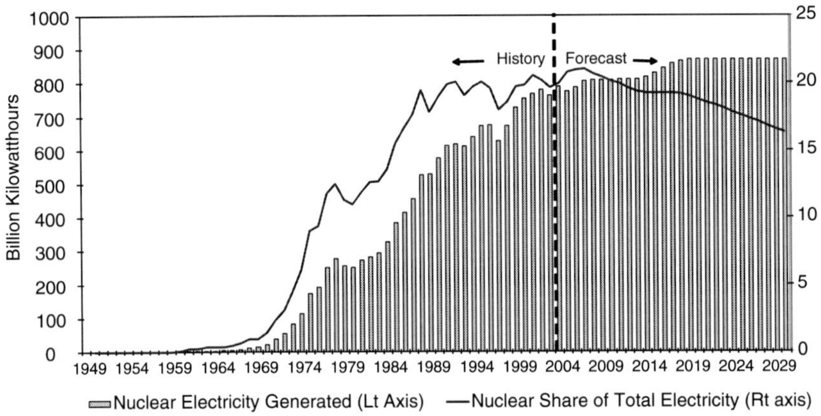

FIGURE 2-13 U.S. nuclear power generation: history and forecast.
SOURCE: EIA, 2006a.

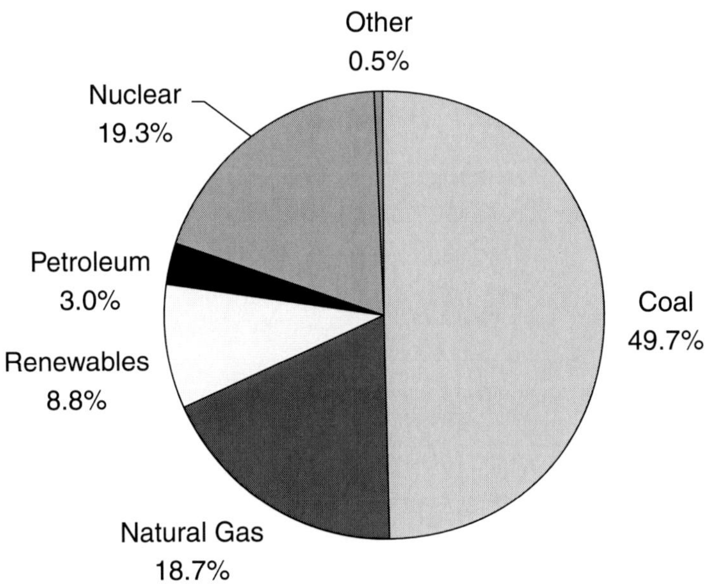

FIGURE 2-14 Generation of electrical power by fuel type, 2005.
SOURCE: EIA, 2006a.

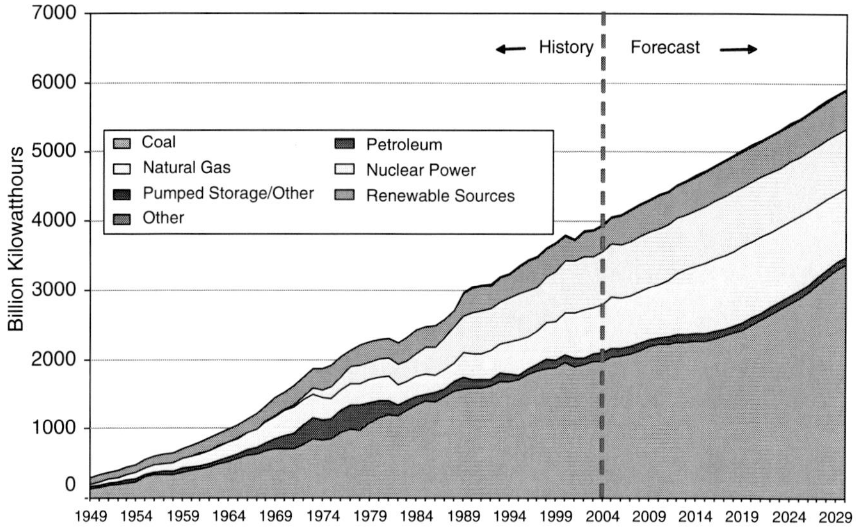

FIGURE 2-15 Net generation of electrical power by fuel type: history and forecast. SOURCE: EIA, 2006a.

- Coal accounts for nearly half of electrical generation.
- Nuclear energy accounts for 20 percent of electrical generation.
- Natural gas accounts for 18 percent.
- Renewables (mostly hydro) account for 9 percent and petroleum accounts for 3 percent.
- Total electrical generation is projected to continue its steady increase, bolstered by an increase in generation from coal (and natural gas to a lesser extent).

Figure 2-16 shows the history and forecast of U.S. electricity sales by sector, 1949 to 2030. It illustrates the following:

- The industrial sector consumed more electricity than any other sector until the mid-1990s, at which time residential consumption exceeded industrial consumption.
- By the late 1990s, commercial electricity consumption also exceeded industrial consumption.
- Forecasts suggest that residential and commercial electricity consumption will continue to increasingly exceed industrial consumption.
- By 2015, commercial consumption is projected to exceed residential consumption, and the difference will increase through 2030.

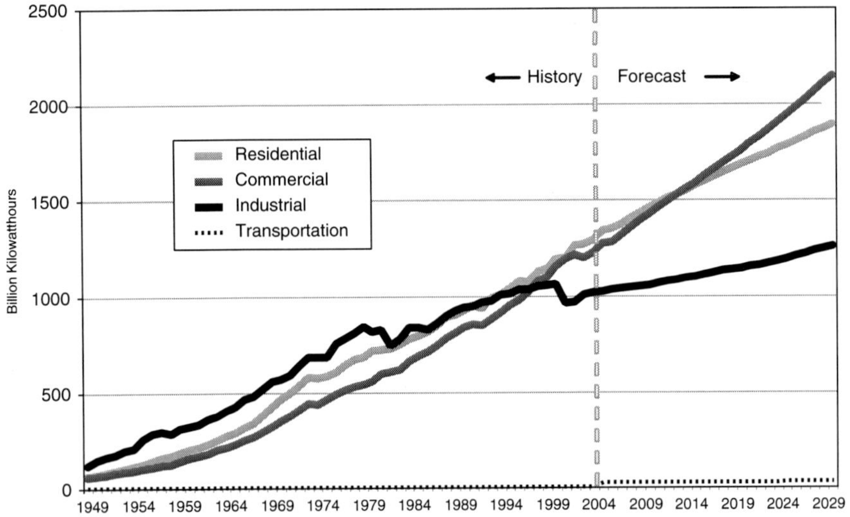

FIGURE 2-16 U.S. electricity sales by sector: history and forecast.
SOURCE: EIA, 2006a.

Figure 2-17 shows the history and forecasts of U.S. energy consumption in the commercial sector by source, and suggests that

- Use of electricity and natural gas will continue to increase at the expense of coal and petroleum.
- System energy losses are, and will continue to be, larger than energy use.

In 2004, over half of commercial-sector energy was lost due to electrical system losses. Electrical energy comprised 24 percent of total consumption, followed by natural gas at 18 percent and petroleum at 4 percent. The shares of coal and renewables were negligible.

Figure 2-18 shows the end use of energy in the U.S. commercial sector in 2004. It indicates that

- The commercial sector consumed 17.5 quads (18.5 EJ).
- The largest single end use is for lighting, followed by space heating, space cooling, and water heating.
- Other uses (such as telecommunications equipment, ATMs, service station equipment, etc.) account for 37 percent of commercial-sector end use.

In 2004, electricity system losses accounted for 46 percent of residential fuel consumption. Natural gas accounted for 24 percent and electricity accounted

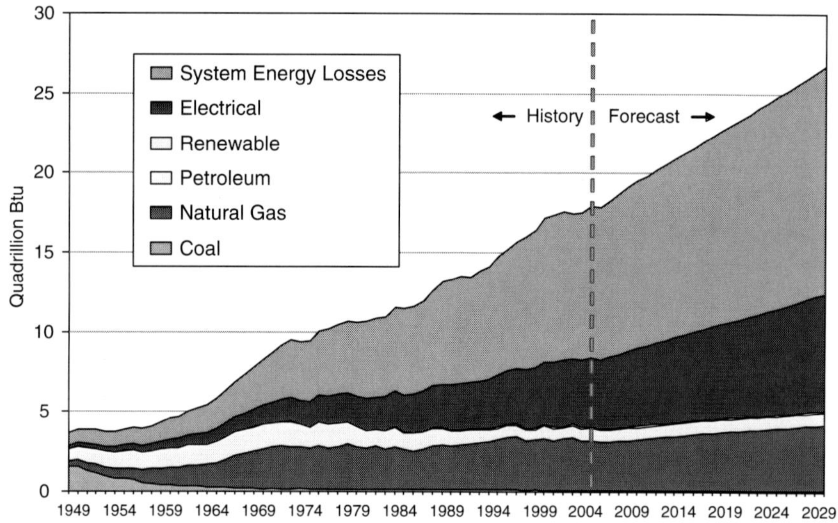

FIGURE 2-17 U.S. commercial-sector energy consumption by source: history and forecast.
SOURCE: EIA, 2006a.

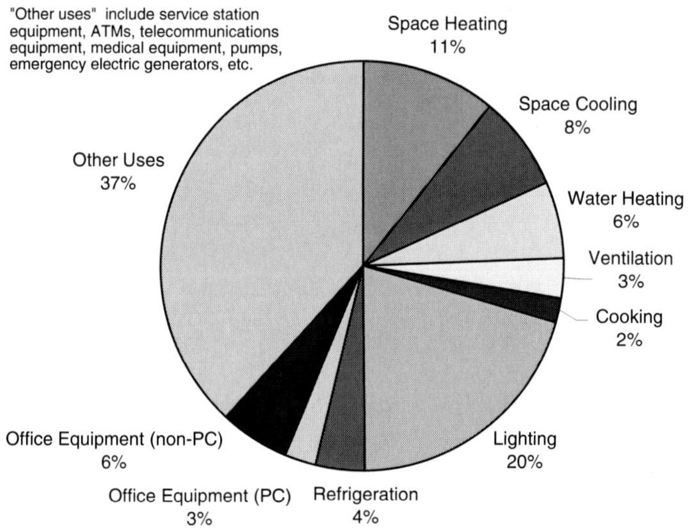

FIGURE 2-18 U.S. commercial-sector energy end use, 2004.
SOURCE: EIA, 2005.

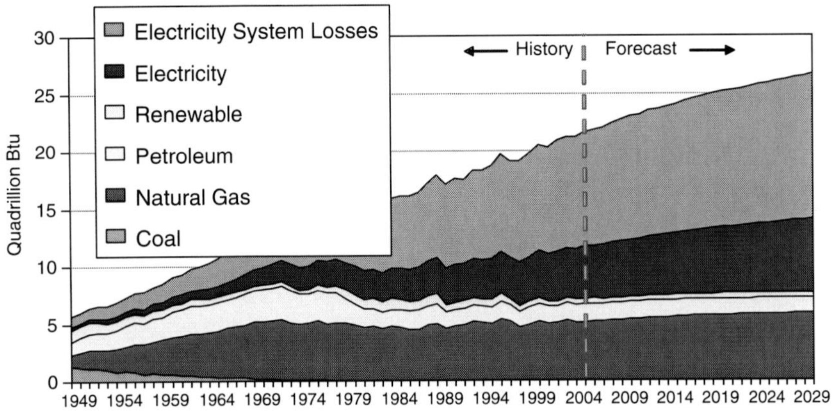

FIGURE 2-19 U.S. residential-sector fuel use by type: history and forecast.
SOURCE: EIA, 2006a.

for 21 percent. Petroleum accounted for 7 percent and renewables accounted for 2 percent. Figure 2-19 shows the historical and projected residential fuel consumption by type. It illustrates that

- Residential-sector fuel consumption increased from 6 quads (6.3 EJ) in 1949 to 22 quads (23.2 EJ) in 2004 and is forecast to increase to 27 quads (28.5 EJ) in 2030.
- Electricity system losses account for about as much as all fuels consumed and are forecast to continue to do so.
- The percent use of the various fuels has remained relatively constant and is forecast to continue to do so.

Petroleum use has dominated the transportation sector and is projected to do so through 2030. Although natural gas and renewable fuels (e.g., biodiesel) make up a share of total consumption, their contributions are a small fraction and forecasts for the future vary widely. In 2004, the U.S. transportation sector used 28.1 quads (29.6 EJ), and petroleum accounted for 96 percent of this.

China

Research from several agencies shows that by 2020, China's primary energy demand could range between 2,440 and 2,900 Mtce (73.9-87.9 EJ), which will roughly double its demand from 2000 (1,300 Mtce) (Figure 2-20). Table 2-1 summarizes the research of IEA, the Asia Pacific Energy Research Centre, under the aegis of Asia-Pacific Economic Cooperation forum, and the joint work of

TABLE 2-1 Comparison on Forecasts of China's Primary Commercial Energy Demand in 2020

Agency	Year	Base Year	Forecasting Method	Demand in 2010 (Mtce)				Demand in 2020 (Mtce)			
				Total	Coal	Oil	NG	Total	Coal	Oil	NG
IEA	2002	2000	Sectoral analysis	1860	1220	480	81	2438	1512	650	146
APERC	2002	1999	Reference scenario	2059	1090	469	99	2781	1414	710	196
ERI, NDRC	2003	2000	Reference scenario	2068	1365	524	108	2896	1788	795	193

SOURCES: IEA, 2002; APEC, 2002; ERI, 2002.

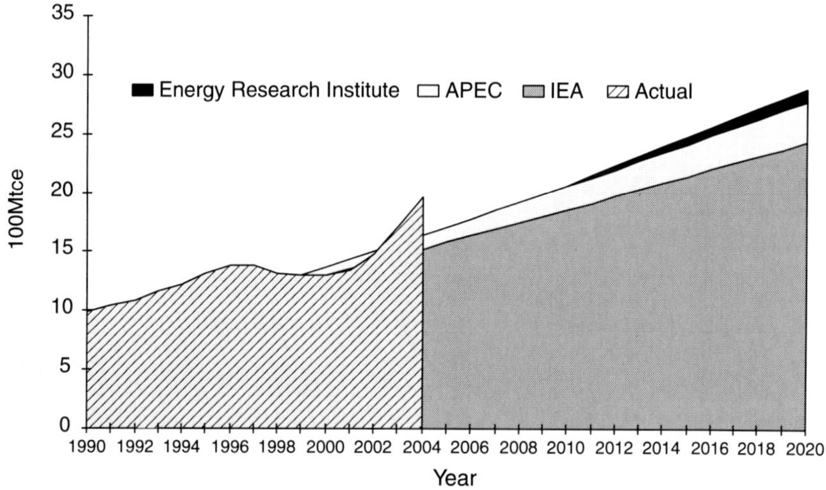

FIGURE 2-20 Forecast of commercial energy consumption in China.
SOURCE: Values before 2004: "China Statistical Year Book"; all others are projections by the respective agencies.

BOX 2-5
The Challenges of Energy Forecasting

There are many limitations in developing long-term forecasts of energy supply and demand. For example, as illustrated in the figure below, natural gas supply can be difficult to project and can shift dramatically from year to year. The Department of Energy's Energy Information Administration (EIA) analyzes various issues that impact U.S. energy markets, such as energy prices, technological advances, changes in public policy, and economic growth (EIA, 2007).

Improvements in technology impact energy supply and demand forecasts. Using advanced technologies reduces production costs and decreases the amount of natural resources being consumed. Energy prices also have the ability to impact supply and demand forecasting by increasing or decreasing the energy resources that are available to consumers. Additionally, policy decisions made by governments and regulating organizations can affect the oil supply, changing energy projections. Long-term energy projections do not consider the impact of these trends. These forecasts are of limited value if outside factors are not examined as part of future trends in energy supply.

In China, these forecasts are made even more difficult by the fact that separate agencies make independent projections. While the National Development and Reform Commission's Energy Research Institute (ERI) makes overall energy consumption projections, the more detailed forecasts on supply and demand are carried out by agencies such as the China Coal Association, or the China Petroleum Sector. This obviously leads to difficulties in cross-referencing information.

Energy forecasters also tend to underestimate the impact of unmodeled "surprises," a key example in the United States being the response to the 1973 oil embargo and the resultant gains in energy efficiency (Craig et al., 2002). While forecasters attempt to capture social trends (e.g., increasing concern over global warming), predicting technological breakthroughs or events which bring about behavioral change is not an easy task. Given all of these challenges, long-term energy forecasts are nonetheless useful tools for energy planners, and additionally, they are illustrative examples of prevailing perceptions and trends.

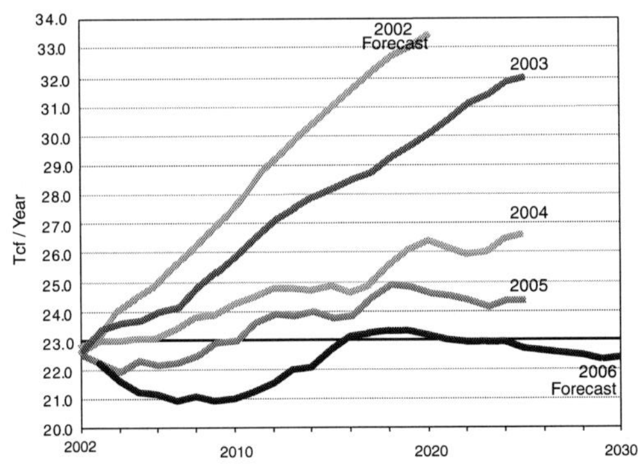

FIGURE Box 2-5 U.S. natural gas supply forecast, 2002-2006.
SOURCE: EIA.

China's Energy Research Institute and the NDRC. The figure in Box 2-5 depicts these various projections over time, and also illustrates the difficulty and potential inaccuracies with energy forecasts.

Coal

Coal consumption has been increasing rapidly, as Figure 2-21 indicates. In 2000, 1.32 billion tons (40 EJ) of coal were consumed; this number increased to 2.17 billion tons (65.6 EJ) by 2005. Figure 2-22 shows the forecast for coal consumption through 2020. The general trends forecast are as follows:

- Industrial coal consumption will continue its rapid rise (particularly for the electric power industry).
- Residential consumption will remain relatively constant.
- Proportions are projected to decline gradually (particularly in urban areas).
- As mentioned earlier, imports are projected to increase, and China could soon be a net importer of coal.

In 2005, industry accounted for 93.5 percent of coal consumption (though this includes electric power generation as an industry). Residential coal consumption totaled 4 percent, ranking second (see Figure 2-23).

As has been noted, much of the coal consumed by industry is used to generate electricity. In 2003, the coal used to generate power or heat supply totaled 876 million tons (26.5 EJ), or 53.5 percent of total consumption. Steel-making,

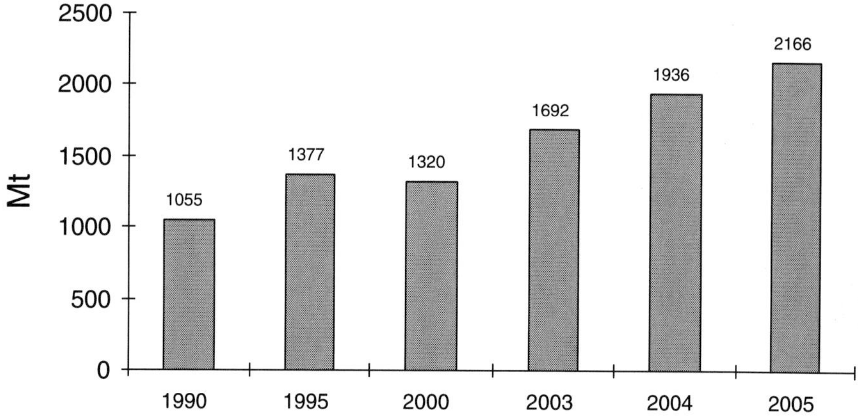

FIGURE 2-21 Coal consumption in recent years.
SOURCE: NBS, 2006b.

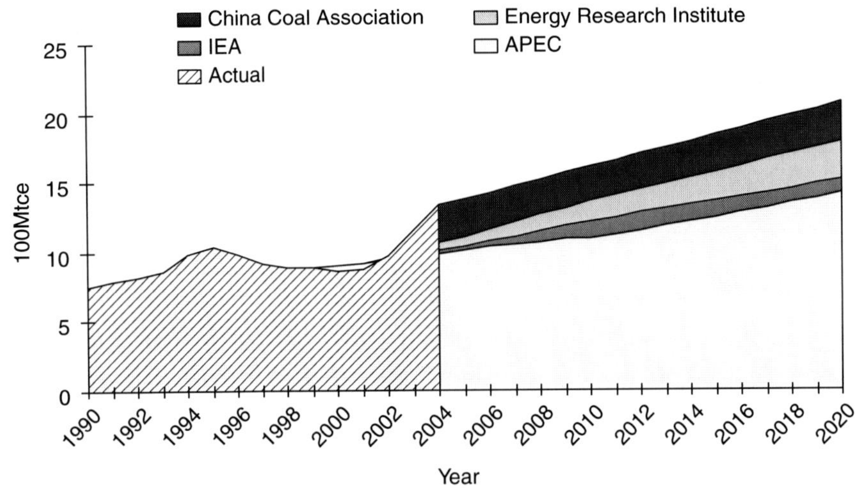

FIGURE 2-22 Forecast of coal consumption in China.
SOURCE: Values before 2004: "China Statistical Year Book," all others are projections by the respective agencies.

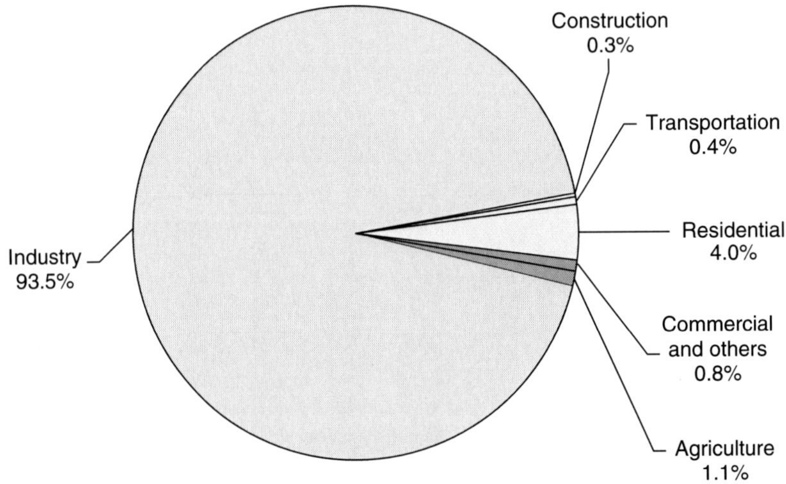

FIGURE 2-23 Coal consumption by sector, 2005.
SOURCE: NBS, 2006b.

building materials, and the chemical industry were the next largest consumers, proportionally consuming 11 percent, 11 percent, and 5 percent, respectively.

China was a net exporter of coal in 2006, exporting more than 60 Mt (1.8 EJ), while importing slightly more than 30 Mt (0.9 EJ). While detailed projections are not currently available, it is expected that coal imports will surpass exports by 2020 at the latest, though most demand will still be met by domestic production.

Petroleum

Petroleum consumption has increased rapidly, from 224 Mt (10 EJ) in 2000, to 325 Mt (14.6 EJ) in 2005, as shown in Figure 2-24. Petroleum consumption by sector is illustrated in Figure 2-25.

Consumption of gasoline, kerosene, and diesel fuel has been increasing. These petroleum products are mainly used in transportation, industry, and commerce. About 45.7 percent of gasoline produced is used in transportation, 41.5 percent in industry and commerce. Diesel fuel is mainly used in transportation (41.5 percent) and industry (21.8 percent). Kerosene is mainly used in civilian shipping and transportation, which consumes two-thirds of the total produced. In recent years the quantity of residual fuel oil consumption has increased slightly. In 2003, about 77 percent of residual fuel oil was used in industry, and 22 percent was used in transportation. Much of this increased demand and consumption for petroleum-based fuels is coming from the transportation sector, which correlates to the similarly steep rise in automobiles in China (see Figure 2-26). In 2003, there were 23.8 million automobiles in China, an increase of nearly 300 percent

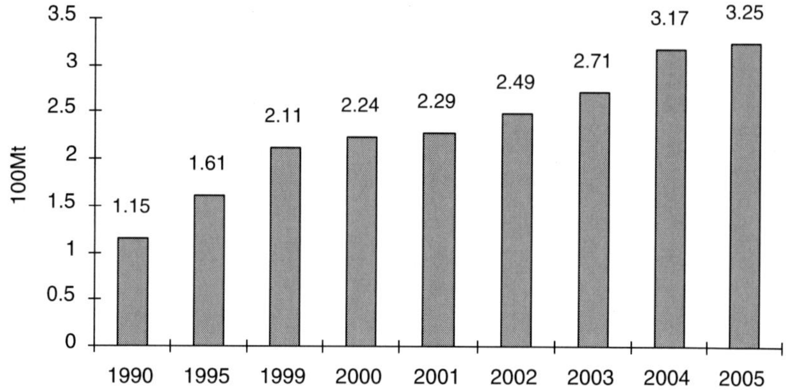

FIGURE 2-24 Petroleum consumption in recent years.
SOURCES: NBS, 2005b, 2006b.

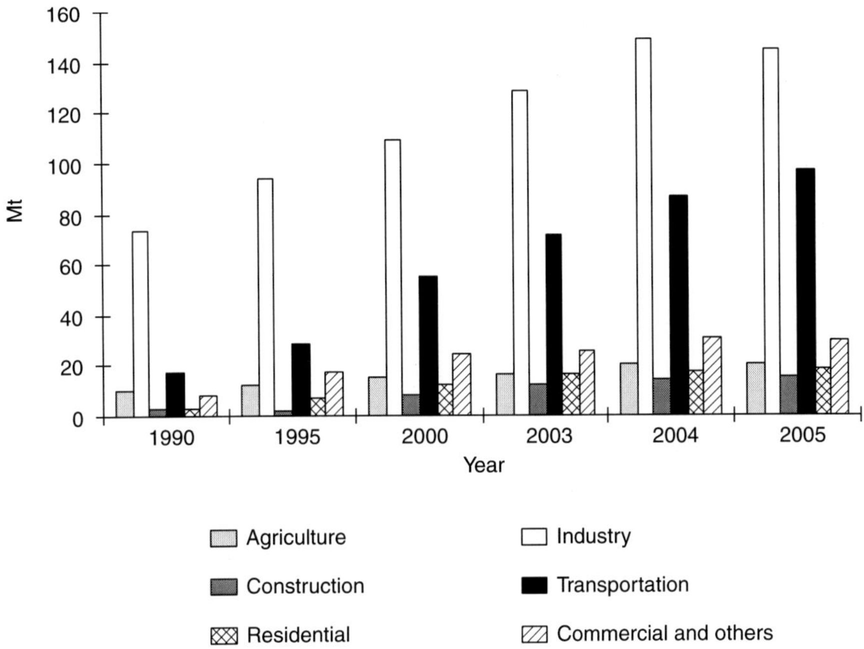

FIGURE 2-25 Petroleum consumption by sector.
SOURCE: NBS, 2006b.

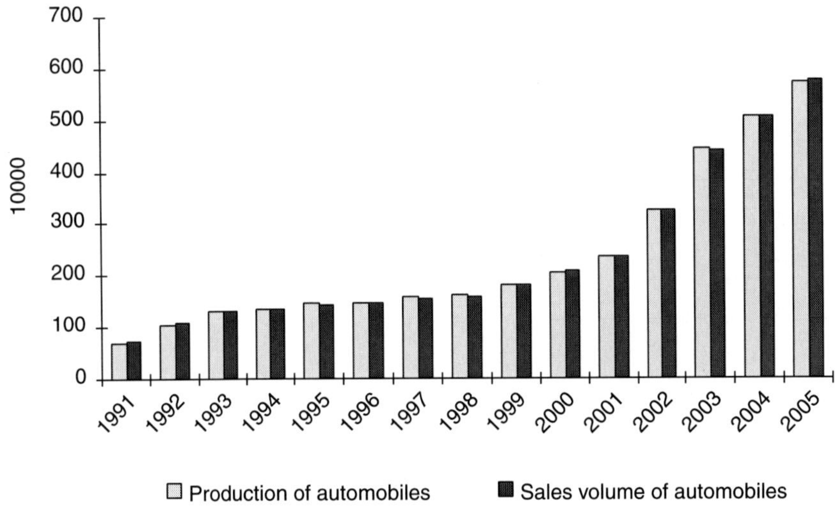

FIGURE 2-26 Energy demand from automobiles.
SOURCE: CATARC, 2005.

from 1991. If China continues at this pace, it could have a passenger car fleet the size of the U.S. fleet by 2030 (Wang et al., 2005).

As China's level of development continues to rise, its consumption of petroleum is also projected to rise (see Figure 2-27), thus widening the gap between domestic production and demand. In order to address the issue of energy security, the Chinese government established a petroleum reserve system, as well as oil and gas support bases overseas. Additionally, in 2005, the government increased research for petroleum substitution strategies.

Based on the preliminary results of this research, the government has decided to emphasize substituting CTL and biofuels for petroleum-based fuels. At present, the NDRC is coordinating research between related government departments and research institutions on a petroleum substitution strategy. Construction began in August 2004 on a demonstration factory to produce CTL in Inner Mongolia province. The Shenhua direct coal liquefaction project is scheduled to be completed in 2007, with commercial demonstration planned to begin in 2008. The demonstration scale is planned to be one million tons/yr (about 20,000 bpd), and the eventual full production scale is planned to be 5 million tons/yr (about 100,000 bpd). In early 2006, Shanxi province began constructing a demonstration factory to produce 160,000 tons annually through indirect coal liquefaction. In 2006 construction also began on a larger (1 million tons annually) indirect coal liquefaction plant using domestically-developed technology; it is being built in

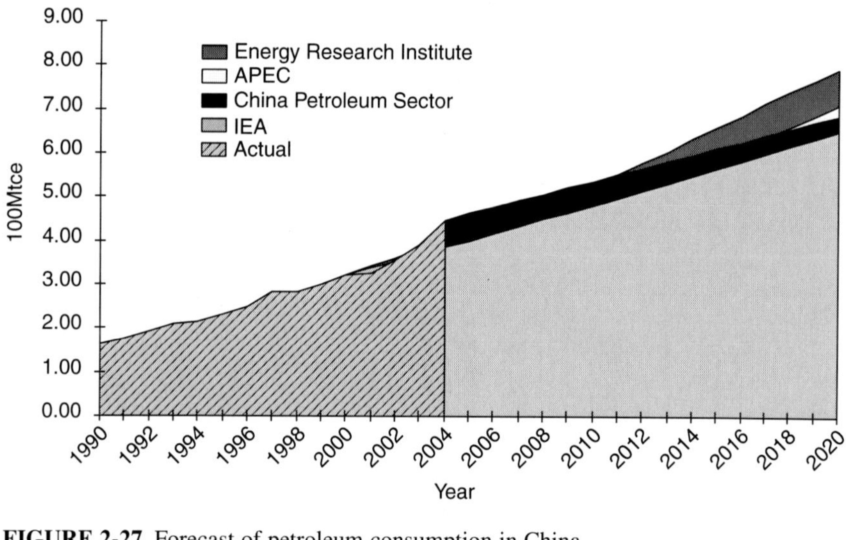

FIGURE 2-27 Forecast of petroleum consumption in China.
SOURCE: Values before 2004: "China Statistical Year Book." Values after 2004: prediction results.

Shanxi province and will be finished in 2010. The overall Chinese goal is to be producing 50 million tons/yr (2.25 EJ) of CTL by 2020. According to the Chinese, this is a key part of their long-term strategy of limiting oil imports to no more than 50 percent of their total requirements. Additionally, nine provinces have launched demonstrations of biofuel use, both in the commercial and transportation sectors. Similarly, research is under way on substituting dimethyl ether for liquefied petroleum gasoline and for developing and utilizing biodiesel.

Natural Gas

Natural gas consumption has increased from 24.5 billion m^3 (0.98 EJ) in 2000 to 47.9 billion m^3 (1.9 EJ) in 2005 (NBS, 2006a). About 74 percent of natural gas was used in industry (including power generation, heat supply, and chemical production), while 16.6 percent was used in the residential sector. Figure 2-28 predicts that natural gas consumption will continue to increase, and although the total consumption may increase four-fold by 2020, its relative contribution to total energy consumption will not be substantial. Most of the increased consumption will result from residential use, in switching from coal to natural gas-based heating systems.

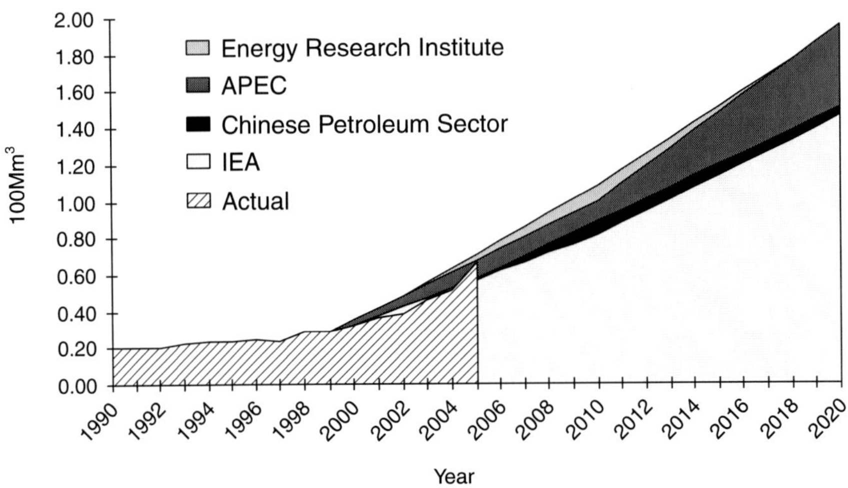

FIGURE 2-28 Forecast of natural gas consumption in China.
SOURCE: Values before 2004: "China Statistical Year Book." Values after 2004: prediction results.

Energy Consumption by Sector

Electricity consumption has increased rapidly over the past 10 years, as shown in Figure 2-29. Industry has been the main consumer of electric power, consuming about 74 percent of total electric power generated. Residential use was the second largest sector, at a distant 11 percent of the total. China's rapid economic development, particularly the trend of heavy industry development, has caused electricity supply shortages. As per capita GDP has increased, so too has residential consumption, a trend which will continue into the future. Industrial consumption of electricity will also increase, but by closing older, inefficient facilities and by upgrading technologies and techniques, industry's share of total consumption will decrease. Figure 2-30 shows projections for electricity consumption through 2020.

According to the NDRC, overall energy consumption per 10,000 RMB gross domestic product has been decreasing at a rate of 4 percent per year from 1991 to 2002, saving 70 Mtce (2.12 EJ) of energy overall. For that period, coal consumption per unit power production decreased by 11.2 percent, steel consumption per ton decreased by 29.6 percent, and cement consumption per ton decreased by 21.9 percent; and, as a result, the gap with the advanced countries has been narrowed (NDRC, 2005). However, energy consumption per GDP remains rather high in comparison to developed countries.

Finally, Figure 2-31 shows energy consumption by sector for 2005. It illustrates that industry is by far the dominant sector in terms of consumption. Residen-

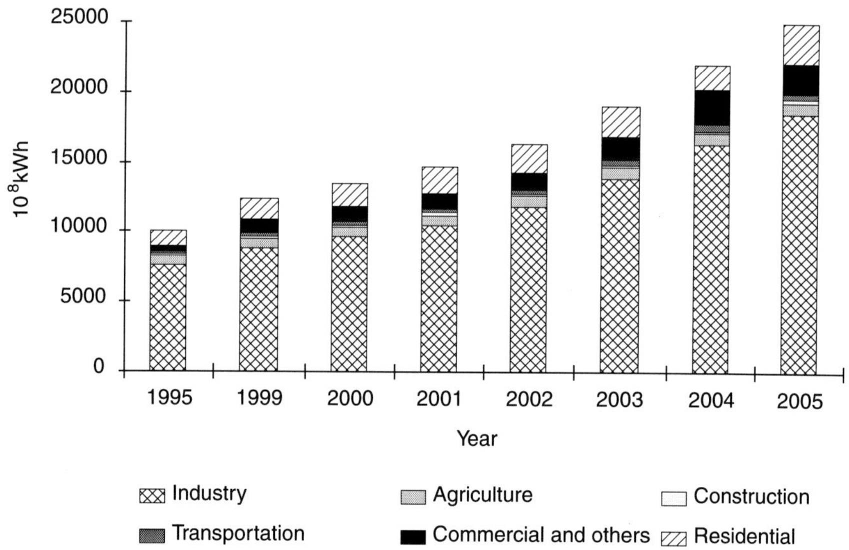

FIGURE 2-29 Electricity consumption by sector.
SOURCE: NBS, 2006a.

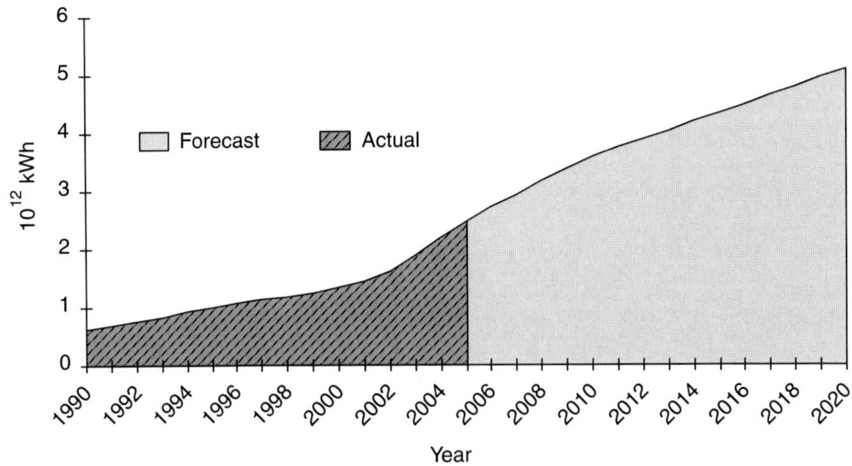

FIGURE 2-30 Forecast of electricity consumption in China.
SOURCE: Values before 2005: "China Statistical Year Book." Values after 2005: prediction results.

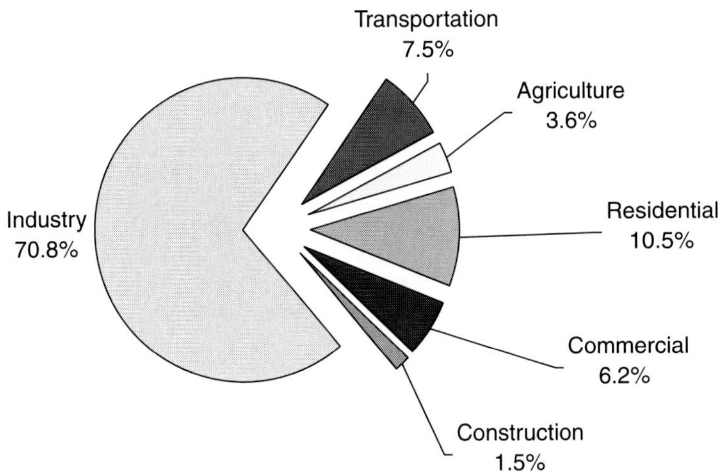

FIGURE 2-31 Energy consumption by sector for 2005.
SOURCE: NBS, 2005a.

tial consumption accounts for just over 11 percent, followed by the transportation sector at 7.5 percent. The transportation sector is the most likely sector to gain in share of total consumption over the next 20-30 years. This has obvious ramifications for China's energy policy, security, as well as concern over emissions.

REFERENCES

Akimoto, H., T. Ohara, J. Kurokawa, and N. Horii. 2006. Verification of energy consumption in China during 1996-2003 by using satellite observational data. Atmospheric Environment 4:7663-7667.

APEC (Asia-Pacific Economic Cooperation). 2002. APEC Energy Demand and Supply Outlook 2002.

Bartis, J.T. 2007. "Policy Issues for Coal-to-Liquid Development," Testimony before the Committee on Energy and Natural Resources, U.S. Senate. RAND Corporation, May.

CATARC (China Automotive Technology and Research Center). 2005. Automotive Industry of China. Tianjin.

Craig, P.P., A. Gadgil, and J.G. Koomey. 2002. What can history teach us? A retrospective of long-term energy forecasts for the United States. Annual Review of Energy and the Environment 27:83-118.

DOE (U.S. Department of Energy). 2006. Energy Policy Act 2005, Section 1837: National Security Review of International Energy Requirements, Washington, D.C.: U.S. Department of Energy.

EIA (Energy Information Administration). 2006a. Annual Energy Outlook—2006. Washington, D.C.: U.S. Department of Energy.

EIA. 2006b. International Energy Outlook—2006. Washington, D.C.: U.S. Department of Energy.

EIA. 2006c. Eliminating MTBE in Gasoline in 2006. Washington, D.C.: U.S. Department of Energy.

EIA. 2007. Energy Outlook—2007. Washington, D.C.: U.S. Department of Energy.

EPA (U.S. Environmental Protection Agency). 2002. Clean Alternative Fuels: Fischer-Tropsch. Washington, D.C.: U.S. Environmental Protection Agency.

ERI (Energy Research Institute). 2002. Scenario Analysis of Energy Demand and Supply in China for 2002. National Development and Reform Commission.

Flalka, J.J. and R. Gold. 2004. Fears of Terrorism Crush Plans for Liquefied-Gas Terminals. The Wall Street Journal, May 14.

Freerks, R.L. 2007. Emissions and Environmental Performance of Coal-to Liquids Fischer-Tropsch Fuels. Rentech, Inc.

IEA (International Energy Agency). 2002. China Energy Outlook 2002. Paris: International Energy Agency.

Liu, J. 2002. Study of China's Resources-Utilization Strategy. Beijing: China Agriculture Press.

MLR (Ministry of Land Resources). 2006. Third National Oil and Gas Resources Investigation.

NBS (National Bureau of Statistics—China). 2005a. China Statistical Yearbook. 2005. Beijing: China Statistics Press.

NBS. 2005b. China Energy Statistical Yearbook. 2005. Beijing: China Statistics Press.

NBS. 2006a. China Statistical Yearbook. 2006. Beijing: China Statistics Press.

NBS. 2006b. China Energy Statistical Yearbook. 2006. Beijing: China Statistics Press.

NCC (National Coal Council). 2006. Coal: America's Energy Future. Washington, D.C.: National Coal Council.

NDRC (National Development and Reform Commission). 2005. China Medium and Long Term Energy Conservation Plan.

NPC (National Petroleum Council). 1999. Meeting the Challenges of the Nation's Growing Natural Gas Demand.

NRC (National Research Council). 2001. Energy Research at DOE: Was It Worth It? Energy Efficiency and Fossil Energy Research 1978 to 2000. Washington, D.C.: National Academy Press.

Sinton, J.E. and Fridley, D.G. 2003. Comments on recent energy statistics from China. Sinosphere 6(2):6-12.

SSEB (Southern States Energy Board). 2006. American Energy Security: Building a Bridge to Energy Independence and a Sustainable Energy Future. Norcross, Georgia.

Tu, J. 2006. China's Botched Coal Statistics. China Brief, Vol. 6, Issue 21. Washington, D.C.: The Jamestown Foundation.

Wang, M., Jiang, Y., He, D. and Yang, H. 2005. Toward a Sustainable Future: Energy, Environment and Transportation in China. W. Zhou and J. Szyliowicz, eds. Beijing: China Communications Press.

3

Air Pollution: Sources, Impacts, and Effects

Concern about the effects of air pollution has existed for centuries. In 14th-century England, regulations were introduced regarding the burning of sea coal, and violators were tortured for producing foul odors. In the United States, the first air pollution regulations also dealt with coal, and in the 19th century, coal and smoke ordinances were passed in Chicago, St. Louis, and Cincinnati. In the 20th century, concern about air pollution in the United States can be traced to severe pollution episodes, such as the 1948 episode in Donora, Pennsylvania (near Pittsburgh), which resulted in nearly 7,000 illnesses and 20 deaths. Although rare, episodes such as those in Donora dramatized the acute health effects of air pollutants. Following the passage of the 1970 Clean Air Act Amendments that required the setting of National Ambient Air Quality Standards (NAAQS)—the imposition of emission standards for hazardous air pollutants and control on mobile source emissions—the United States embarked on a process of improving air quality, while maintaining a growing but changing economy.

Beginning in the 1970s, China started to pay attention to pollution caused from coal combustion. China has experienced rapid economic growth of up to 7-8 percent of GDP per year since the mid-1980s. Within that period, the Law of the People's Republic of China on the Prevention and Control of Atmospheric Pollution was passed by the Committee of People's Congress Council in September of 1987, with amendments in 1995 and in 2000. This short period of fast economic growth has led to higher living standards, but has also caused severe problems in environmental pollution. In recent years the Chinese government has made major efforts to reduce emissions, which have partly compensated for the rapid growth in energy consumption and urbanization. Improving air quality has become an urgent task.

Air quality management efforts in the United States are guided by the NAAQS, which are set based on health and welfare effects of the major air pollutants: particulate matter (PM), ozone (O_3), carbon monoxide (CO), sulfur dioxide (SO_2), nitrogen dioxide (NO_2), and lead. Standards for PM have been established for two size categories: PM of less than 10 microns in aerodynamic diameter (PM_{10}) and PM of less than 2.5 microns in diameter. Likewise, air quality management efforts in China are guided by National Air Quality Standards, which have been established for SO_2, NO_2, and PM_{10}. The current health-based NAAQS are listed in Table 3-1, along with air quality standards adopted by China (Table 3-2) and guidelines established by the World Health Organization (WHO) (Table 3-3). In addition to efforts focused on these major air pollutants, both countries are also concerned with toxic air pollutants that are less ubiquitous. These toxic air pollutants (or hazardous air pollutants, as they are called in the Clean Air Act) include benzene and other aromatic compounds from motor vehicles, fuels, and other combustion sources, and mercury from solid waste and coal combustion.

China has developed an Air Pollution Index (API) based on its air quality standards, and cities use this tool to report their air quality (see Chapter 4, Table 4-1). Like their national standards, the API contains classes (I-V) which allow cities to comply at various levels. Cities are encouraged to achieve Class II standards

TABLE 3-1 U.S. National Ambient Air Quality Standards (NAAQS)

Pollutant	Primary Stds.	Averaging Times	Secondary Stds.
Carbon monoxide	9 ppm (10 mg/m^3)	8-hour	None
	35 ppm (40 mg/m^3)	1-hour	None
Lead	1.5 μg/m^3	Quarterly average	Same as primary
Nitrogen dioxide	0.053 ppm (100 μg/m^3)	Annual (arithmetic mean)	Same as primary
Particulate matter (PM_{10})	Revoked	Annual (arith. mean)	
	150 μg/m^3	24-hour	
Particulate matter ($PM_{2.5}$)	15.0 μg/m^3	Annual (arith. mean)	Same as primary
	35 μg/m^3	24-hour	
Ozone	0.08 ppm (171 μg/m^3)	8-hour	Same as primary
	0.12 ppm (257 μg/m^3)	1-hour (Applies only in limited areas)	Same as primary
Sulfur oxides	0.03 ppm (78 μg/m^3)	Annual (arith. mean)	—
	0.14 ppm (364 μg/m^3)	24-hour	—
	—	3-hour	0.5 ppm (1300 μg/m^3)

TABLE 3-2 China's Ambient Air Quality Standards (GB 3095-1996) (μg/m^3 unless otherwise noted)

Pollutant	Class I Standard	Class II Standard	Class III Standard	Averaging Times
SO_2	20	60	100	Annual
	50	150	250	Daily
	150	500	700	1-hour
TSP	80	200	300	Annual
	120	300	500	Daily
PM_{10}	40	100	150	Annual
	50	150	250	Daily
NO_x	50	50	100	Annual
	100	100	150	Daily
	150	150	300	1-hour
NO_2	40	40	80	Annual
	80	80	120	Daily
	120	120	240	1-hour
CO (mg/m^3)	4.00	4.00	6.00	Daily
	10.00	10.00	20.00	1-hour
O_3	120	160	200	1-hour

TABLE 3-3 World Health Organization (WHO) Air Quality Guidelines (in μg/m^3)

Pollutant	Interim target 1	Interim target 2	Interim target 3	Standard	Averaging Times
[a]$PM_{2.5}$	35	25	15	**10**	Annual
	75	50	37.5	**25**	24-hour
PM_{10}	70	50	30	**20**	Annual
	150	100	75	**50**	24-hour
O_3	160	—	—	**100**	8-hour
NO_2	—	—	—	**40**	Annual
	—	—	—	**200**	1-hour
SO_2	125	50	—	**20**	24-hour
	—	—	—	**500**	10-minute

[a]Preferred guideline, according to WHO.
SOURCE: WHO, 2006.

or better, although as of 2006, nearly 48 percent of urban residents lived in cities not meeting these standards (SEPA, 2007). China is not the only country with varying classes of air quality standards, and WHO recently adopted interim targets to accompany its own guidelines, in order to facilitate implementation in more polluted areas (WHO, 2006). A benefit of the interim targets, and the rationale for China's classes of standards, is that they allow governments to consider local circumstances in developing an approach to balance health risks, technological feasibility, economic considerations, and other factors (WHO, 2006). However, by virtue of its formulation in China, in which air quality is reported by class and dominant pollutant (see Chapter 9, Table 9-5), the API has some drawbacks as well. Since the API reflects a comprehensive index of all pollutants, and air quality rankings reflect the lowest level of compliance, cities have tended to focus on a particular pollutant (e.g., PM_{10}) in order to improve their overall rankings. While this clearly provides benefits associated with the reduction of the particular pollutant, it also works against efforts to adopt a multipollutant reduction strategy. It also may lead governments to overlook increasing trends in concentrations of pollutants which are still satisfying a certain criteria; emissions targets in China are established with a target API rank in mind, providing a disincentive to limit emissions other than for the predominant pollutant.

This chapter first discusses the effects of air pollution in the United States and China, and then reviews trends in their air pollutant emissions and concentrations, and ends with a discussion of key source-receptor relationships.

AIR POLLUTION EFFECTS

United States

Health Effects of Air Pollution

In the United States, air pollution is regulated because of concerns about its impact on human health, visibility, and the environment (NRC, 2004). Economic analysis of air pollution control efforts in the United States indicate that historically, the benefits have far outweighed the costs. In 1997, the U.S. Environmental Protection Agency (EPA) estimated that the costs of control measures undertaken in the United States from 1970 through 1990 totaled nearly \$500 billion (1990 dollars). The benefits accruing from the emissions reductions over that same period totaled more than \$20 trillion (1990 dollars), outweighing the costs by a ratio of 40 to 1 (EPA, 1997). Control programs adopted more recently still show highly favorable benefit-to-cost balances. In 2005, the EPA adopted the Clean Air Interstate Rule, requiring an additional 70 percent reduction in SO_2 emissions and a 60 percent reduction in NO_x emissions from large stationary sources by 2015. EPA estimates the cost of these controls will be about \$3 billion per year in 2015, while the annual benefits that year will be about \$90 billion (EPA, 2005). Analyzing the health benefits of any regulation requires flexible, innovative, and multidisciplinary participation and guidance from scientific experts (NRC, 2002).

Over the past 15 years, as the concentrations of CO, SO_2, NO_2, and Pb have declined, the focus of health studies and control efforts has increasingly turned to PM and O_3 as the most important major air pollutant species of concern. Correspondingly, the primary focus of this section is on the current understanding of the health effects of PM and O_3 in the United States. This section also discusses the risks of selected toxic air pollutants, which are generally less ubiquitous than the major pollutants mentioned above, but which may also have profound health implications where significant exposures occur (see also Box 3-1).

There have been several recent reviews of the improved understanding of the health effects of exposure to $PM_{2.5}$ (Lippmann et al., 2003; EPA, 2004; Pope and Dockery, 2006). Epidemiological studies have linked exposure to $PM_{2.5}$ to a range of adverse respiratory effects. Significantly, both short- and long-term $PM_{2.5}$ exposures are linked to a heightened risk of premature mortality. There is roughly a 10 percent increase in adult mortality rates for every 10 $\mu g/m^3$ of annual-average $PM_{2.5}$, a 0.25-1 percent increase per 10 $\mu g/m^3$ 24-hour average PM_{10}, and 0.2-0.8 percent increase per 10 $\mu g/m^3$ increase in 1-hour peak ozone (Pope et al., 2002; Cohen et al., 2004; WHO, 2006; Smith et al., 2004; HEI, 2006; Ostro et al., 2006). Some short-term studies have also linked "coarse" particles in the size range from 2.5 to 10 microns ($PM_{10-2.5}$) to premature mortality, but the results for this size fraction are less consistent than for $PM_{2.5}$ (Brunekreef and Forsberg, 2005). An important change in focus for fine particles ($PM_{2.5}$) has been from effects on the respiratory system to their role in cardiovascular effects. It is now hypothesized that much of the mortality that is associated with particle exposure results is related to effects on the cardiovascular system (EPA, 2004). This change in focus has led to substantial improvements in the understanding of pathways by which particulate pollutants might induce significant cardiac effects.

In terms of morbidity effects, there may be a greater role for larger particles, including particles in the 2.5 to 10 μm size range ($PM_{10-2.5}$). For example, some epidemiological studies have found that PM_{10} and $PM_{10-2.5}$ have greater effects than $PM_{2.5}$, particularly on emergency room visits (Brunekreef and Forsberg, 2005). Nonetheless, the health impacts of $PM_{2.5}$ are both better known and larger than for other pollutants, thus countries are beginning to focus more attention on $PM_{2.5}$ concentrations. Burden of disease estimates are helpful in assessing priorities to manage health-related pollutants. Cohen et al. (2005) estimate that outdoor ambient $PM_{2.5}$ pollution causes 3 percent of mortality from cardiopulmonary disease, roughly 5 percent of mortality from various cancers, and about 1 percent of mortality from acute respiratory infections in children 5 years and under, worldwide. This burden is unevenly distributed, though, occurring predominantly in developing countries (and 65 percent in Asia alone). Owing to uncertainties, including the fact that only mortality was considered in the assessment, Cohen et al. suggest that the impact of urban air pollution on burden of disease is likely underestimated.

Two age groups, older adults and the very young, are potentially at greater risk for PM-related effects. Epidemiologic studies have generally not shown striking differences between adult age groups. However, some epidemiologic studies have suggested that serious health effects, such as premature mortality, are greater among older populations (e.g., Dockery et al., 1993; Pope et al., 1995). Epidemiologic evidence has reported associations with emergency hospital admissions for respiratory illness and asthma-related symptoms in children. In the United States, approximately 22 million people, or 11 percent of the population, have received a diagnosis of heart disease, about 20 percent of the population have hypertension, and about 9 percent of adults and 11 percent of children in the United States have been diagnosed with asthma. In addition, about 26 percent of the U.S. population are under 18 years of age, and about 12 percent are 65 years of age or older.

To put the estimates of premature mortality impacts of PM into context, overall health statistics indicate that there are approximately 2.5 million deaths from all causes per year in the U.S. population, with about 100,000 deaths from chronic lower respiratory diseases (Kochanek et al., 2004). EPA estimates the cost of meeting the revised 24-hour $PM_{2.5}$ standards at \$5.4 billion in 2020. This estimate includes the costs of purchasing and installing controls for reducing pollution to meet the standard. It also estimates that meeting the revised 24-hour standard will result in health and visibility benefits ranging from \$9 billion to \$76 billion a year by 2020. These costs and benefits are in addition to the estimated costs (\$7 billion) and benefits (\$20-160 billion per year by 2015) associated with meeting the 1997 standards for fine PM (EPA, 2006b).

Epidemiological and clinical evidence links short-term exposure to elevated ozone levels to respiratory symptoms and illness, with epidemiological studies also showing a positive association with emergency room visits and hospital admissions (EPA, 2007b). There is also evidence for an association between elevated O_3 concentrations and premature mortality (Bell et al., 2004; Gryparis et al., 2004). There are limited health effects that can be ascribed to the other criteria pollutants, given the apparent large influence of PM and ozone. Although EPA does make estimates in the relevant criteria documents, it appears that there are relatively few deaths or serious illnesses arising from exposure to these other pollutants at the levels currently present in the United States.

The last assessment of the effects of hazardous air pollutants was performed based on emissions and concentration estimates for 1999. From a national perspective, benzene is the most significant air toxic for which cancer risk could be estimated, contributing 25 percent of the average individual cancer risk identified in this assessment (EPA, 2006a). Based on EPA's national emissions inventory, the key sources for benzene are on-road (49 percent) and non-road mobile sources (19 percent). EPA projects that on-road and non-road mobile source benzene emissions will decrease by about 60 percent between 1999 and 2020, as a result of motor vehicle standards, fuel controls, standards for non-road engines and equipment, and motor vehicle inspection and maintenance programs. Most of these programs reduce benzene simultaneously with other volatile organic compounds. Diesel PM, which often contains benzene and other toxic pollutants, is currently regulated in California, where estimates attribute 70 percent of cancers resulting from ambient air pollution to diesel PM (SCAQMD, 2000). EPA lists diesel PM as a possible carcinogen, but has not adopted a unit risk factor.

Of the 40 air toxics showing the potential for acute respiratory effects, acrolein is the most significant, contributing 91 percent of the nationwide average non-cancer hazard identified in EPA's assessment (SCAQMD, 2000). Note that the health information and exposure data for acrolein include much more uncertainty than those for benzene. Based on the national emissions inventory, the key sources for acrolein are open burning, prescribed fires and wildfires (61 percent), and on-road (14 percent) and non-road (11 percent) mobile sources. The apparent dominance of acrolein as a non-cancer "risk driver" in both the 1996 and 1999 national-scale air toxics assessments has led to efforts to develop an effective monitoring test method for this pollutant. EPA projects acrolein emissions from on-road sources will be reduced by 53 percent between 1996 and 2020, as a result of existing motor vehicle standards and fuel controls.

EPA's national air toxics assessment estimates that in most of the country, people have a lifetime cancer risk from breathing air toxics between 1 and 25 in one million (SCAQMD, 2000). This means that out of one million people, between 1 and 25 have an increased likelihood of contracting cancer as a result of breathing air toxics from outdoor sources, if they were exposed to 1999 levels over the course of their lifetime. The assessment estimates that most urban locations in the United States have air toxics lifetime cancer risk greater than 25 in a million. Risks in transportation corridors and in some other locations are greater than 50 in a million. In contrast, one out of every three Americans (330,000 in a million) will contract cancer during a lifetime, when all causes (including exposure to air toxics) are taken into account. Based on these results, the risk of contracting cancer is increased less than 1 percent due to the inhalation of air toxics from outdoor sources. As another comparison, the national risk of contracting cancer from radon exposure is on the order of 1 in 500 (2,000 in a million). Note that risks from human-caused air toxics are commonly viewed differently, because they are involuntary and subject to control, than are risks that are naturally occurring or voluntarily assumed.

Mercury is a toxic metal that is widely distributed around the globe due to emissions from both anthropogenic and natural sources. Exposure to high levels of mercury is associated with neurologic and kidney damage. For most people, the greatest health risk from mercury is posed by the consumption of fish contaminated with methyl mercury, which can bioaccumulate up the food chain. Impacts on fetal development due to maternal exposure are of special concern. In the United States, the Center for Disease Control and Prevention's National Health and Nutrition Examination Survey found that approximately 6 percent of childbearing-aged women had blood mercury levels at or above the reference dose of 5.8 μg/L, an estimated level assumed to be without appreciable harm (CDC, 2004). Most states in the United States have extensive advisories recommending that people limit consumption of fish caught in local waters, in order to avoid excessive mercury intake. In response to concerns about mercury contamination, international efforts are under way to reduce mercury use in manufacturing. In

BOX 3-1
Emissions, Exposure, and Intake Fraction

In addition to information on ambient concentrations, it is also useful to have an understanding of human exposure and intake fraction (*iF*) when considering human health impacts (Bennet et al., 2002). Intake fraction refers to the mass intake of pollutants by people during a given amount of time, relative to the mass of emissions released into the environment. This concept is useful in helping to set priorities in cities—emission sources such as cement mixers in urban areas exhibit high intake fractions, and therefore are more likely to impact human health than are sources which are located further from population centers, or which disperse their pollutants more effectively (e.g., through taller smokestacks). Mobile sources, specifically automobiles, also contribute disproportionately to human exposure, due to their prevalence in population centers, emitting at street level (Laden et al., 2000; Marshall et al., 2005). Of course, simply relocating or redistributing these emissions, while ostensibly reducing the risk in a given urban population, can potentially transfer this risk to other regions. Therefore, decisions to relocate or redistribute sources of emissions must be considered in light of downwind populations and the regional airshed.

Studies in the United States and in European cities have underscored the large contribution of mobile sources (Laden et al., 2000; Schwartz et al., 2002; Hoek et al., 2002; Maynard et al. 2007) to air pollution-related mortality. Recent studies in China have indicated that, although the electrical power generation sector is responsible for the bulk of SO_2 and TSP emissions, its *iF* value is generally much lower than are three other key industries: mineral production, chemical, and metallurgy (Wang et al., 2006; Ho and Nielsen, 2007). Improved understanding of the relationship between pollution sources and human exposure will further aid cities in assessing their risks and in developing appropriate strategies to reduce air pollution.

the United States, control technology has been required to significantly reduce mercury emissions from medical waste incinerators and from municipal solid waste combustors, and control requirements have recently been adopted for coal-fired power plants.

Environmental Effects

The welfare effects of greatest interest include visibility degradation, impacts on crop production and ecosystem health, and materials damage. During the 1970s and 1980s, recognition that lakes and streams in the eastern United States were becoming acidified due to atmospheric deposition led to requirements for power plants to significantly reduce emissions of sulfur and nitrogen oxides (NRC, 2004). Recovery of some lakes and streams in the northeastern United States is now being observed as a consequence of these emissions reductions, but full recovery will require additional controls and, in some places, will take decades (NAPAP, 2005). Deposition of reactive nitrogen in the form of ammonia and ammonium and nitrate ions is receiving increasing attention in the United States, due to concerns about eutrophication of coastal zones and over-fertilization of sensitive terrestrial ecosystems (NRC, 2004).

Visibility is most affected by airborne particulate matter, particularly PM with particle diameters between 0.1 and 1.0 μm. The U.S. goal for visibility is to restore visibility in protected national parks and wilderness areas to natural levels by 2064. In conjunction with the National Park Service, other federal land managers, and state organizations, the EPA has supported visibility monitoring in national parks and wilderness areas since 1988. The monitoring network was originally established at 20 sites, but it has now been expanded to 110 sites that represent all but one (Bering Sea) of the 156 mandatory Federal Class I areas across the country. Annual average visibility conditions (reflecting light extinction due to both anthropogenic and non-anthropogenic sources) vary regionally across the United States. The rural east generally has higher levels of impairment than do remote sites in the west, with the exception of urban-influenced sites such as San Gorgonio Wilderness (California) and Point Reyes National Seashore (California), which have annual average levels comparable to certain sites in the northeast (EPA, 2004). Higher visibility impairment levels in the east are due to generally higher concentrations of anthropogenic fine particles, particularly sulfates, and higher average relative humidity levels. In fact, sulfates account for 60-86 percent of the haziness in eastern sites.

Regional trends in Class I area visibility are updated and presented in the EPA's National Air Quality and Emissions Trends Report (EPA, 2001). Eastern trends for the 20 percent haziest days from 1992-1999 showed about a 16 percent improvement. However, visibility in the east remains significantly impaired, with an average visual range of approximately 20 km on the 20 percent haziest days. In western Class I areas, aggregate trends showed little change during 1990-1999 for the 20 percent haziest days. Average visual range on the 20 percent haziest days in western Class I areas is approximately 100 km.

PM does produce some effects on crops and forested ecosystems by the deposition of reactive nitrogen species. At this time, there is limited information on the extent of these effects and their trends. The larger influence on crop production, forest growth, and other indicators of ecosystem health and function is from ozone. Detrimental effects on vegetation include reduction in agricultural and commercial forest yields, reduced growth and increased plant susceptibility to disease, and potential long-term effects on forests and natural ecosystems. Plants are significantly more sensitive to ozone than are people. Murphy et al. (1999) evaluated benefits to eight major crops associated with several scenarios concerning the reduction or elimination of O_3 precursor emissions from motor vehicles in the United States. Their analysis reported a \$2.8 billion to \$5.8 billion (1990 dollars) benefit from the complete elimination of O_3 exposures from all sources (i.e., ambient O_3 reduced to a background level assumed to be 0.025 to 0.027 ppm). The EPA is currently considering setting "secondary" welfare-based NAAQS for ozone to address the effects of this pollutant on crops and other vegetation.

There is damage to materials from acidic species (gaseous and particulate) and ozone as well as from black carbon particles. With the decreases in the emissions that have occurred over the past 35 years, materials effects in the United States have become less important than in the past, and do not currently figure strongly in decisions regarding the setting of secondary air quality standards.

China

Health Effects

Air pollution and its impact on people's health and the environment is a matter of great concern in China. Heavy reliance on coal in power production and a rapidly growing car fleet, usually in combination with outdated technologies and poor maintenance, have led to concentrations of air pollutants far exceeding the limits of both national air quality standards and the air quality guidelines recommended by the WHO (2006). Nearly all of China's rural residents and a shrinking fraction of urban residents use solid fuels (biomass and coal) for household cooking and heating. As a result, by the use of global meta-analyses of epidemiological studies, it is estimated that indoor air pollution from solid fuel use in China alone is responsible for ~420,000 premature deaths annually, more than the estimated 300,000 attributed to urban outdoor air pollution in the country (Zhang and Smith, 2005).

The major air pollutants monitored in China are suspended particulates, SO_2, and NO_x. Health end-points studied in China in association with air pollution include changes in mortality of all causes, of respiratory disease, cardiovascular disease, and cerebrovascular disease, and morbidity, as well as the number of outpatient and emergency room visits. Increases in respiratory and other clinical symptoms and decrease in lung functions and immune functions are also studied. However, in comparison with air monitoring data, data on the effects of human health are limited. Aunan and Pan (2004) specifically summarized the relationships between PM_{10} and SO_2 and mortality, hospital admissions, and chronic respiratory symptoms and diseases. They expressed the exposure-response functions in terms of percentage change (per unit of exposure), rather than as absolute numbers. They derived the following coefficients for acute effects: a 0.03 percent (standard error [S.E.] 0.01) and a 0.04 percent (S.E. 0.01) increase in all-cause mortality per $\mu g/m^3$ of PM_{10} and SO_2, respectively; a 0.04 percent (S.E. 0.01) increase in cardiovascular deaths per $\mu g/m^3$ for both PM_{10} and SO_2; and a 0.06 percent (S.E. 0.02) and a 0.10 percent (S.E. 0.02) increase in respiratory deaths per $\mu g/m^3$ of PM_{10} and SO_2, respectively. For hospital admissions due to cardiovascular diseases, the obtained coefficients are 0.07 percent (S.E. 0.02) and 0.19 percent (S.E. 0.03) for PM_{10} and SO_2, respectively, whereas the coefficients for hospital admissions due to respiratory diseases are 0.12 percent (S.E. 0.02) and 0.15 percent (S.E. 0.03) for PM_{10} and SO_2, respectively. Exposure-response functions for the impact of long-term PM_{10} levels on the prevalence of chronic respiratory symptoms and diseases are derived from the results of cross-sectional questionnaire surveys, and indicate a 0.31 percent (S.E. 0.01) increase per $\mu g/m^3$ in adults and 0.44 percent (S.E. 0.02) per $\mu g/m^3$ in children. With some exceptions, Chinese studies report somewhat lower exposure-response coefficients, as compared to studies in Europe and the United States.

Increasing China's already severe air pollution will substantially increase the incidence of respiratory diseases throughout the country, as air pollution is estimated to be the primary cause of nearly 50 percent of all respiratory ailments (Brunekreef and Holgate, 2002). According to the United Nations Environmental Programme statistics (UNEP, 1999), soot and particle pollution from the burning of coal causes approximately 50,000 deaths per year in China, while some 400,000 people suffer from chronic bronchitis annually in the country's 11 largest urban areas. The United Nations Development Programme estimated that the death rate from lung cancer in severely polluted areas of China was 4.7-8.8 times higher than in areas with good air quality (UNDP, 2002). Extrapolating from 1995 emission levels and trends, the World Bank estimated that by 2020, China will need to spend approximately \$390 billion—or about 13 percent of projected GDP—to pay for the healthcare costs that will accrue solely from the burning of coal (World Bank, 1997). Zaozhuang, a coal-dependent eastern city, was estimated to be spending 10 percent of its GDP on air pollution-related health damages in 2000; and without additional controls, this share could rise to 16 percent by 2020 (Wang and Mauzerall, 2004). In Shanghai, which has been taking measures to diversify its fuel mix, it was estimated that health impacts due to particulate air pollution in urban areas in 2001, totaled \$625 million, accounting for over 1 percent of the city's GDP (Kan and Chen, 2004). The Health Effects Institute recently made available a compendium of epidemiologic studies of air pollution in Asia from 1980-2006, including 69 studies from mainland China (HEI, 2006).

Visibility

There is limited research about the impact of air pollution on visibility in China. Qiu and Yang (2000) analyzed the visibility trends in northern China from 1980 to 1994 based on 0.74 μm aerosol optical depths at five meteorological observations. In Zhengzhou (Figure 3-1) and Geermu, visibility showed an improving trend during this period and a possible reason for this is vertical distribution shifts of aerosol particles up in the troposphere. Visibility at Urumqi, Harbin, Beijing, and Zhengzhou in winter is impaired. At Harbin, the visibility range in summer is about twice that in winter. Cheung et al. (2005) utilized a formula developed

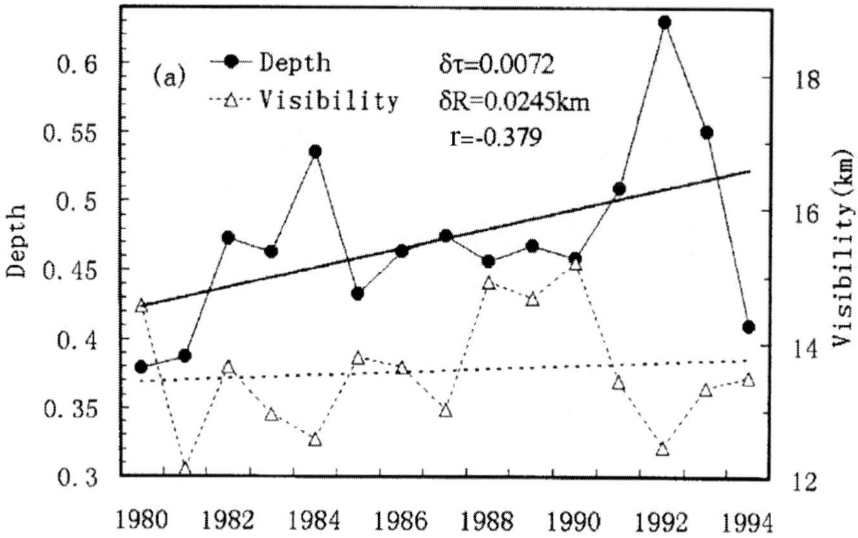

FIGURE 3-1 Variation characteristics of year-averaged aerosol optical depth and visibility over Zhengzhou during 1980-1994.
SOURCE: Qiu and Yang, 2000.

in the U.S. Interagency Monitoring to Protect Visual Environment (IMPROVE) study to calculate the contribution of organic matter and sulfate in $PM_{2.5}$ to the light extinction in the Pearl River Delta region, China. The fine sulfate is a larger contributor with humid climates. Sun et al. (2006) characterized the chemical compositions of $PM_{2.5}$ and PM_{10} in haze-fog episodes in Beijing. The serious air pollution in haze-fog episodes was strongly correlated with meteorological conditions and with the emissions of pollutants from anthropogenic sources.

Cultural Relics

There are more than 30 World Cultural Heritage sites in China and most of them are suffering damage from air pollution. From 2002 to 2005, the State Administration of Cultural Heritage initiated a project to investigate the corrosion status of cultural collections inside most of the museums in China. About 10 percent of collections were damaged seriously by indoor air pollution. Outdoor cultural relics, like grottos and carved stones, are eroded predominantly by acid rain. However, very little research has focused on indoor air pollution and environmental assessment in China's museums. For example, Christoforou et al. (1996) developed computer-based models to simulate the air flow into the caves and particle deposition within the Yungang caves. It was found that horizontal surfaces within caves 6 and 9 at Yungang would become completely covered by a full monolayer of particles in less than half a year, under the April conditions studied there, and will be soiled even more rapidly under annual average conditions. Cao et al. (2005a) investigated the indoor air quality in Emperor Qin's Terra-Cotta Museum in Xi'an, China, during summer 2004. The average levels of indoor $PM_{2.5}$ and total suspended particles (TSPs) were 108.4 and 172.4 μg/m^3. Sulfate ((32.4±6.2) percent), organics ((27.7±8.0) percent), and geological material ((12.5±3.4) percent) dominated indoor $PM_{2.5}$, followed by ammonium ((8.9±2.8) percent), nitrate ((7.0±2.9) percent), and elemental carbon (EC, (3.9±1.5) percent). High concentrations of acidic aerosols will erode the terra-cotta warriors and horses, especially in the summer season with high temperature (30°C) and relative humidity (70 percent) and with undesirable solar radiation inside the museum.

Agriculture

Effects of air pollution on crops have been studied in China since the 1970s. Cao (1989) reported the effects of short-term exposure of three sensitive groups of crops to SO_2 and hydrofluoric acid (HF). The sensitive species are reported to have suffered 5 percent injury at 880±1430 mg/m^3 SO_2 and 12±48 mg/m^3 HF during an 8-hour exposure.

Black carbon is the aerosol most responsible for reducing atmospheric transparency and visibility by so much in China that agricultural productivity is reduced by an estimated 10-20 percent (Chameides et al., 1999), with additional loss from black carbon deposited on plant leaves (Bergin et al., 2001). Chameides et al. (1999) estimated that regional haze in China is currently depressing optimal yields of ~70 percent of the crops grown in China by at least 5-30 percent.

Chang and Hu (1996) reported that the average yield reduction for vegetables in Chongqing is 24 percent. Shu et al. (1993) estimated the annual cost of forest damage by acid rain in Guangxi province to be $80 million. Ou et al. (1996) estimated the economic loss due to acid rain damages to crops and materials in the Xiamen area to be about $6 million, which equals about 1 percent of the GDP for the area. Chang and Hu (1996) reported that the annual damage from air pollution in Chongqing in 1993 was about $220 million, which is 4.4 percent of the GDP.

A recent study on the impacts of ozone on agriculture in China found that reductions in crop yields in 1990 were less than 3 percent for most grain crops, but that predictions for 2020 suggested that crop losses for soybeans and spring wheat might reach 20 and 30 percent, respectively (Aunan et al., 2000). Wang and Mauzerall (2004) estimated that due to O_3 concentrations in 1990, China lost 1-9 percent of their yield of wheat, rice, and corn and 23-27 percent of their yield of soybeans, with an associated value of 1990 US$3.5 billion. In 2020, assuming no change in agricultural production practices, grain loss due to increased levels of O_3 pollution is projected to increase to 2-16 percent for wheat, rice, and corn, and 28-35 percent for soybeans; the associated economic costs are expected to increase 82 percent by 2020, compared to 1990 costs.

Acid Deposition

Acid rain emerged as an important environmental problem in China in the late 1970s. The region with the most serious acid rain problem in China is in the southwest; approximately 90 percent of the monitoring stations with a mean pH of less than 5.6 are located south of the Yangtze River (Zhao et al., 1988). The deposition of sulfur is in some places higher than what was reported from the "black triangle" in central Europe in the early 1980s (Larssen et al., 2006). Since 1989, an acid rain network including 88 stations has been operated by the China Meteorological Administration. Based on the observational data, China's acid rain problem experienced two stages in the last two decades. The first stage was characterized by rapid economic development and increasing acid deposition through the mid-1990s, followed by a period of relative stability which persists today. Although no further worsening has been observed in recent years, the situation is far from being ameliorated (Ding, 2004).

EMISSIONS AMOUNTS AND TRENDS

United States

In recent years, concern about air pollution in the United States has evolved to a focus on chronic exposures to non-lethal concentrations of air pollutants. Since the late 1960s and early 1970s, emission controls on industrial and vehicular sources of air pollution have resulted in dramatic improvements in some facets of air quality. For example, in the 1960s, most industrialized urban areas in the eastern United States exceeded air quality standards for ambient concentrations of sulfur dioxide. But, after three decades of control efforts, directed largely at industrial sources, there are now no areas in the United States that exceed the NAAQS for SO_2.

Carbon Monoxide (CO)

There are now only a few areas in the United States that continue to exceed NAAQS for carbon monoxide (NRC, 2003), due largely to control strategies that reduced CO emissions (per kilometer traveled) from motor vehicles by more than 95 percent.

Figure 3-2 shows the emissions of CO from 1983 to 2002.

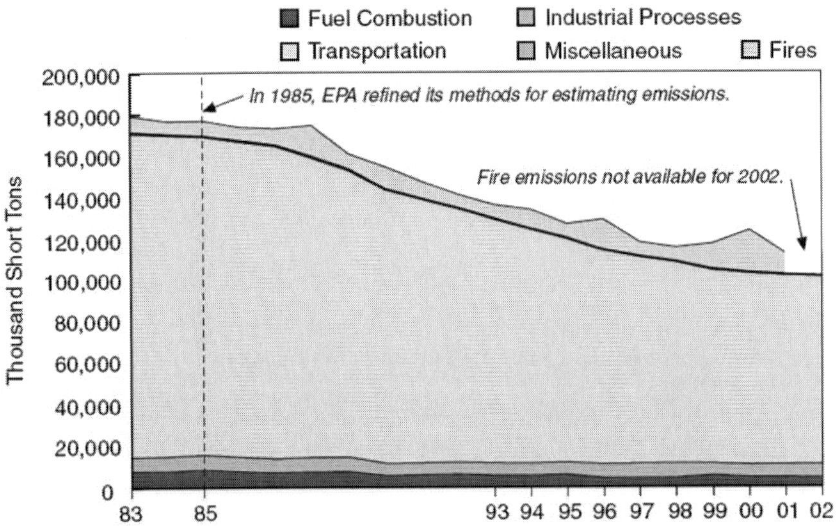

FIGURE 3-2 CO emissions by source, 1983-2002.
SOURCE: EPA, 2006b.

Nitrogen Oxides (NO_x)

Oxides of nitrogen arise from combustion sources including mobile and stationary sources. Although reductions of NO_x have resulted in meeting the NAAQS, further emission reductions have been mandated to reduce ozone and particulate matter concentrations. Figure 3-3 shows the EPA's estimates of reductions in NO_x emissions over the past 20 years. Some recent studies have suggested that these estimates may overestimate the amount of reductions achieved from mobile sources in recent years (Parrish, 2006).

Sulfur Dioxide (SO_2)

For SO_2, the major sources include the combustion of coal and oil, smelting of non-ferrous metals, petroleum refining, and other industrial processes. The overall trends in estimated SO_2 emissions are shown in Figure 3-4. Sulfur dioxide emissions are thought to be relatively well known, because they are dominated by large stationary sources with direct emissions monitors. As with NO_x, emissions of SO_2 are generally higher in the eastern United States than in the less densely populated western half of the country.

It should be noted that reductions in NO_x and SO_2 have occurred while the total generation of electricity has increased. Similarly, since 1970, aggregate emissions traditionally associated with vehicles have significantly decreased (with

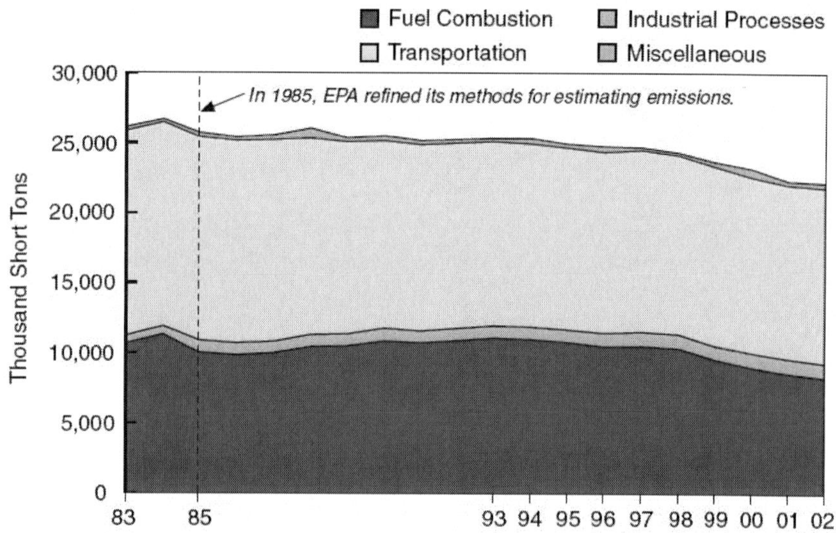

FIGURE 3-3 NO_x emissions by source, 1983-2002.
SOURCE: EPA, 2006b.

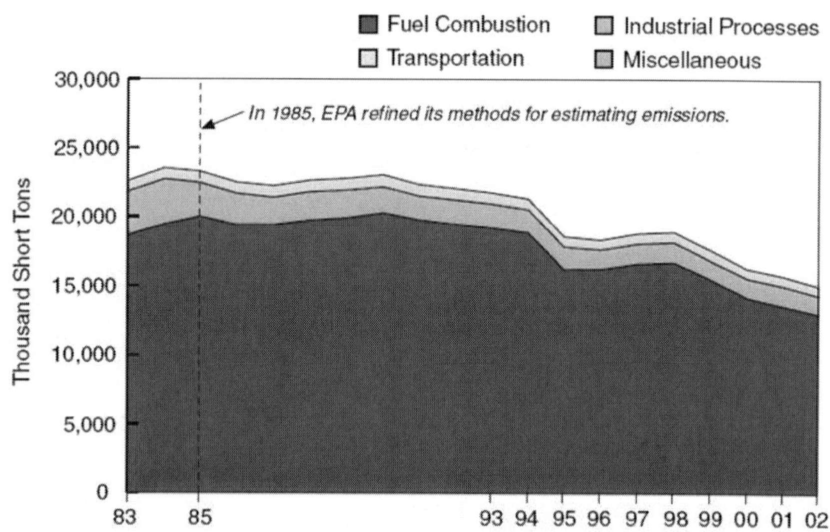

FIGURE 3-4 SO_2 emissions by source, 1983-2002.
SOURCE: EPA, 2006b.

the exception of NO_x), even as vehicle miles traveled (VMT) have increased by approximately 149 percent. NO_x emissions from vehicles increased between 1970 and 1999 by 16 percent, due mainly to emissions from light-duty trucks and heavy-duty vehicles. However, as recent trends show, vehicle travel is having a smaller and smaller impact on emissions as a result of stricter engine and fuel standards, even with additional growth in VMT.

PM_{10}

At this time, relatively few locations in the United States are in violation of the PM_{10} NAAQS. Emissions trends for PM_{10} are shown in Figure 3-5. The trends data reflect emissions from fuel combustion, transportation, and industrial production, but do not include emissions of PM_{10} from "fugitive" sources, including windblown dust and agricultural activities.

$PM_{2.5}$

With the promulgation of a NAAQS for $PM_{2.5}$ in 1997, concern for fine particles and their emission became a higher priority. Figure 3-6 presents the trends in the primary $PM_{2.5}$ emissions where there is a truncated time series, since it was

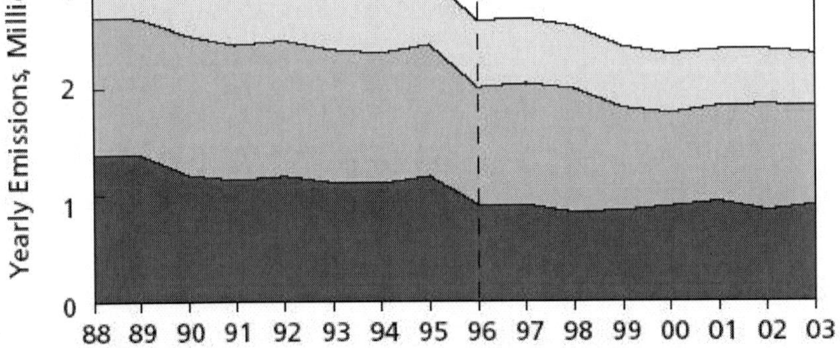

FIGURE 3-5 PM_{10} emissions by source, 1988-2003.
SOURCE: EPA, 2006b.

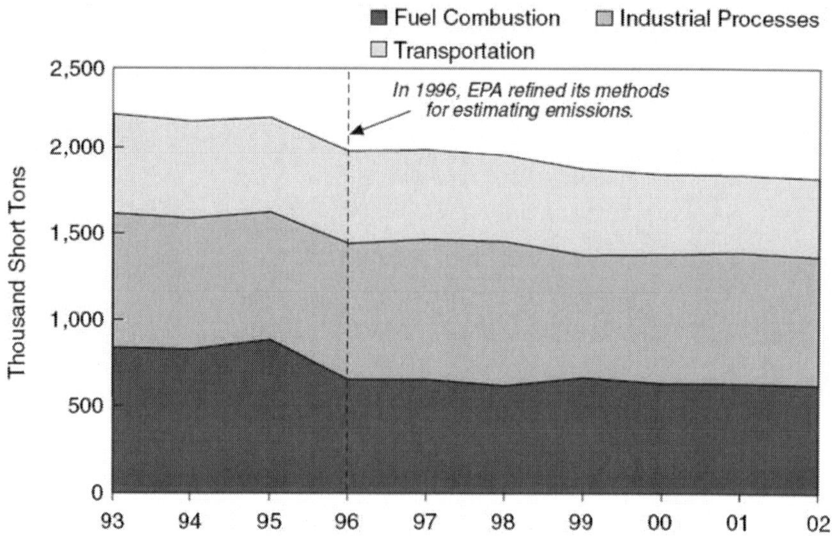

FIGURE 3-6 $PM_{2.5}$ emissions by source, 1993-2002.
SOURCE: EPA, 2006b.

not until 1993 that this size fraction was separately inventoried. Most of the $PM_{2.5}$ is secondary in nature and thus emissions of SO_2, NO_x, and VOC (volatile organic compounds) that serve as the precursors to the secondary fine particles are also important to understanding fine particle concentrations (Turpin and Huntzicker, 1995; Cabada et al., 2004). Assessing $PM_{2.5}$ sources is a complicated issue, as the fractions (primary versus secondary) vary by pollution sources, meteorology, and local geography (Seinfeld and Pankow, 2003; Kanakidou et al., 2005; Pun and Seigneur, 2007). Moreover, reductions in precursors will not necessarily lead to reductions in $PM_{2.5}$ (Gaydos et al., 2005; Vayenas et al., 2005; Robinson et al., 2007), and therefore accurate emissions inventories for all precursors are critical, as are their inclusion in monitoring and modeling, in order to analyze and characterize $PM_{2.5}$ formation.

Lead (Pb)

The final criteria pollutant for which there are direct emissions into the atmosphere is lead in total suspended particles. The trend for Pb in TSP is presented in Figure 3-7. It can be seen that there was a very sharp drop in lead emissions during the early 1980s, as lead was phased out of gasoline. Complete removal was achieved by 1988, except for Alaska where leaded fuel ceased to be sold in 1991. Today, the principal remaining lead emissions arise from primary and

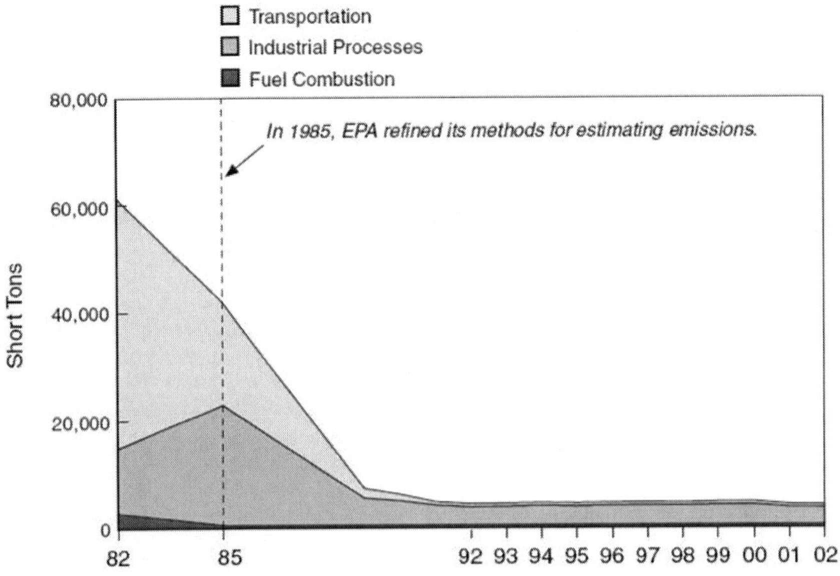

FIGURE 3-7 Lead emissions by source, 1982-2002.
SOURCE: EPA, 2006b.

secondary lead smelting and from other non-ferrous metal processes, rather than from gasoline combustion (see Box 3-2).

Mercury

Mercury is regulated in the United States as a hazardous air pollutant. The EPA estimates that from 1990 to 1999, anthropogenic emissions of mercury in the United States declined from 220 short tons per year to 115 short tons per year, due largely to reductions in emissions from medical waste incinerators and municipal waste combustors (EPA, 2007a). As of 1999, utility coal boilers were the largest source of mercury emissions, producing about 45 short tons per year of emissions in the United States. The Clean Air Mercury Rule adopted in 2005 would limit emissions from this sector to 38 short tons in 2010 and to 15 short tons in 2018. More rapid reductions may be achieved in response to state-level regulations.

Volatile Organic Compounds (VOCs)

VOCs are regulated in the United States as an important precursor of ozone, and because some VOCs are toxic in their own right. Emissions of VOCs have

BOX 3-2
Mobile Source Emissions in the United States

Emissions from motor vehicles are grouped into two main categories: (1) major gaseous and particulate air pollutants, which can be found in relatively high amounts in the atmosphere; and (2) air toxics, which usually are found in smaller concentrations, but are carcinogens, causing cancers and other adverse human health effects at very low levels. The major gaseous and particulate pollutants to which motor vehicles contribute include carbon monoxide (CO); ozone (O_3)—through its atmospheric precursors volatile organic compounds (VOCs) and nitrogen oxides (NO_x); fine particulate matter, PM_{10} and $PM_{2.5}$; and nitrogen dioxide (NO_2). Transportation sources account for a small share of total SO_x emissions, but emissions from vehicles increased from 1970 to 1999, largely due to increased emissions from non-road vehicles. The toxics emitted from motor vehicles include aldehydes (acetaldehyde, formaldehyde, etc.), benzene, 1,3-butadiene, and a large family of substances known as polycyclic organic matter (including polycyclic aromatic hydrocarbons, or PAHs).

On-road vehicles are the dominant source of emissions from the transportation sector. Light-duty gasoline-powered vehicles (cars and trucks) account for the predominant share of CO, NO_x, and VOC emissions from on-road vehicles, but, in the United States, their share of all criteria pollutants has been decreasing over time.

declined due to a variety of regulatory actions, including more stringent controls on exhaust emissions and evaporative fuel losses from vehicles, better control of emissions in the refining and distribution of motor vehicle fuels, and the imposition of controls on a variety of industrial and mobile emissions sources under regulatory programs designed to address toxic air pollution. The trends in estimated VOC emissions are shown in Figure 3-8.

China

China has conducted emission control since the 1980s, beginning later than in the United States. At first, emission control was focused on dust emissions and then on soot emissions, because of economic limitations. The central government has encouraged desulfurization measures for power plants and industrial emissions, but little action was taken prior to 2000. More stringent policies, such as an enhanced levy system, have been implemented by the government in recent years. The reduction of SO_2 emissions and improvement of ambient air quality for SO_2 has been observed in some big cities, such as Beijing.

China currently has nationwide statistics of emission amounts only on three major pollutants: SO_2, soot, and industrial dust. Industrial soot emission refers

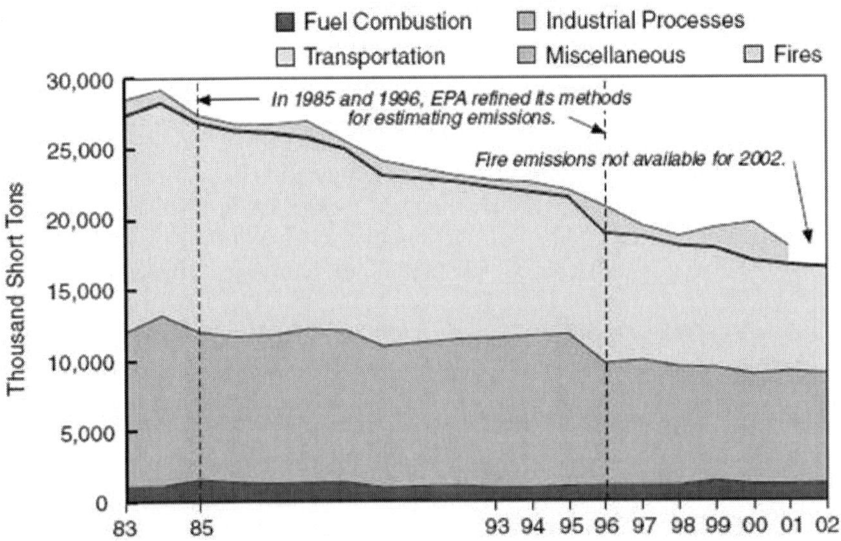

FIGURE 3-8 VOC emissions by source, 1983-2002.
SOURCE: EPA, 2006b.

to PM produced as a result of fuel combustion (and includes power plants as an industrial source). Other soot refers to fuel combustion from all other social and economic activities and operation of public facilities. It is calculated on the basis of estimated coal consumption by households and others. Industrial dust refers to the volume of PM emitted by industrial processes and suspended in the air for a given period of time, such as dust from refractory material of iron and steel works, dust from coke-screening systems and sintering machines of coke plants, dust from lime kilns, and dust from cement production in building material enterprises, but excluding soot and dust emitted from power plants. In this section, most historical data on emissions only go back to 1997. While data exist for years prior to 1997, those data do not capture emissions from townships and counties and, therefore, are less useful when compared to more recent data.

As discussed below, during 1994-2005, emissions of major air pollutants in China increased year after year. Among them, the emission of air pollutants from industrial sources had a dramatic increase. This is closely related to the sustained growth of the national economy and the mode of economic growth in China. Meanwhile, emission of air pollutants from residential sources was under fairly good control, resulting in a steadily declining trend of the emissions of sulfur dioxide and soot. This is related to the gradual implementation of clean energy strategies and integrated pollution prevention and control programs in cities. But exhaust emissions of automobiles were rising rapidly and are becoming a major source of urban atmospheric pollution. See Box 3-3 at the end of this section for more information on mobile sources in China.

SO_2

In recent years, owing to sustained economic growth, particularly in heavy industrial sectors such as steel, cement, and aluminum, China's industrial energy consumption, and consequently SO_2 emissions, have steadily increased (Figure 3-9). Over this same period, residential emissions have remained relatively stable.

Streets and Waldhoff (2000) estimated that SO_2 emissions are projected to increase from 25.2 mt in 1995 to 30.6 mt in 2020, provided emission controls are implemented on major power plants; if this does not happen, emissions could increase to as much as 60.7 mt by 2020. Emissions are concentrated in the populated and industrialized areas of China: the northeastern plain, the east central and southeastern provinces, and the Sichuan Basin (Streets and Waldhoff, 2000).

In 2004, over 52 percent of all China's SO_2 emissions came from just nine provinces: Shandong, Hebei, Shanxi, Guizhou, Sichuan, Henan, Jiangsu, Inner Mongolia, and Guangdong. Shandong province emitted the largest amount of SO_2 from industrial sources, approximately 8.2 percent of China's total industrial SO_2 emissions. This is a result of energy-intensive heavy industrial activity, but also of high-sulfur content in the local coal (SSB, 2006). Guizhou province reported the largest amount of SO_2 emissions from residential sources, approximately 19.7 percent of China's total residential SO_2 emissions, due to the prevalence of coal combustion for cooking and heating, as well as high-sulfur content (up to 2.5 percent) in the local coal (Ministry of Commerce, 2007; Xiao and Liu, 2004).

An examination of the five most heavily polluting industrial sectors[1] from 2000-2004 reveals that their SO_2 emissions comprised, on average, more than 80 percent of China's total, while their economic contributions averaged less than one-third of the total (SEPA, 2005a).

Since 2004, SO_2 emissions from coal-fired power plants are estimated at 9.3 Mt in China, accounting for 49 percent of total industrial emissions. The top five provinces for power plant SO_2 emission are Shandong, Hebei, Henan, Inner Mongolia, and Shanxi, which emitted 33.8 percent of total SO_2 emissions from power plants in China. Of 1,196 coal-fired power plants included in the statistics on emissions, only 425 have installed desulfurization facilities; they only reduced 1.57 million tons of SO_2 emissions, which is much below their desulfurization capacity. Thus, China still has significant room for reducing emissions of SO_2.

[1]Production/supply of electric power, gas, and water; non-metal mineral production; smelting/pressing of ferrous metals; raw chemicals and chemical products; and smelting/pressing of non-ferrous metals.

BOX 3-3
Mobile Sources in China

Between 2000 and 2004 the passenger car fleet in China more than doubled and the diesel truck fleet has also grown rapidly: heavy-duty vehicles more than tripled between 1998 and 2002, and light-duty diesel trucks doubled in the same time frame. The motorcycle fleet, which includes many highly polluting two-stroke vehicles, also doubled over the same time period (He et al., 2005). Rapid urbanization and the improvement of the standard of living are expected to stimulate the purchase and use of motor vehicles, which will have a significant impact on urban air pollution.

In general, mobile sources are currently contributing 45-60 percent of NO_x emissions, 40-90 percent of VOC emissions, and about 80-90 percent of CO emissions in typical Chinese cities (Wang et al., 2005). More important than the relative contribution of vehicles and other sources to overall emissions is the contribution of motor vehicle emissions to personal exposure; because vehicles emit at ground level near the breathing zones of people, their emissions are frequently more important than is reflected by the total tonnage emitted. In particular, in urban centers, along roadsides, and especially in urban street canyons in crowded central business districts, mobile sources can contribute 2 to 10 times as much pollution as in general background situations.

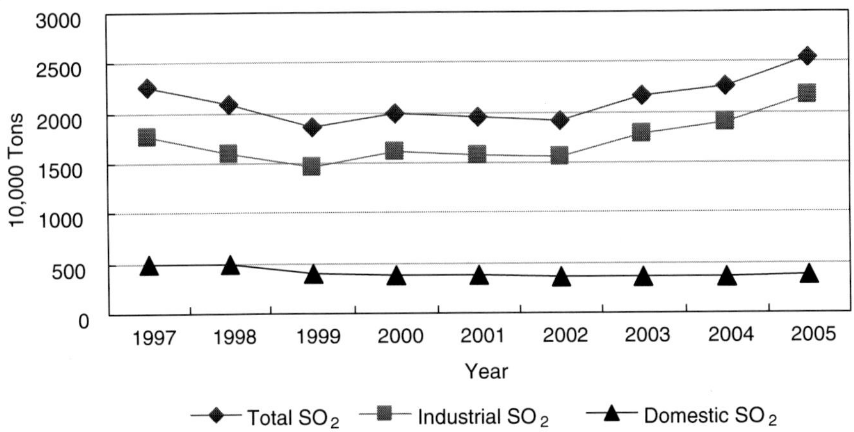

FIGURE 3-9 Trends of emission amounts of SO_2 in China, 1997-2005.
SOURCE: SEPA, 2001a, 2005b, 2006.

Soot

As shown in Figure 3-10, industrial soot emissions decreased remarkably in the last decade of the 20th century. But, these emissions, like SO_2 emissions, have increased steadily in recent years.

The provinces where emissions exceeded 600,000 tons in order were Shanxi, Sichuan, Henan, Hebei, and Inner Mongolia. These five provinces emitted 37.5 percent of total soot emissions in China. The major emission sectors are electric power, non-metal mineral products, and smelting and pressing of ferrous metals, which accounted for 67 percent of the total emissions from industry; among them electric power contributed 44.2 percent.

Industrial Dust

The emission of industrial dust has not changed much since the year 2000. The trend is decreasing (see Figure 3-10), probably due to the effectiveness of dust removal by industries. Figure 3-11 shows the emissions of industrial dust by province in 2004. The provinces where emissions exceeded 600,000 tons per year were Hunan, Hebei, Henan, and Shanxi, in that order. Their emissions contributed 31.4 percent of the total industrial emissions. The non-metal mineral sector emitted 70.2 percent of the total dust emission from China's industries and the smelting and pressing of ferrous metals sector contributed 15.2 percent.

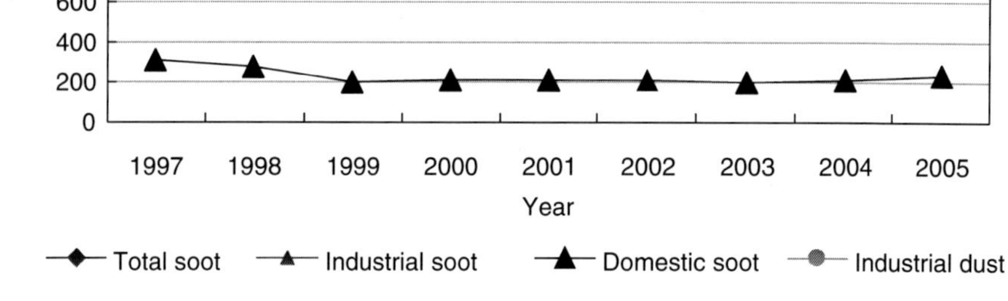

FIGURE 3-10 Trends of emission amounts of soot and industrial dust in China, 1997-2005.

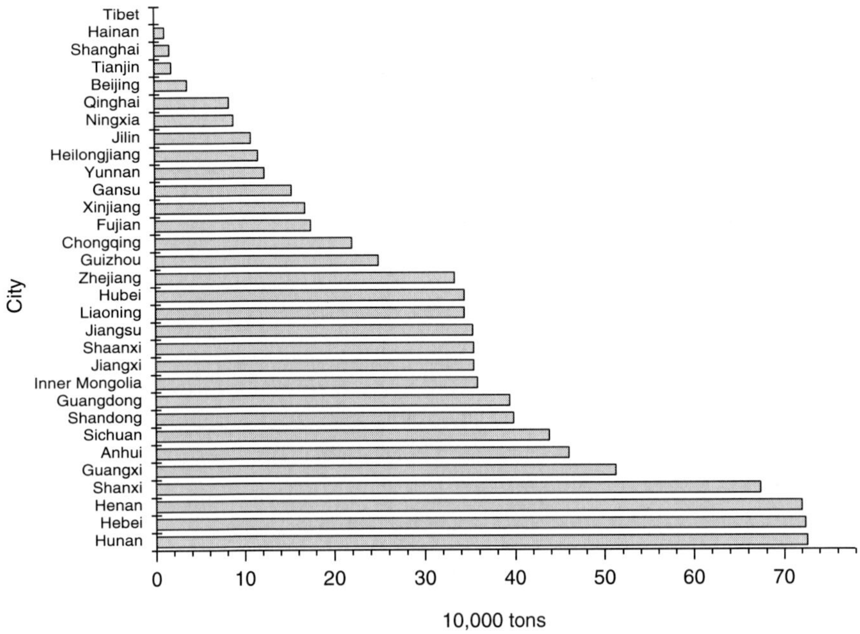

FIGURE 3-11 The distribution of industrial dust emission amount by province in 2004.
SOURCE: SEPA, 2005a.

Mercury

Researchers at Tsinghua University and Argonne National Laboratory have collaborated to develop multiyear inventories of mercury emissions in China, covering the period from 1995 to 2003. They estimate that mercury emissions from anthropogenic sources in China totaled about 700 tons in 2003, up from about 550 tons in 1995. The major sources of anthropogenic mercury emissions were non-ferrous metal smelting and coal combustion, with the latter contributing about 260 tons in 2003, mainly from the industrial and power sectors. While emissions rose nationwide, Wu et al. (2006) concluded that mercury emissions in some provinces had declined, for example in Liaoning Province, due to reduced metal smelting, and in Beijing Province due to an increased use of pollution control technology in the power sector.

AMBIENT CONCENTRATIONS

United States

In the United States, ambient concentrations are measured for the major criteria pollutants, CO, NO_2, SO_2, O_3, PM, and lead in total suspended particles, as well as for selected VOCs and other air toxics. Approximately 25 years of data are available from 1980 through 2005.

Carbon Monoxide (CO)

Carbon monoxide pollution is primarily related to emissions from motor vehicles. In the United States, improved catalytic converter technology coupled with improved fuel quality has led to a steady decline in ambient concentrations as shown in Figure 3-12.

Nitrogen Dioxide (NO_2)

The NAAQS for oxides of nitrogen are designated for NO_2. The trend in ambient NO_2 concentrations in the United States is shown in Figure 3-13. The NAAQS is currently met at all of the locations where NO_2 is monitored. The highest concentrations are in southern and central California.

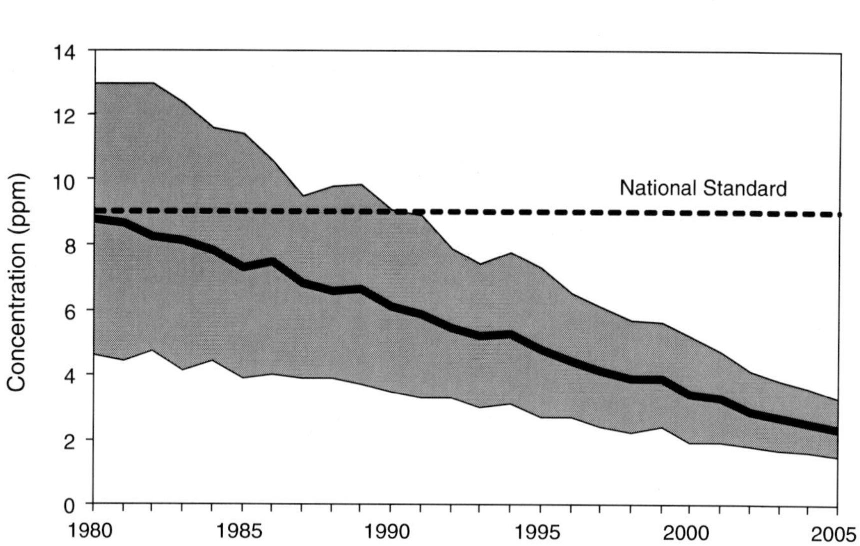

FIGURE 3-12 CO concentrations, 1980-2005, based on annual second maximum 8-hour average. National trend based on 152 sites.
SOURCE: EPA, 2007b.

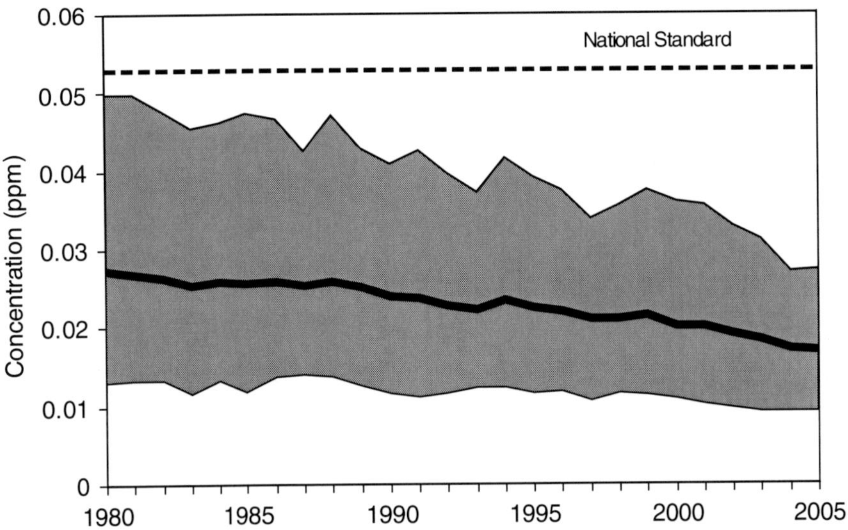

FIGURE 3-13 NO_2 concentrations, 1980-2005, based on annual arithmetic average. National trend based on 88 sites.
SOURCE: EPA, 2007b.

Sulfur Dioxide (SO_2)

The ambient concentrations of SO_2 have also declined over the past two decades (Figure 3-14). The major drop in ground-level SO_2 occurred in the 1970s, and all sites in the United States now meet the NAAQS for SO_2.

PM_{10}

As discussed previously, airborne particulate matter is characterized as PM_{10} and $PM_{2.5}$. PM_{10} concentrations have declined in the United States since a standard was imposed for this size class in 1987, as shown in Figure 3-15.

$PM_{2.5}$

Data for $PM_{2.5}$ have been collected over a much shorter time period, with national trend data available since 1999, as shown in Figure 3-16. The most significant concentrations are in southern and central California and in many of the large urban areas of the midwestern and eastern United States (e.g., Pittsburgh).

The United States has also been making measurements of fine particle composition. Beginning in 1988, the IMPROVE program has been making particle

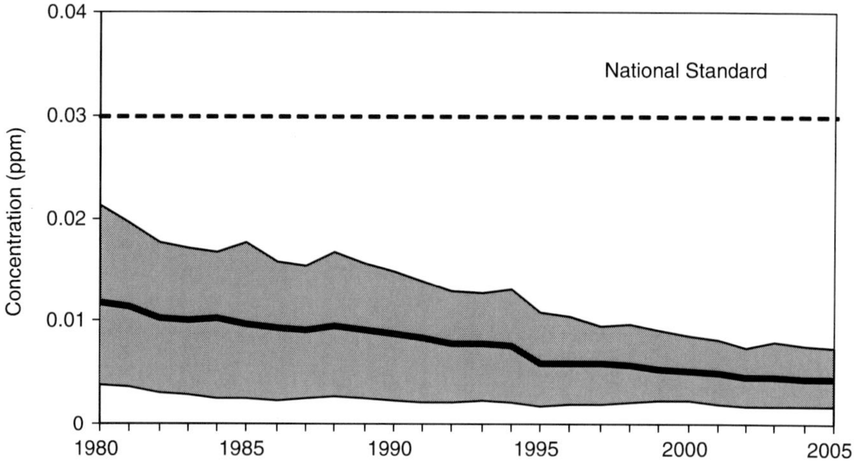

FIGURE 3-14 SO_2 concentrations, 1980-2005, based on annual arithmetic average. National trend based on 163 sites.
SOURCE: EPA, 2007b.

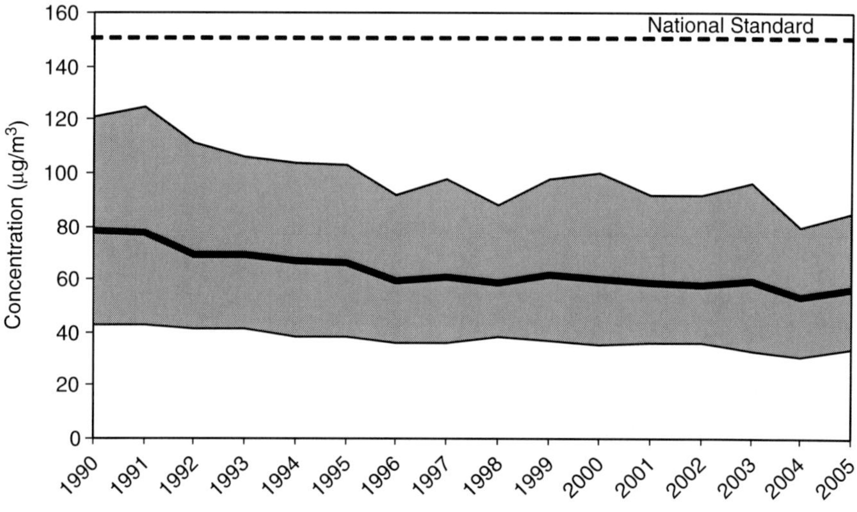

FIGURE 3-15 PM_{10} concentrations, 1990-2005, based on seasonally weighted annual average. National trend based on 435 sites.
SOURCE: EPA, 2007b.

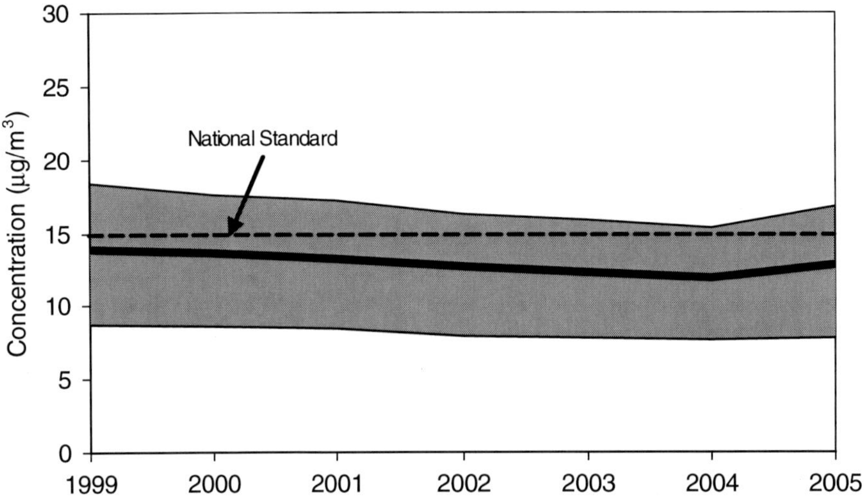

FIGURE 3-16 $PM_{2.5}$ concentrations, 1999-2005, based on seasonally weighted annual average. National trend based on 658 sites.
SOURCE: EPA, 2007b.

composition measurements in national parks, forests, and other areas where visibility degradation is a concern. In 2001, the IMPROVE network, which had focused primarily on the west, added sites in the eastern and central United States. Data from this network indicate that in the eastern United States, the rural $PM_{2.5}$ is dominated by sulfate with some nitrate and carbon, while the western particle compositions are more dominated by nitrate in California and carbonaceous aerosol in much of the rest of the region. A second composition monitoring network, the Speciation Trends Network, was initiated by the EPA in 2000 to 2001 to provide composition data from urban sites across the United States. The composition patterns in these urban locations are presented in Figure 3-17. Sulfate and carbonaceous particles are prominent in urban aerosols measured at the eastern STN sites. Urban sites in California show a relatively large influence from nitrate compared to eastern cities.

Lead (Pb)

Lead in total suspended particulate decreased sharply in the 1980s, as the use of leaded gasoline declined. Figure 3-18 presents the trends in airborne Pb concentration.

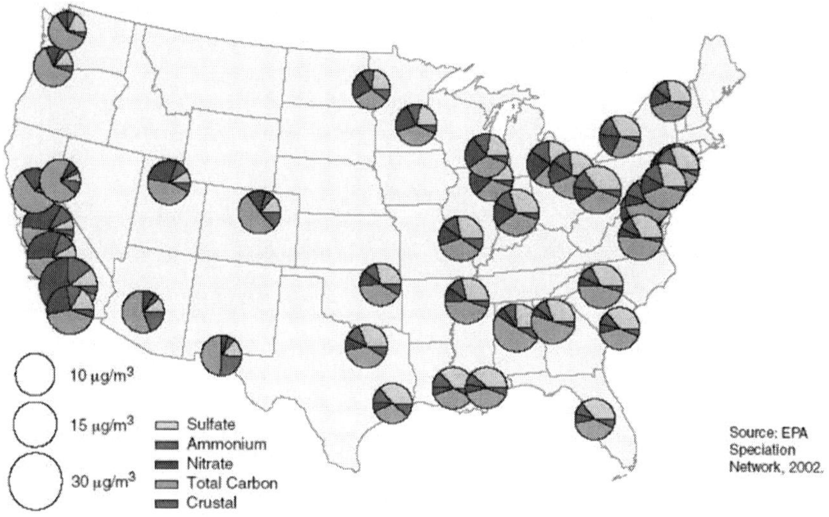

FIGURE 3-17 Annual average $PM_{2.5}$ concentrations ($\mu g/m^3$) and particle type in urban areas, 2002.

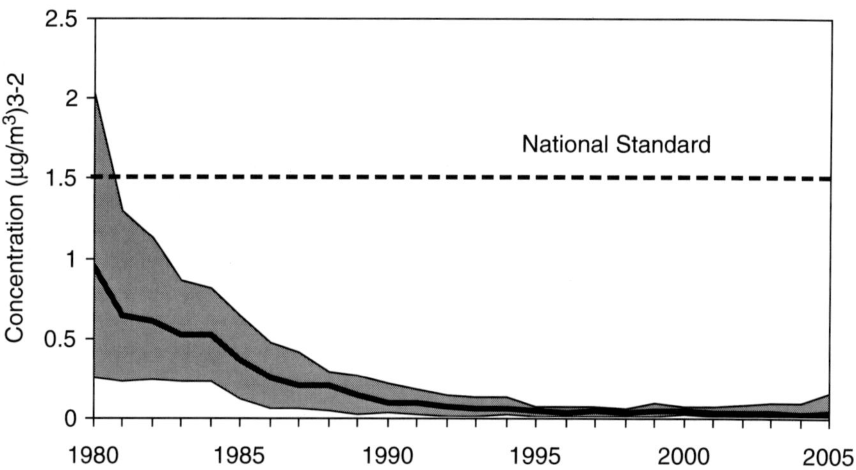

FIGURE 3-18 Lead concentrations, 1980-2005, based on annual maximum quarterly average. National trend based on 16 sites.
SOURCE: EPA, 2007b.

Ozone

Ozone is a secondary pollutant formed in the atmosphere from reactions of nitrogen oxides and VOCs. The trends in ozone concentrations are shown in Figure 3-19, and the geographical distribution of measured ozone concentrations are given in Figure 3-20. Considering country-wide trends, ozone concentrations have declined modestly since 1980. However, many parts of the United States do not meet the current ozone NAAQS.

Until the early 1990s, much of the focus in ozone control was on the reduction of VOC concentrations. The 1990 Clean Air Act Amendments required areas in moderate to severe non-attainment of the ozone standard to make additional measurement of the VOC precursors, as part of the Photochemical Assessment Monitoring Program Stations network. The trends in VOC values from 1995 to 2001 are shown in Figure 3-21.

It can be seen that there was a greater decrease in VOC concentrations than in the commensurate levels of ozone. In many locations, it is now recognized that control of NO_x is a critical pathway to ozone control. However, it is clear that for many areas it will be difficult to achieve the current 8-hour ozone standard, and the current value is under review in consideration of recommendations from the Clean Air Scientific Advisory Committee to reduce the 8-hour standard concentration to no greater than 70 ppb.

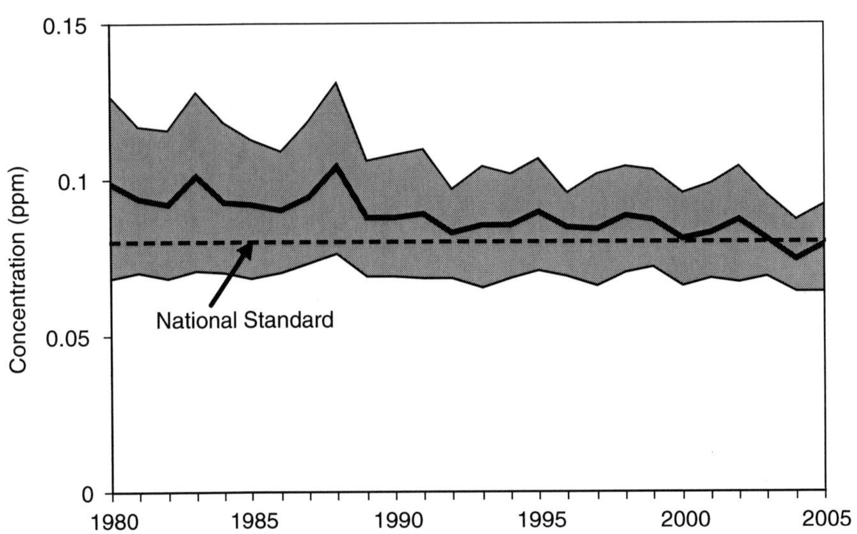

FIGURE 3-19 Ozone concentrations, 1980-2005, based on annual fourth maximum 8-hour average. National trend based on 286 sites.
SOURCE: EPA, 2007b.

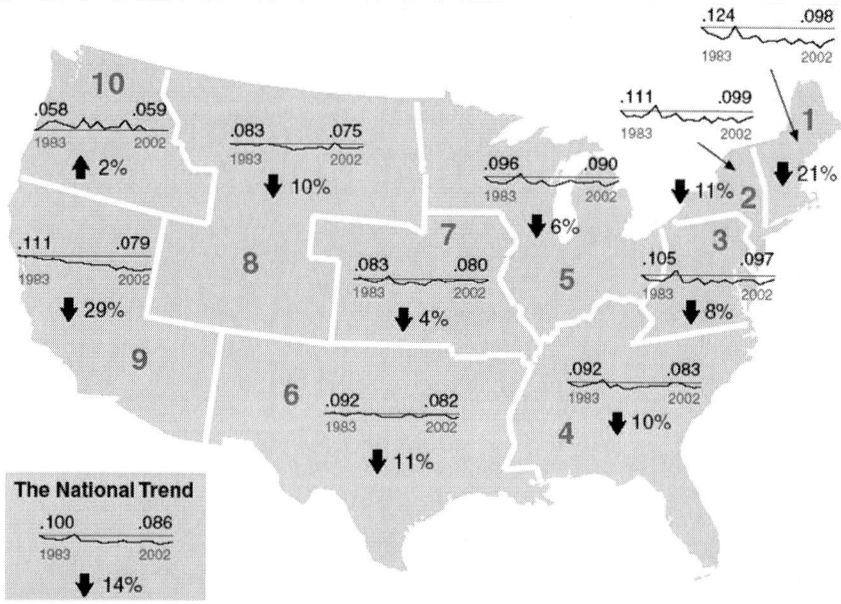

FIGURE 3-20 Trend in 8-hour O_3 levels, 1983-2002, averaged across EPA regions, based on annual fourth maximum 8-hour average.
SOURCE: EPA, 2003.

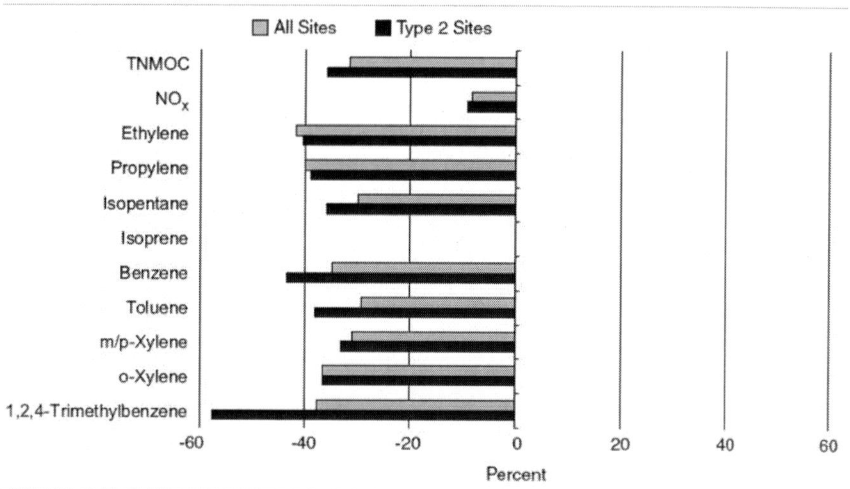

FIGURE 3-21 Median percent change (1995-2001) at PAMS monitors for selected VOC species.
SOURCE: EPA, 2003.

China

In China, ambient concentrations are measured for the major criteria pollutants, CO, NO_x, SO_2, TSP/ PM_{10}, O_3, and Pb. China issued its first ambient air quality standards in 1982 and revised them in 1996 and 2000. The National Air Quality Standard of China divides the standard levels into three grades. For protecting residential health, it is required that the ambient air quality must achieve the Grade II level. Monitoring is the responsibility of local environmental protection bureaus, which typically report air quality in terms of the standard(s) achieved (Ref. to Chapter 4 of this report). Among the criteria pollutants, only four are routinely monitored for the whole country. Monitoring of ozone and lead is not required for every city, so they are not included in the national air quality statistics.

Results shown in Figure 3-22 from the routine monitoring network of 360 cities for the period from 1999-2005 reveal that the air quality has improved, but nearly 40 percent of urban areas do not meet the Grade II air quality standards.

China has high levels of SO_2 and TSP due in large part to coal use. Meanwhile, the number of motor vehicles has increased substantially since the mid-1980s, primarily in urban areas and in city clusters, leading to increased emissions of NO_x, VOCs, and particulates, and causing higher levels of ozone in the summer in urban areas—and much higher levels of inhalable particulates (PM_{10}

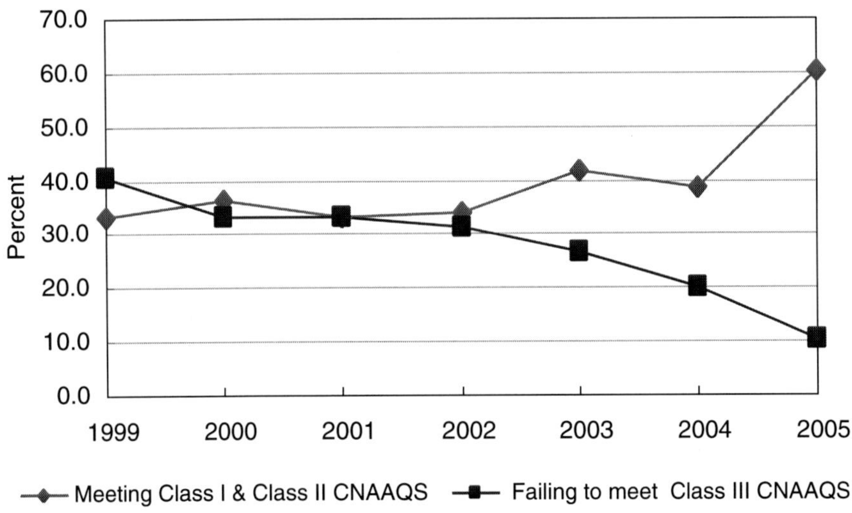

FIGURE 3-22 The percentage of 360 cities achieving different levels of air quality standards from 1999-2005.
SOURCES: SEPA, 2000, 2001a, 2001b, 2002, 2003, 2004, 2005a, 2005b, 2006.

and $PM_{2.5}$) throughout the country. In general, urban air quality across China basically remains stable against a background of rapid economic growth and accelerated urbanization. In addition, the air quality of some cities enjoyed some improvement.

CO

Carbon monoxide was observed in high concentration in China's large cities. Recently, strict emission standards and improved catalytic converter technology have led to a steady decline in ambient concentrations as shown in Figure 3-23.

NO_x /NO_2

In China, the concentration limit for NO_x was replaced by a limit for NO_2 in 2000. As shown in Figure 3-24, after the year 2000, the ambient air concentration of NO_2 usually did not exceed the Grade II standard. Ambient concentrations were relatively high in some major cities, such as Guangzhou, Beijing, Shanghai, Hangzhou, Harbin, Urumqi, Nanjing, Chengdu, and Wuhan.

It is important to note that, following the revision in 2000 to measure NO_2 instead of NO_x, most monitoring sites were not adjusted, and thus their locations, often at street level, are not suitable to accurately measure the true NO_x concentrations in the urban atmosphere. NO_x concentrations are generally higher than the NO_2 measurements reveal, but are no longer being reported, and the result has

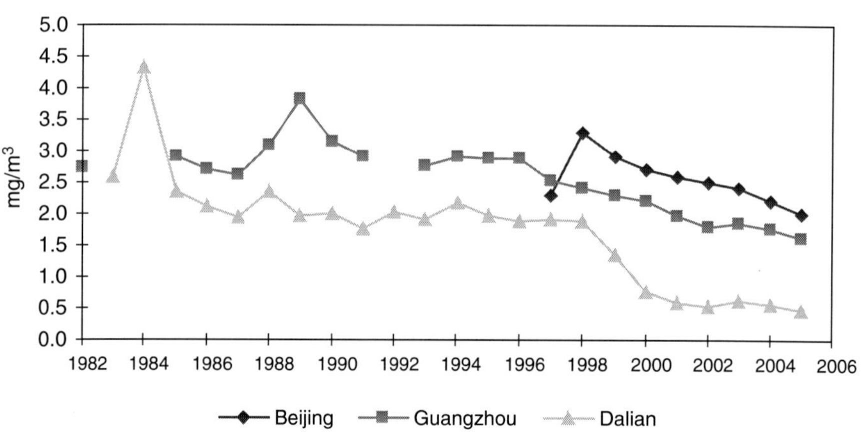

FIGURE 3-23 Carbon monoxide ambient concentration trend in three large cities in China.
SOURCE: Beijing, Guangzhou, and Dalian EPB annual reports.

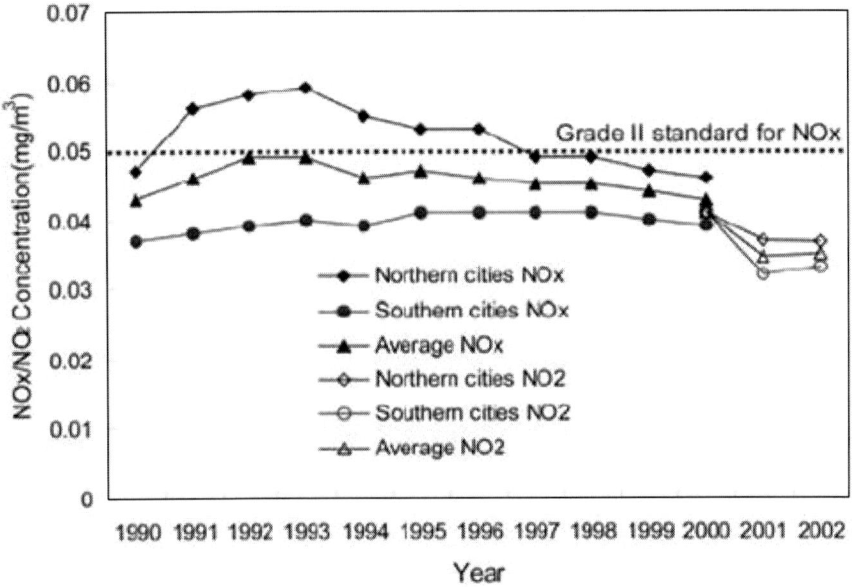

FIGURE 3-24 Average annual urban NO_x concentrations in China from 1990-2002.
NOTE: NO_x and NO_2 concentrations in 2000 are not exact because of the lack of the data of some cities.
SOURCE: Hao and Wang, 2005.

been that many cities, while ostensibly satisfying Grade II standards for NO_2, are in reality suffering from NO_x pollution. Furthermore, these cities have focused on other criteria pollutants which exceed the Grade II standard and have paid much less attention to NO_x emissions reduction strategies.

Recent satellite observations have also provided a new insight into China's air pollution. In reviewing NO_2 column data from two satellites, GOME and SCIAMACHY, from the period 1996-2002, a significant increase on the order of 50 percent was observed over China, which suggested a larger and more rapid increase in NO_2 emissions than local and national inventories might suggest (Richter et al., 2005; Irie et al., 2005). Wang et al. (2007) used data from the Dutch-Finnish Ozone Monitoring Instrument to show that aggressive measures taken by Beijing to restrict motor vehicle traffic during the 2006 Sino-African Summit resulted in NO_x emission reductions of more than 40 percent.

SO_2

SO_2 is a major air pollutant in China because nearly 70 percent of the fuel used is coal. Routine monitoring of SO_2 started in the 1970s. Figure 3-25 shows

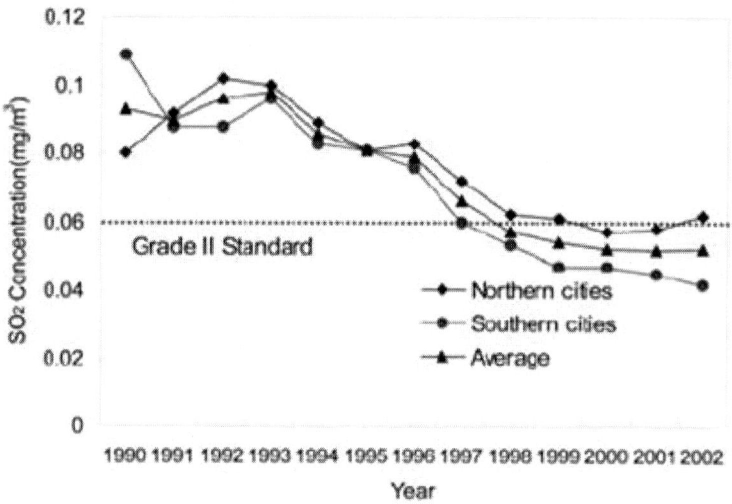

FIGURE 3-25 Average annual urban SO_2 concentrations in China from 1990-2002.
SOURCE: Hao and Wang, 2005.

that in the northern cities and southern cities, the average concentration of SO_2 decreased dramatically from 1990 to 2000, at which point average concentrations began to rise again (see also Figure 3-26). The higher SO_2 levels in the northern cities are the result of a combination of factors. Most regions of China experience higher SO_2 concentrations in the winter as a result of increased heating needs, as coal continues to be the dominant source. Northern cities have a greater heating load, but they also have a large concentration of energy-intensive industries which contribute to higher year-round SO_2 emissions.

TSP/PM$_{10}$

Figure 3-27 shows that concentrations of TSP have declined across China since 1990. This has been accomplished as a result of improved regulation, the installation of pollution controls (e.g., fabric filters), and the shuttering or relocating of certain industries. Still, many cities continue to face challenges in reducing PM concentrations.

Since 2000, high concentrations of PM_{10} have been the most frequent cause of Class II violations in China. In Beijing, the annual average level of PM_{10} fluctuated around 160 μg/m^3 from 2000 to 2004 (Beijing EPB, 2006). Megacities such as Beijing, Shanghai, and Guangzhou are frequently among the cities of the world with the highest levels of airborne PM (UNEP, 2002). Large areas of China are exposed to high levels of particulate pollution.

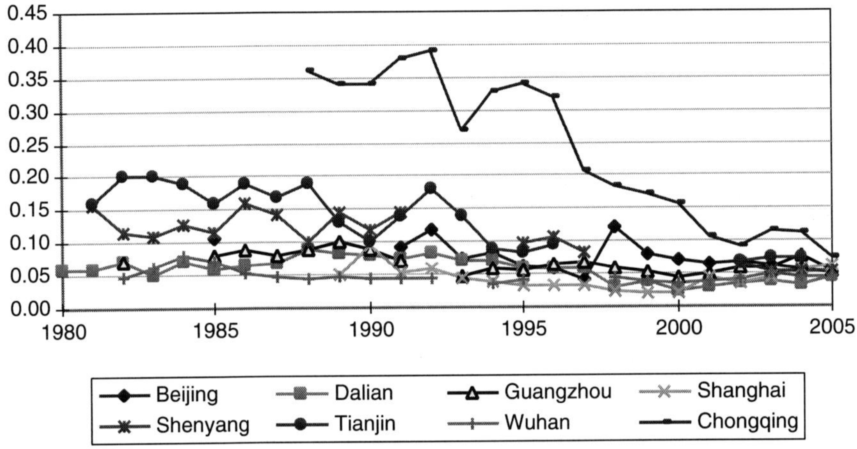

FIGURE 3-26 Average annual SO_2 concentrations in several large cities, China, 1980-2005 (mg/m^3).

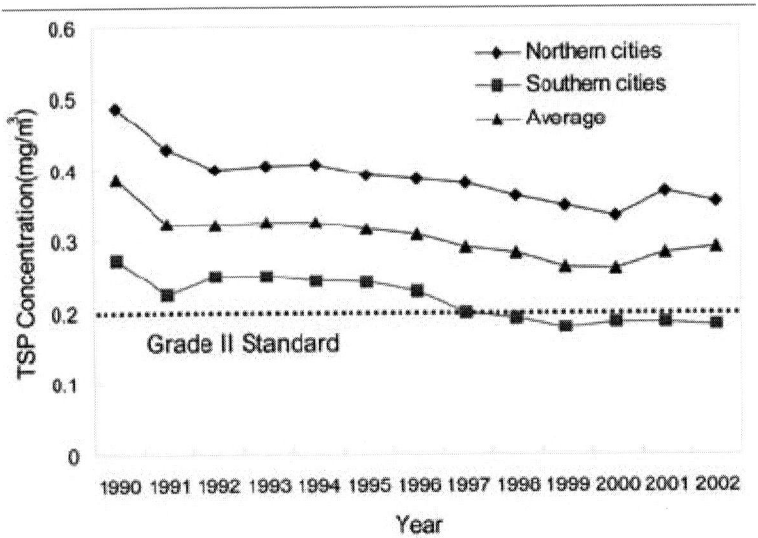

FIGURE 3-27 Average annual urban TSP concentration in China from 1990-2002.
SOURCE: Hao and Wang, 2005.

$PM_{2.5}$

China has not yet set standards for $PM_{2.5}$. Unlike other air pollutants mentioned before, all the information reported here for $PM_{2.5}$ comes from individual research studies. In China, much work has been done to characterize ambient concentration, chemical composition, size distribution, optical properties, seasonal variation, horizontal and vertical profiles, transport, and source-receptor relationships. Ambient air quality measurements of $PM_{2.5}$ have been made in megacities, such as Beijing, Shanghai, and Guangzhou, and at the regional scale, for example, in the Pearl River Delta region and theYangtze River Delta region. Results show that $PM_{2.5}$ has very high concentration levels, sometimes close to100 μg/m^3. The ratio of $PM_{2.5}$ to PM_{10} is about 50-70 percent. Thus, $PM_{2.5}$ is an important air pollutant in urban areas and is especially important in regional pollution (Figure 3-28).

Ozone

There are only a few cities in China routinely monitoring for ozone, including Beijing and Lanzhou in Gansu province. In fact, high concentrations of

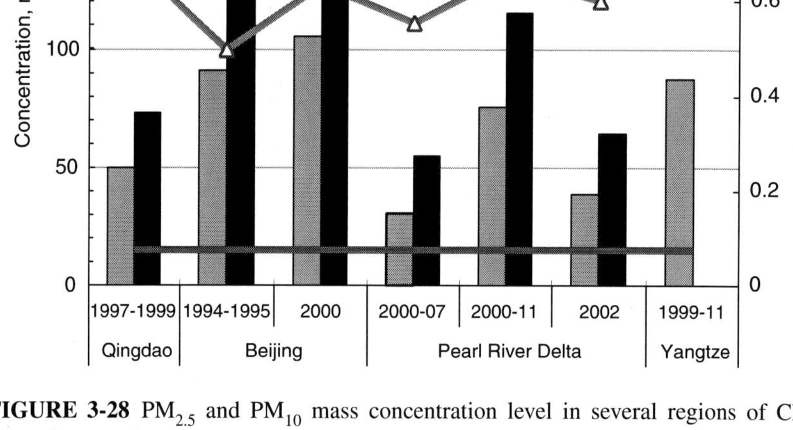

FIGURE 3-28 $PM_{2.5}$ and PM_{10} mass concentration level in several regions of China (annual average).
SOURCE: Peking University.

ground-level ozone have been observed for many years in several of China's urban areas.

In the mid-1970s, photochemical smog first appeared in the Xigu petroleum industry district in Lanzhou. In 1986, Beijing also experienced photochemical smog in the summer. There the O_3 level gradually increased later in the day exceeding Class II of the CNAAQS. Researchers at Peking University measuring the diurnal variations of episodic ground-level ozone found that concentrations have increased sharply since the 1990s, and often exceed 200 ppb (Figure 3-29). Systematic monitoring data collected by the Beijing Municipal Environment Monitoring Center showed that the hours of O_3 concentration exceeding Grade II and the days of exceeding Grade II are still very high (see Figure 3-30).

More recently, some southern cities, especially coastal cities, have faced the threat of photochemical smog in the summer and fall, even on the regional scale.

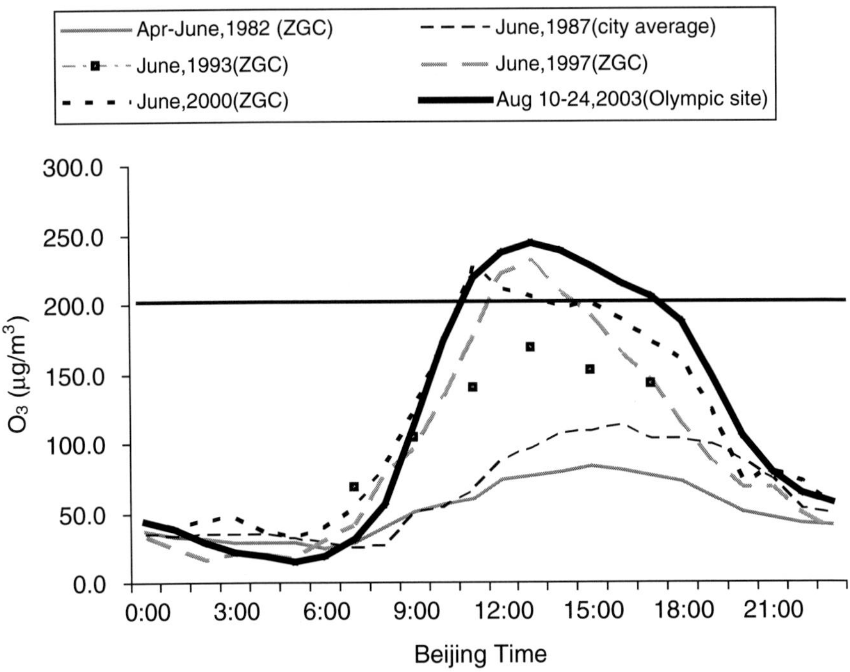

FIGURE 3-29 The diurnal variation and trends in the episodic concentrations of ambient O_3 measured in Zhongguancun (ZGC), Beijing (1982-2003), a northwest suburb of the city, about 20 km from Tian'anmen square. The 2008 Olympic Games site is about 4 km north of ZGC. The yellow line indicates the 1-hour average O_3 concentration at grade II, according to the national ambient air quality standards of China (2000 amendment to GB3095-1996).
SOURCE: Shao et al., 2006.

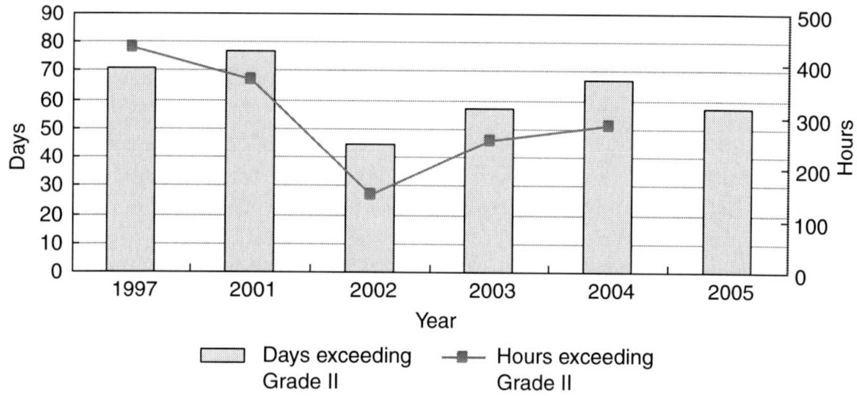

FIGURE 3-30 The hours and days of ozone concentration exceeding NAAQS Grade II in different years.
SOURCE: Beijing EPB, 2006.

The Pearl River Delta region, Guangzhou, and surrounding areas have frequently experienced high O_3 concentrations (Zhang et al., 1998). A similar study in the Yangtze River Delta region showed that high ozone concentrations are also often found at sites some distance removed from urbanized or industrial regions (Shao et al., 2006).

SOURCE RECEPTOR RELATIONSHIPS

In the United States, emissions reductions planning efforts are often made at the state level in State Implementation Plans (SIPs). SIPs are designed to reduce concentrations of ambient pollutants to below current NAAQS, and must make a showing that the plan will achieve the desired goal. To effectively and efficiently meet this objective, a quantitative understanding of source impacts at receptor locations is required. This information then guides policy makers at various levels within government in developing the most efficient and cost-effective approaches for attaining the health- and welfare-based air quality standards. Historically, emissions reduction strategies are guided by various types of statistical and mathematical modeling to identify and to "quantitatively" estimate a value with an associated uncertainty, and to apportion the source impacts at a given receptor location(s).

Two general approaches are used to estimate source contributions quantitatively at a receptor location(s). These are source-oriented and receptor-oriented methods (NRC, 2004). Source-oriented approaches start from the source of the emissions and use models or modeling systems to describe the transport, transformation, and fate of those emissions from the source to the receptor. Receptor-oriented approaches begin with ambient concentrations measured at a receptor site(s) and use statistical or similar approaches to estimate source contributions at the site(s). Both source-oriented and receptor-oriented approaches have limitations, but they complement each other, so the use of both is advised to obtain the best estimates of the impact of sources or source types at receptor locations.

Source-Oriented Approaches

Source-oriented approaches start from the emissions source of the pollutant and work forward in time (rather then starting from the receptor and working backward in time) to estimate the contribution of a source(s) at a receptor location(s). Source-oriented approaches use advanced mathematical models or modeling systems to describe the fate of those emissions between the source and the receptor (Russell, 2007; Seigneur and Moran, 2004, and references within). The most advanced source-oriented modeling systems, called chemical transport models (CTMs), consist of three major components: an emission model, a meteorological model, and an atmospheric process model (chemistry model). Each model may be composed of several modules. Ambient concentration data are typically not incorporated directly into CTMs (except to provide initial and boundary conditions) but are used for model performance evaluation (Seigneur et al., 2000; Seigneur and Moran, 2004; Russell, 2007). Morris et al. (2004) reported on a comparison of the Community Multiscale Air Quality (CMAQ) model and the CAMx model, which are state-of-the-art models that are widely used in the United States, using high-time-resolution data. Sulfate is well reproduced; organic carbon (OC), using a factor to convert OC to organic material of 1.8, is reproduced reasonably well; while nitrate and EC are overestimated by both models.

Most importantly, CTM can be used to predict changes in PM mass and in components observed at a receptor, due to future or predicted changes in emissions (e.g., emissions reductions recommended through a SIP). Source-oriented models also effectively link source and receptor for secondary PM air pollutants (e.g., sulfate, nitrate, SOA) (Russell, 2007) but are not as effective for tracking primary species. Thus, there is synergy between source- and receptor-oriented models, with the application of both providing the most accurate picture of the impact of source emissions at receptor locations.

Receptor-Oriented Approaches

Receptor approaches are observationally based and may involve simple analyses such as time-series analysis, or correlation between or among pollutants; or they may involve more complex multivariate approaches usually referred to as receptor modeling (Hopke, 1991, 2003; Seinfeld and Pandis, 1998; Brook et al., 2004; and references within these publications). In all cases, receptor approaches use ambient concentrations collected at a receptor(s) and possibly other variables and work backwards to the source to estimate source contributions to ambient PM loadings at the receptor location. Receptor methods primarily describe the current situation, since they are observationally based, and therefore, are not used in predicting changes in PM concentrations due to changes in emissions. While receptor models can separate primary from secondary components, they are used most effectively to link primary species observed at a receptor to source types or categories (source apportionment), or individual sources (e.g., a specific emitter) (source attribution), and to quantify (value with an uncertainty estimate) the source contribution at the receptor. Secondary components are usually grouped by compound (e.g., ammonium sulfate, ammonium nitrate), but quantitative separation into source categories is usually not obtained. Source markers or tracers (i.e., usually multiple markers used to identify a given source type) are used to identify primary sources, and these may include inorganic and/or organic species.

The three most widely used receptor models are CMB (Friedlander, 1973; Watson et al., 1984), PMF (Paatero, 1997), and UNMIX (Lewis et al., 2003). All three are based on the general mass balance equation and require ambient concentration measurements. PMF and UNMIX require only ambient concentrations and the factors developed are interpreted by the investigator as specific source types (e.g., motor vehicle, soil dust, etc.). Specific source information is not required for PMF and UNMIX. CMB requires the assumption that the sources are known and that source profiles (i.e., mass fractions of individual chemical species comprising the emissions) are available for each source. The EPA has supported the development of these three models for use in the regulatory process, and they are widely used in SIPs being developed across the country.[2]

Model Integration

In most planning efforts, it is useful to employ both source- and receptor-oriented models to be able to cross-compare results. Receptor models can help to identify problems with the source modeling (Core et al., 1982), while source models can provide predictions of the effects of specific control actions. In both cases, efforts have to be made to provide the critical input data. In the case of source models, the biggest issue is obtaining good emissions inventories. It is often a problem to identify all of the sources and to characterize their emissions. In the case of receptor models, there needs to be a program of ambient monitoring for a sufficient number of chemical species over a sufficiently long time frame—so that it provides an appropriate basis for the receptor model applications.

[2]Current information about these models is available at <http://www.epa.gov/scram001/receptorindex.htm>.

Applications in China

It is important to understand the contribution of each emission source of air pollutants to ambient concentrations, to establish effective measures for risk reduction. Source-oriented and receptor-oriented models have been used to investigate source apportionment and source/receptor relationships for different air pollutants in the urban atmosphere in China since the 1980s (Wang, 1985; Zhao et al., 1991). Most of the methods described above were adapted to differentiate the primary pollutants, such as mineral dust and fugitive dust.

Application of source-oriented models in China has been limited until relatively recently, due to incomplete emission inventory data. However, some researchers have used international emission data, local meteorological fields, and advanced models to simulate the spatial and temporal distribution of pollutants in the context of international field studies, and more recently to examine local or regional control options. More recently, Chinese researchers have begun to incorporate local emissions inventory data into source-oriented modeling studies (e.g., Chen et al., 2007).

For example, M.G. Zhang et al. (Zhang, 2004; Zhang et al., 2005, 2006) applied CMAQ (Models-3 Community Multi-scale Air Quality modeling system) coupled with the Regional Atmospheric Modeling System to East Asia to analyze the production and transport processes of organic carbon (OC), black carbon (BC), and sulfur compounds in the spring of 2001, when two large field campaigns, TRACE-P (TRAnsport and Chemical Evolution over the Pacific) and ACE-Asia (Aerosol Characterization Experiment—Asia), were being conducted over a broad area covering northeastern Asia and the western Pacific (Figure 3-31).

Wang et al. (2005) used the STEM-2K1 atmospheric chemistry and transport model with MM5 meteorological fields, anthropogenic emissions estimates from Streets et al. (2003), and biogenic emissions from the GEIA inventory to compare contributions from transportation, power generation, and industry to concentrations of ozone, SO_2, NO_x, and CO in Guangdong Province in March 2001. They concluded that the transportation sector was the largest contributor to ozone levels, and found that in their simulations, ozone formation in urban areas was limited by VOC emissions, whereas ozone formation in rural areas was NO_x-limited.

Chen et al. (2007) applied the CMAQ model with MM5 meteorological fields to investigate the contributions of transport from surrounding provinces to PM_{10} pollution in Beijing. The study used county-level emissions inventories for primary PM from the environmental protection administrations of Beijing and the surrounding provinces, and examined simulated and measured PM_{10} concentrations for 4 months in 2002. Observed PM_{10} concentrations were reproduced relatively well, except for the month of April, when concentrations were underpredicted, because the simulations did not account for extreme sandstorm events. The modeling analysis indicated that transboundary pollution contributes significantly to PM_{10} concentrations in Beijing, with especially high contributions when pollution levels in Beijing are elevated.

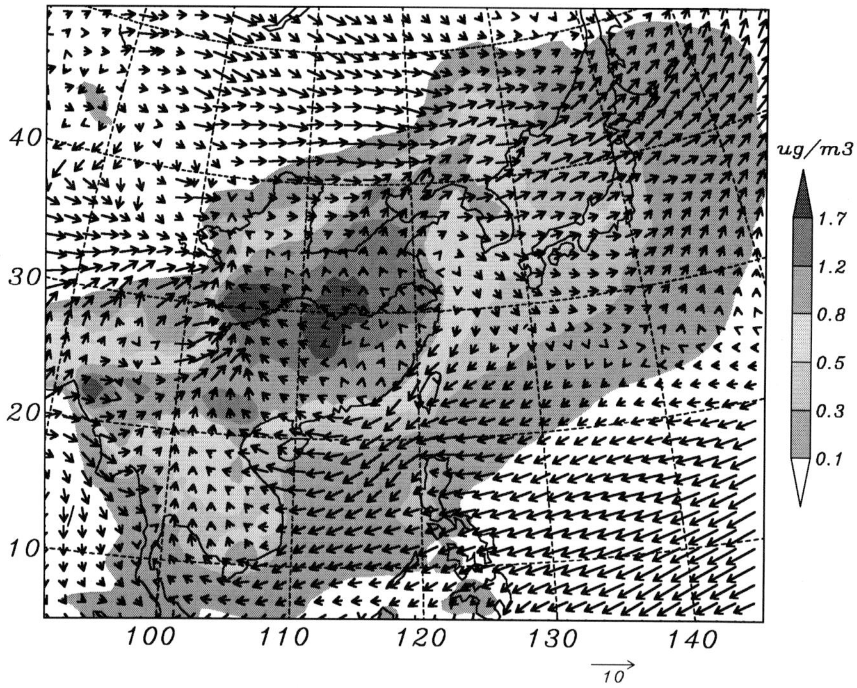

FIGURE 3-31 Horizontal distributions of average black carbon aerosol concentrations and wind fields for the lowest model layer in the period of March-April 2001.

Compared with the application of source-oriented models, receptor-oriented models have been utilized more extensively in China. Particulate pollution is the most serious pollutant in Chinese cities, so models like principle component analysis, absolutely principle component analysis, positive matrix factorization, and chemical mass balance (CMB) have been important analysis tools; and, since the 1980s, these tools have been used in more than 20 cities that have no emissions inventories (Wang, 1985; Zhao et al., 1991; Chen et al., 1994; X.Y. Zhang et al., 2001; Y.H. Zhang et al., 2004). Earlier studies focused on TSP. After requirements shifted to PM_{10} control in 1996, more studies investigated the source apportionment of PM_{10}. Recent studies in Beijing and Hong Kong were aimed at the source apportionment of $PM_{2.5}$ (Ho et al., 2006, Song et al., 2006a, 2006b). At the same time, source apportionment studies have been extended to specific pollutants like polycyclic aromatic hydrocarbons (PAHs), OC, and EC (Qi et al., 2002; Cao et al., 2005b; Peng et al., 2005; Wan et al., 2007). New methods including genetic algorithms, neural networks, fuzzy set theory, and nested CMB were developed and utilized in source apportionment studies (Feng et al., 2002; Li et al., 2000, 2003; Li and Ding 2005).

Eleven typical source apportionment studies for PM in Chinese cities are summarized in Appendix C and show that major sources of TSP include coal combustion dust, fugitive (soil) dust, and construction dust. These sources also contribute significantly to PM_{10} in some cities, especially in northern China. In Hong Kong, a developed city without extensive construction activity or coal utilization, receptor model results suggest that secondary aerosol and motor vehicle exhaust are relatively important sources of PM_{10} as well as of $PM_{2.5}$.

While receptor models are useful for estimating source contributions when accurate emissions inventories are not available, better information is needed to support their use. For example, local source profiles need to be developed step by step; long-term and systematic sampling should be implemented, chemical analysis methods should be compared nationally and internationally, and the source apportionment results should be reconciled with the results from source modeling and from emissions inventories.

REFERENCES

Aunan, K. and X.C. Pan. 2004. Exposure-response functions for health effects of ambient air pollution applicable for China—a meta-analysis. Science of the Total Environment 329:3-16.

Aunan, K., T.K. Berntsen, and H.M. Seip. 2000. Surface ozone in China and its possible impact on agricultural crop yields. Ambio 29:294-301. Stockholm: The Royal Swedish Academy of Sciences.

Beijing EPB (Beijing Municipal Environmental Protection Bureau). 2006. Report on the State of the Environment in Beijing 2001-2005.

Bell, M.L., A. McDermott, S.L. Zeger, J.M. Samet, and F. Dominici. 2004. Ozone and short-term mortality in 95 U.S. urban communities. Journal of the American Medical Association 292:2372-2378.

Bergin, M.H., R. Greenwald, J. Xu, Y. Berta, and W.L. Chameides. 2001. Influence of aerosol dry deposition on photosynthetically active radiation available to plants: A case study in the Yangtze delta region of China. Geophysical Research Letters 28:3605-3608.

Brook, J.R., T..F Dann, and R. Vet. 2004. Time-Integrated Particle Measurements: Status in Canada. European Monitoring and Evaluation Program (EMEP) Workshop on Particulate Matter Measurement and Modeling. New Orleans, Louisiana.

Brunekreef, B. and B. Forsberg. 2005. Epidemiological evidence of effects of coarse airborne particles on health. European Respiratory Journal 26:309-318.

Brunekreef, B. and S.T. Holgate. 2002. Air pollution and health. Lancet 360:1233-1242.

Cabada, J., S. Pandis, R. Subramanian, A. Robinson, A. Polidori, and B. Turpin. 2004. Estimating the secondary organic aerosol contribution to $PM_{2.5}$ using the EC tracer method. Aerosol Science and Technology 38(Suppl)1:140-155.

Cao, H. 1989. Air pollution and its effects on plants in China. Journal of Applied Ecology 26:763-773.

Cao J.J., B. Rong, S.C. Lee, J.C. Chow, K.F. Ho, J.G. Watson, Z.S. An, K. Fung, S.X. Liu, and C.S. Zhu. 2005a. Composition of indoor aerosols at the Emperor Qin's Terra-cotta Museum, Xi'an, China during summer, 2004. China Particuology 3(3):170-175.

Cao, J.J., F. Wu, J.C. Chow, S.C. Lee, Y. Li, S.W. Chen, Z.S. An, K.K. Fung, J.G. Watson, C.S. Zhu, and S.X. Liu. 2005b. Characterization and source apportionment of atmospheric organic and elemental carbon during fall and winter of 2003 in Xi'an, China. Atmospheric Chemistry and Physics 3127-3137.

CDC (Centers for Disease Control and Prevention). 2004. Blood mercury levels in young children and childbearing-aged women—United States, 1999-2002. Morbidity and Mortality Weekly Report 53(43):1018-1020.

Chameides, W. L., H. Yu, S.C. Liu, M. Bergin, X. Zhou, L. Mearns, G. Wang, C.S. Kiang, R.D. Saylor, C. Luo, Y. Huang, A. Steiner, and F. Giorgi. 1999. Case study of the effects of atmospheric aerosols and regional haze on agriculture: An opportunity to enhance crop yields in China through emission controls? Proceedings of the National Academy of Sciences 96:13626-13633.

Chang, Y. and T. Hu. 1996. The Economic Cost of Damage from Atmospheric Pollution in Chongqing. Report of Chongqing Environmental Protection Bureau.

Chen, D.S., S.Y. Cheng, L. Liu, T. Chen, and X.R. Guo. 2007. An integrated MM5-CMAQ modeling approach for assessing trans-boundary PM_{10} contribution to the host city of 2008 Olympic summer games—Beijing, China. Atmos. Environ. 41:1273-1250.

Chen, S.L., Y.B. Zheng, Q. Zhao, et al. 1994. Source apportionment of TSP in Chongqing. Chongqing Environmental Sciences 16(4):30-32 (In Chinese).

Cheung, H.C., T. Wang, K. Baumann, and H. Guo. 2005. Influence of regional pollution outflow on the concentrations of fine particulate matter and visibility in the coastal area of southern China. Atmospheric Environment 39:6463-6474.

Christoforou, C.S., L.G. Salmon, and G. Cass. 1996. Fate of atmospheric particles within the Buddhist cave temples at Yungang China. Environmental Science and Technology 30:3425-3434.

Cohen, A.J., H.R. Anderson, B. Ostro, K.D. Pandey, M. Krzyzanowski, N. Künzli, K. Gutschmidt, C.A. Pope III, I. Romieu, J.M. Samet, and K.R. Smith. 2004. Mortality impacts of urban air pollution, in M. Ezzati, A.D. Rodgers, A.D. Lopez, C.J.L. Murray (eds.), Comparative Quantification of Health Risks: Global and Regional Burden of Disease due to Selected Major Risk Factors. Geneva: World Health Organization, Vol. 2, pp. 1353-1433.

Cohen, A.J., H.R. Anderson, B. Ostra, K.D. Pandev, M. Krzyzanowski, N. Künzli, K. Gutschmidt, A. Pope, I. Romieu, J.M. Samet, and K. Smith. 2005. The global burden of disease due to outdoor air pollution. Journal of Toxicology and Environmental Health Part A 68:1-7.

Core, J.E., P.L. Hanrahan, and W.M. Cox. 1982. Particulate dispersion model evaluation: A new approach using receptor models. Journal of the Air Pollution Control Association 1142-1147.

Ding, G. 2004. Database from the acid rain network of China Meteorological Adminstration and its preliminary analyses. Journal of Applied Meteorological Science 15(Z1):85-94 (in Chinese).

Dockery, D.W., C.A. Pope III, X. Xu, J.D. Spengler, J.H. Ware, M.E. Fay, B.G. Ferris, Jr., and F.E. Speizer. 1993. An association between air pollution and mortality in six U.S. cities. New England Journal of Medicine 329:1753-1759.

EPA (U.S. Environmental Protection Agency). 1997. The Benefits and Costs of the Clean Air Act, 1970 to 1990. Washington, D.C.: U.S. Environmental Protection Agency.

EPA. 2001. National Air Quality and Emissions Trends Report, 1999. Washington, D.C.: U.S. Environmental Protection Agency.

EPA. 2003. Latest Findings on National Air Quality: 2002 Status and Trends. Washington, D.C.: U.S. Environmental Protection Agency.

EPA. 2004. Air Quality Criteria for Particulate Matter, October. Washington, D.C.: U.S. Environmental Protection Agency.

EPA. 2005. Regulatory Impact Analysis for the Final Clean Air Interstate Rule, EPA-452/R-05-002. Washington, D.C.: Office of Air and Radiation.

EPA. 2006a. National-Scale Air Toxics Assessment For 1999: Estimated Emissions, Concentrations and Risk. Washington, D.C.: Office of Air and Radiation.

EPA. 2006b. Regulatory Impact Analysis Of EPA's Final Revisions to the National Ambient Air Quality Standards for Particle Pollution (Particulate Matter). Washington, D.C.: Office of Air and Radiation.

EPA. 2007a. Controlling Power Plant Emissions: Emissions Progress, http://www.epa.gov/mercury/control_emissions/emissions.htm.

EPA. 2007b. Review of National Ambient Air Quality Standards for Ozone. Final Staff Paper, January.

Feng, Y.C., Z.P. Bai, and T. Zhu. 2002. Principle and application of nested chemical mass balance. Environmental Science 23(Suppl):103-108 (in Chinese).

Friedlander, S.K. 1973. Chemical element balances and identification of air pollution sources. Environmental Science and Technology 7:235-240.

Gaydos, T.M., C.O. Stanier, and S.N. Pandis. 2005. Modeling of in situ ultrafine atmospheric particle formation in the eastern United States. Journal of Geophysical Research 110:D07S12.

Gryparis, A., B. Forsberg, K. Katsouyanni, A. Analitis, G. Touloumi, J. Schwartz, E. Samoli, S. Medina, H.R. Anderson, E.M. Niciu, H.E. Wichmann, B. Kriz, M. Kosnik, J. Skorkovsky, J.M. Vonk, and Z. Dörtbudak. 2004. Acute effects of ozone on mortality from the 'Air Pollution and Health: A European Approach' project. American Journal of Respiratory and Critical Care Medicine 170:1080-1087.

Hao, J.M. and L.T. Wang. 2005. Improving urban air quality in China: Beijing case study. Journal of the Air and Waste Management Association 55(9):1298-1305.

He, K. et al. 2005. Oil consumption and CO_2 emissions in China's road transport: Current status, future trends, and policy implications. Energy Policy 33:1499-1507.

HEI (Health Effects Institute). 2006. PAPA-SAN: Public Health and Air Pollution in Asia—Science Access on the Net, http://www.healtheffects.org/Asia/papasan-overview.htm.

Ho, K.F., J.J. Cao, S.C. Lee, and C.K. Chan. 2006. Source apportionment of $PM_{2.5}$ in urban area of Hong Kong. Journal of Hazardous Materials B138:73-85.

Ho, M.S., and C.P. Nielsen, eds. 2007. Clearing the Air: The Health and Economic Damages of Air Pollution in China. Cambridge, Mass.: MIT Press.

Hoek, G., B. Brunekreef, S. Goldbohm, P. Fischer, and P.A. van den Brandt. 2002. Association between mortality and indicators of traffic-related air pollution in the Netherlands: A cohort study. Lancet 360:1203-1209.

Hopke, P.K., ed. 1991. Receptor Modeling for Air Quality Management. Amsterdam: Elsevier Science Publishers.

Hopke, P.K. 2003. Recent Developments in Receptor Modeling. Journal of Chemometrics 17:255-265.

Irie, H., K. Sudo, H. Akimoto, A. Richter, J.P. Burrows, T. Wagner, M. Wenig, S. Beirle, Y. Kondo, V.P. Sinyakov, and F. Goutail. 2005. Evaluation of long-term tropospheric NO_2 data obtained by GOME over East Asia in 1996-2002. Geophysical Research Letters 32(11):L11810.

Kan, H.D. and B.H. Chen. 2004. Particulate air pollution in urban areas of Shanghai China: Health-based economic assessment. Science of the Total Environment 322:71-79.

Kanakidou, M., J.H. Seinfeld, S.N. Pandis, I. Barnes, F.J. Dentener, M.C. Facchini, R. Van Dingenen, B. Ervens, A. Nenes, C.J. Nielsen, E. Swietlicki, J.P. Putaud, Y. Balkanski, S. Fuzzi, J. Horth, G.K. Moortgat, R. Winterhalter, C.E.L. Myhre, K. Tsigaridis, E. Vignati, E.G. Stephanou, and J. Wilson. 2005. Organic aerosol and global climate modeling: A review. Atmospheric Chemistry and Physics, 5:1053-1123.

Kochanek, K.D., S.L. Murphy, R.N. Anderson, and C. Scott. 2004. Deaths: Final data for 2002. National Vital Statisitics Reports; vol. 53, no. 5. Hyattsville, MD: National Center for Health Statistics.

Laden, F., L.M. Neas, D.W. Dockery, and J. Schwartz. 2000. Association of fine particulate matter from different sources with daily mortality in six U.S. cities. Environmental Health Perspectives 108(10).

Larssen, T., E. Lydersen, D.G. Tang, Y. He, J.X. Gao, H.Y. Liu, L. Duan, H.M. Seip, R.D. Vogt, J. Mulder, M. Shao, Y.H. Wang, H. Shang, X.S. Zhang, S. Solberg, W. Aas, T. Okland, O. Eilertsen, V. Angell, Q.R. Liu, D.W. Zhao, R.J. Xiang, J.S. Xiao, and J.H. Luo. 2006. Acid rain in China. Environmental Science & Technology 40(2):418-425.

Lewis, C.W., G.A. Norris, T.L. Conner, and R.C. Henry. 2003. Source apportionment of Phoenix $PM_{2.5}$ aerosol with the unmix receptor model. Journal of the Air Waste Management Association 53:325-338.

Li, Z.Y. and H.K. Ding. 2005. Application of BP neural network to sources apportionment of atmospheric particulates. Environmental Monitoring in China 21(2):74-76 (in Chinese).

Li, Z.Y. and L.H. Peng. 2000. Source apportionment of atmospheric particulates based on genetic algorithm. Research of Environmental Sciences 13(6):19-21 (in Chinese).

Li, Z.Y., C.J. Ni, and J. Ding. 2003. Application of rough sets theory to sources apportionment of atmospheric particulates. Journal of Sichuan University 35(4):112-114 (in Chinese).

Lippmann, M., M. Frampton, J. Schwartz, D. Dockery, R. Schlesinger, P. Koutrakis, J. Froines, A. Nel, J. Finkelstein, J. Godleski, J. Kaufman, J. Koenig, T. Larson, D. Luchtel, L.-J.S. Liu, G. Oberdörster, A. Peters, J. Sarnat, C. Sioutas, H. Suh, J. Sullivan, M. Utell, E. Wichmann, and J. Zelikoff. 2003. The U.S. Environmental Protection Agency Particulate Matter Health Effects Research Centers Program: A midcourse report of status, progress, and plans. Environmental Health Perspectives 111(8).

Marshall, J.D., S.K. Teoh, and W.W. Nazaroff. 2005. Intake fraction of nonreactive vehicle emissions in US urban areas. Atmospheric Environment 39(7):1363-1371.

Maynard, D., B.A. Coull, A. Gryparis, and J. Schwartz. 2007. Mortality risk associated with short-term exposure to traffic particles and sulfates. Environmental Health Perspectives 115(5):751-755.

Ministry of Commerce. 2007. Guizhou: Planned investments of over 200 million Yuan to address issues of residential coal use, May 30.

Morris, R.E., B. Lau, T.W. Tesche, D. McNally, C. Loomis, G. Stella, G. Tonnesen, and Z. Wang. 2004. VISTAS Emissions and Air Quality Modeling—Phase I Task 4cd Report: Model Performance Evaluation and Model Sensitivity Tests for Three Phase I Episodes, August.

Murphy, J.J., M.A. Deluki, D.R. McCubbin, and H.J. Kim. 1999. The cost of crop damage caused by ozone air pollution from motor vehicles. Journal of Environmental Management 55:273-289.

NAPAP (National Acid Precipitation Assessment Program). 2005. National Acid Precipitation Assessment Program Report to Congress: An Integrated Assessment.

NRC (National Research Council). 2002. Estimating Public Health Benefits of Air Pollution Regulations. Washington, D.C.: National Academy Press.

NRC. 2003. Managing Carbon Monoxide in Meteorological and Topographical Problem Areas. Washington, D.C.: The National Academies Press.

NRC. 2004. Air Quality Management in the United States. Washington, D.C.: The National Academies Press.

Ostro, B., H. Tran, and J. Levy. 2006. The Health Benefits of Reduced Tropospheric Ozone in California. Journal of the Air and Waste Management Association 56:1007-1021.

Ou, S., L. Pan, and M. Zhuang. 1996. Estimation of the effects and economic loss caused by acid deposition in Xiamen area. Res. Environ. Sci. 9(5):57-59 (in Chinese).

Paatero, P. 1997. Least squares formulation of robust, non-negative factor analysis. Chemometrics and Intelligent Laboratory Systems 37:23-35.

Parrish, D.D. 2006. Critical evaluation of US on-road vehicle emission inventories. Atmospheric Environment 40:2288-2300.

Peng, L., J. Li, and T. Zhu. 2005. Carbon isotope characteristics of polycyclic aromatic hydrocarbons in atmospheric particles and their source analysis in urban area of Zhengzhou City. China Environmental Science 25(1):106-109 (in Chinese).

Pope, C.A. and D.W. Dockery. 2006. Health effects of fine particulate air pollution: Lines that connect. Journal of Air and Waste Management Association 56(6).

Pope, C.A. III, M.J. Thun, and M.M. Namboordiri. 1995. Particulate air pollution as a predictor of mortality in a prospective study of U.S. adults. American Journal of Respiratory and Critical Care Medicine 151:669-674.

Pope, C.A. III, R.T. Burnett, M.J. Thun, E.E. Calle, D. Krewski, K. Ito, and G.D. Thurston. 2002. Lung cancer, cardiopulmonary mortality, and long-term exposure to fine particulate air pollution. Journal of the American Medical Association 287:1132-1141.

Pun, B.K. and C. Seigneur. 2007. Investigative modeling of new pathways for secondary organic aerosol formation. Atmospheric Chemistry and Physics 7:2199-2216.

Qi, S.H., G.Y. Sheng, J.M. Fu, Z.S. Wang, and Y.S. Min. 2002. Source apportionment experiment of polycyclic aromatic hydrocarbons (PAHs) in aerosols from Macao. China Environmental Science 22(2):118-122 (in Chinese).

Qian, Y. and F. Giorgi. 2000. Regional climatic effects of anthropogenic aerosols? The case of Southwestern China. Geophysical Research Letters 27:3521-3524.

Qian, Y., C.B. Fu, R.M. Hu, and Z.F. Wang. 1996. Effects of industrial SO_2 emission on temperature variation in China and East Asia. Clim. Environ. Res. 2:143-149 (in Chinese).

Qian, Y., L.R. Leung, S.J. Chan, and F. Giorgi. 2003. Regional climate effects of aerosols over China: Modeling and observation. Tellus 55B:914-934.

Qiu, J.H. and L.Q. Yang. 2000. Variation characteristics of atmospheric aerosol optical depths and visibility in North China during 1980-1994. Atmospheric Environment 34:603-609.

Richter, A., P. Burrows, H. Nues, C. Granier, and U. Niemeijer. 2005. Increase in tropospheric nitrogen dioxide over China observed from space. Nature 437:129-130.

Robinson, A.L., N.M. Donahue, M.K. Shrivastava, E.A. Weitkamp, A.M. Sage, A.P. Grieshop, T.E. Lane, J.R. Pierce, and S.N. Pandis. 2007. Rethinking organic aerosols: Semivolatile emissions and photochemical aging. Science 315(5816):1259-1262.

Russell, A.G. 2007. EPA supersites program-related emissions-based particulate matter modeling: Applications and advances. Journal of the Air and Waste Management Association. Accepted for publication.

SCAQMD (South Coast Air Quality Management District). 2000. Multiple Air Toxics Exposure Study (MATES-II) Final Report, March.

Schwartz, J., F. Laden, and A. Zanobetti. 2002. The Concentration—response relation between air pollution and daily deaths. Environmental Health Perspectives 110(10).

Seigneur, C. and M.D. Moran. 2004. Using models to estimate particle concentration. Chapter 8 in Particulate Matter Science for Policy Makers: A NARSTO Assessment, P. McMurry, M. Shepherd, and J. Vickery, eds. Cambridge: Cambridge University Press.

Seigneur, C., B. Pun, P. Pai, J. Louis, P.A. Solomon, C. Emery, R. Morris, M.S. Zahniser, D.R. Worsnop, P. Koutrakis, W. White, and I. Tombach. 2000. Guidance for the performance evaluation of three-dimensional air quality modeling systems for particulate matter and visibility. Journal of the Air and Waste Management Association 50:588-599.

Seinfeld, J.H. and S.N. Pandis. 1998. Atmospheric Chemistry and Physics: From Air Pollution to Global Change. New York: John Wiley and Sons.

Seinfeld, J.H. and J.F. Pankow. 2003. Organic atmospheric particulate material. Annual Review of Physical Chemistry 54:121-140.

SEPA (State Environmental Protection Administration). 2000. Report on the State of the Environment in China 1999.

SEPA. 2001a. Annual Report of Environmental Statistics 2000.

SEPA. 2001b. Report on the State of the Environment in China 2000.

SEPA. 2002. Report on the State of the Environment in China 2001.

SEPA. 2003. Report on the State of the Environment in China 2002.

SEPA. 2004. Report on the State of the Environment in China 2003.

SEPA. 2005a. Annual Report of Environmental Statistics 2004.

SEPA. 2005b. Report on the State of the Environment in China 2004.

SEPA. 2006. Report on the State of the Environment in China 2005.

SEPA. 2007. Report on State of the Environment in China 2006.

Shao, M., X. Tang, Y. Zhang, and W. Li. 2006. City clusters in China: Air and surface water pollution. Frontiers in Ecology and the Environment 4(7):353-361.

Shu, J., H. Cao, Y. Gao, et al. 1993. Ecological criteria for Masson pine effected by acidic precipitation and evaluation of economic loss. China Environ. Sci. 13(4):279-283 (in Chinese).

Smith, K.R., S. Mehta, and M. Maeusezahl-Feuz. 2004. Indoor smoke from household solid fuels, in Comparative Quantification of Health Risks: Global and Regional Burden of Disease due to Selected Major Risk Factors. M. Ezzati, A.D. Rodgers, A.D. Lopez, and C.J.L Murray, eds. Geneva: World Health Organization, vol. 2, 1435-1493.

Song, Y., Y.H. Zhang, S.D. Xie, L.M. Zeng, M. Zheng, L.G. Salmon, M. Shao, and S. Slanina. 2006a. Source apportionment of $PM_{2.5}$ in Beijing by positive matrix factorization. Atmospheric Environment 40:1526-1537.

Song. Y., S.D. Xie, Y.H Zhang, et al. 2006b. Source apportionment of $PM_{2.5}$ in Beijing using principal component analysis/absolute principal component scores and UNMIX. Science of the Total Environment 372:278-286.

SSB (Shandong Province Statistical Bureau). 2006. Shandong Statistical Yearbook 2005.

Streets, D.G. and S.T. Waldhoff. 2000. Present and future emissions of air pollutants in China: SO_2, NO_x and CO. Atmospheric Environment 34:363-374.

Streets, D.G. et al. 2001. Black carbon emissions in China. Atmospheric Environment 35:4281-4296.

Streets, D.G., T.C. Bond, G.R. Carmichael, S.D. Fernandes, Q. Fu, D. He, Z. Klimont, S.M. Nelson, N.Y. Tsai, M.Q. Wang, J.H Woo, and K.F. Yarber. 2003. A year-2000 inventory of gaseous and primary aerosol emissions in Asia to support TRACE-P modeling and analysis. Journal of Geophysics Research 108(D21):8809.

Sun, Y.L., G.S. Zhuang, A.H. Tang, et al. 2006. Chemical characteristics of $PM_{2.5}$ and PM_{10} in haze-fog episodes in Beijing. Environmental Science and Technology 40:3148-3155.

Turpin, B.J. and J.J. Huntzicker. 1995. Identification of secondary organic aerosol episodes and quantitation of primary and secondary organic aerosol concentrations during SCAQS. Atmospheric Environment 29:3527-3544.

UNDP (United Nations Development Programme). 2002. China Report on Human Development 2002. Beijing, China: Financial Economy Press.

UNEP (United Nations Environmental Programme). 1999. Global Environment Outlook 2000. London, UK: Earthscan Publications Ltd.

UNEP. 2002. Global Environmental Outlook 2003. London, UK: Earthscan Publications Ltd.

Vayenas, D.V., S. Takahama, C.I. Davidson, and S.N. Pandis. 2005. Simulation of the thermodynamics and removal processes in the sulfate-ammonia-nitric acid system during winter: Implications for $PM_{2.5}$ control strategies. Journal of Geophysical Research 110:D07S14.

Wan, X.L., J.W. Chen, F.L. Tian, et al. 2007. Source apportionment of PAHs in atmospheric particulates of Dalian: Factor analysis with nonnegative constraints and emission inventory analysis. Atmospheric Environment 40(34):6666-6675.

Wang, M.X. 1985. Using factor analysis to study aerosol sources. Chinese Journal of Atmospheric Sciences 9(1):73-81 (in Chinese).

Wang, S.X., J.M. Hao, M.S. Ho, J. Li, and Y.Q. Lu. 2006. Intake fractions of industrial air pollutants in China: Estimation and application. Science of the Total Environment 354:127-141.

Wang, X., G. Carmichael, D. Chen, Y. Tang, and T. Wang, 2005. Impacts of different emissions sources on air quality during March 2001 in the Pearl River Delta (PRD) region. Atmospheric Environment 39:5227-5241.

Wang, X.P. and D.L.Mauzerall. 2004. Characterizing distributions of surface ozone and its impact on grain production in China, Japan and South Korea: 1990 and 2020. Atmospheric Environment 38:4383-4402.

Wang, Y., M.B. McElroy, K.F. Boersma, H.J. Eskes, and J.P. Veefkind. 2007. Traffic restrictions associated with Sino-African summit: Reductions of NO_x detected from space, Geophysical Research Letters 34:L08814.

Watson, J.G., J.A. Cooper, and J.J. Huntzicker. 1984. The effective variance weighting for least squares calculations applied to the mass balance receptor model. Atmospheric Environment 18:1347-1356.

WHO (World Health Organization). 2006. WHO Air Quality Guidelines for Particulate Matter, Ozone, Nitrogen Dioxide and Sulfur Dioxide: Global Update 2005. World Health Organization, Geneva.

World Bank. 1997. China's Environment in the New Century: Clear Water, Blue Skies. The World Bank, Washington, D.C.

Wu, D., X.X. Tie, C.C. Li, et al. 2005. An extremely low visibility event over the Guangzhou region: A case study. Atmospheric Environment 39:6568-6577.

Wu, J., W.M. Jiang, C.B. Fu, et al. 2004. Simulation of the radiative effect of black carbon aerosols and the regional climate responses over China. Advances in Atmospheric Science 21(4):637-649.

Wu, Y., S.X. Wang, D.G. Streets, J.M. Hao, M. Chan, and J.K. Jiang. 2006. Trends in anthropogenic mercury emissions in China from 1995 to 2003. Environmental Science Technology 40:5312-5318.

Xiao, H.Y. and C.Q. Liu. 2004. Chemical characteristics of water-soluble components in TSP over Guiyang, SW China, 2003. Atmospheric Environment 38(37):6297-6306.

Xu, Q. 2001. Abrupt change of the mid-summer climate in central east China by the influence of atmospheric pollution. Atmospheric Environment 35:5029-5040.

Zhang, D. 2006. Source apportionment of TSP in ambient air in Shuozhou center urban. Shanxi Chemical Industry 26(3):61-63 (in Chinese).

Zhang, J. and K.R. Smith. 2005. Indoor air pollution from household fuel combustion in China: A review. The 10th International Conference on Indoor Air Quality and Climate, 1-26.

Zhang, M.G. 2004. Modeling of organic carbon aerosol distributions over east Asia in the springtime. China Particuology, 2(5):192-195.

Zhang, M.G., Y.F. Xu, R.J. Zhang, et al. 2005. Emission and concentration distributions of black carbon aerosol in East Asia during the spring time. Chinese Journal of Geophysics, 48(1):55-61.

Zhang, M.G., Y.F. Pu, R.J. Zhang, et al. 2006. Simulation of sulfur transport and transformation in East Asia with a comprehensive chemical transport model. Environmental Modelling & Software, 21(6):812-820.

Zhang, X.Y., J.J. Cao, et al. 2001. Particulate pollution control in Xi'an. In: Chinese Urban Air Pollution Control. Chinese Science & Technology Press. 222-293 (in Chinese).

Zhang, Y., K. Shao, X. Tang, and J. Li. 1998. The Study of Urban Photochemical Smog Pollution in China. Acta Scicentiarum Naturalum Universitis Pekinesis 34(2-3):392-400.

Zhang, Y.H., X.L. Zhu, L.M. Zeng, et al., 2004. Source apportionment of fine-particle pollution in Beijing, In: Urbanization, Energy, and Air Pollution in China. National Academy of Engineering and National Research Council. Washington, D.C.: The National Academies Press. Pp. 139-153.

Zhao, D., J. Xiong, Y. Xu, and W.H. Chan. 1988. Acid rain in southwestern China. Atmospheric Environment (1967) 22(2):349-358.

Zhao, D.S., M.X. Wang, et al. 1991. Coal-combustion air pollution and aerosol. Chinese Environmental Press 1-404 (in Chinese).

Zhou, X.J., W.L. Li, and Y.F. Luo. 1998. Numerical simulation of the aerosol radiative forcing and regional climate effect over China. Scientia Atmospherica Sinica 22(4):418-427 (in Chinese).

4

Institutional and Regulatory Frameworks

In both the United States and China, air quality management (AQM) developed as a response to health concerns, and typically began with research before progressing to regulatory actions. Today, the United States has an extensive AQM framework which has been successful in addressing serious air quality problems, but still faces challenges such as developing a more integrated, multipollutant approach, as well as an airshed-based approach (NRC, 2004). China's AQM history is comparatively short, but in recent years, as other state agencies were reduced in size, environmental agencies and organizations have enlarged their mandate. While China is still focusing on strategies to control criteria pollutants (defined below) such as PM_{10} and SO_2, it is at the same time seeking to prevent future pollution, control other criteria pollutants, and help more cities improve air quality to emulate the gains Beijing and a few other key cities have made. The following chapter traces the progression of AQM in each country and explains the roles of current institutions and regulatory frameworks. It will detail the related roles of research institutions and non-governmental organizations (NGOs) and their influence on policy and practice. The chapter concludes with a section on energy policy in the United States and China and its relations to air quality goals.

ORGANIZATION OF AIR QUALITY MANAGEMENT IN THE UNITED STATES AND CHINA

United States

The U.S. government is comprised of three co-equal and independent branches—the legislative branch, the executive branch, and the judicial branch.

All three of these branches of government play critical roles in the development and administration of environmental policy. Environmental laws such as the Clean Air Act (CAA) originate in the U.S. Congress. Under the American political system, only the elected national legislature can impose such broad and far-reaching limitations on economic activity, which could competitively disadvantage particular industrial sectors or regions of the country. The executive branch is responsible for implementing these laws, and develops regulations for this purpose through authority that is delegated from the legislature. The judicial branch performs the role of enforcing laws enacted by the legislative branch, including ensuring that the executive branch acts within its statutory authority. In addition, courts in the United States play a major role as arbiters of disputes between citizens, which sometimes involve liability for environmental damage.

In the United States, "judge-made" or "common law" liability rules represent an important forerunner of, and adjunct to, statutory environmental law or regulation. In particular, through a private lawsuit based on the common law claim of nuisance, an injured party can either seek payment for property damage, or try to enjoin the activity that is causing harm. Alongside an extensive framework of environmental regulation, nuisance suits are still employed today to address damage caused by air or water pollution, especially in situations where government regulation has proven to be inadequate. However, prevailing in a nuisance suit is not easy. Private litigation is costly and time-consuming, and the injured party must generally prove that the defendant engaged in an "unreasonable" activity that "substantially" interfered with the enjoyment of his or her property (Powell, 1992). It may be especially difficult for someone injured by air pollution to show that a particular defendant caused quantifiable harm (Diamond v. General Motors Corp., 1971). This point was driven home in 1969 by a class action lawsuit filed on behalf of seven million residents of the Los Angeles area. The suit named almost 1,300 defendants and simply asked the court to "do something about the air." Therefore, while the right to use private lawsuits to address environmental damage is a vital principle of American law, government regulation is generally viewed as a more efficient mechanism for protecting public health and the environment, with the critical advantage that it can be used proactively to prevent harm before it occurs.

Prior to 1970, state, county, and municipal governments in the United States undertook most air pollution regulation. Municipal smoke ordinances in the United States date back to the 19th century. State and local efforts to address air pollution intensified in many areas after World War II, in response to rapid growth in industrial activity that occurred during the war and continued afterwards. These state and local efforts are exemplified by those efforts undertaken in Pittsburgh and Los Angeles in the 1950s and 1960s, which are discussed in Chapters 8 and 10.

The federal government also began to address air pollution after World War II, but began with a comparatively modest focus on research and technical assistance. The National Air Pollution Control Administration was created within the Department of Health, Education, and Welfare (HEW) in 1955, in response to several air pollution crises that occurred in the previous decade, including a 3-day pollution episode in 1948 in the small town of Donora, Pennsylvania, where pollution from steel mills and stagnant meteorological conditions combined to cause a score of deaths and thousands of illnesses (Davis, 2002). The 1963 CAA[1] required HEW to develop air quality criteria by studying relationships between air pollution levels and health and welfare effects; but these criteria were largely advisory and could be enforced only in cases where pollution directly endangered health or welfare. The 1963 Act also authorized the Secretary of HEW to ask the Justice Department to bring suit against polluters, under specified circumstances. Up until 1970, however, only a single enforcement action was taken (Stewart and Krier, 1978). The federal role in air pollution regulation was gradually strengthened through the 1960s, culminating with the passage of the CAA amendments of 1970 (see timeline, Figure 4-1).

The Environmental Protection Agency

The year 1970 is widely recognized as the watershed year for national environmental regulation in the United States. In addition to the CAA amendments, the U.S. Environmental Protection Agency (EPA) was established as an independent agency. The mission of the EPA is to establish and enforce environmental standards across the air, water, and soil media, conduct research on environmental problems and their solutions, and assist the president's Council on Environmental Quality in developing recommendations for new environmental policy initiatives (Lewis, 1985). The EPA was largely assembled from programs that already existed at other departments, including the National Air Pollution Control Administration at HEW. EPA now has 18,000 employees working nationwide, including 1,245 in the Office of Air and Radiation; these employees focus on national air quality policy (see Figure 4-2). A staff of 605 people in the ten regional offices interacts directly with state air pollution control agencies, to coordinate pollution-control efforts and to enforce federal requirements. When combined with researchers at EPA's Office of Research and Development (ORD), nearly 4,000 EPA employees work specifically on issues of air quality.[2]

The 1970 CAA amendments authorized the new EPA to set National Ambient Air Quality Standards and required states to develop and implement plans to meet them. Standards were required to be nationally uniform and to protect public health with an adequate margin of safety. The Act also authorized EPA to set emissions standards for large new stationary sources and for new motor vehicles. The CAA amendments of 1970 established a "cooperative" relationship between states and the federal government, in placing primary responsibility on the states

[1]Public Law 88-206, 77 Stat, 392 (1963).

[2]As of November 2006.

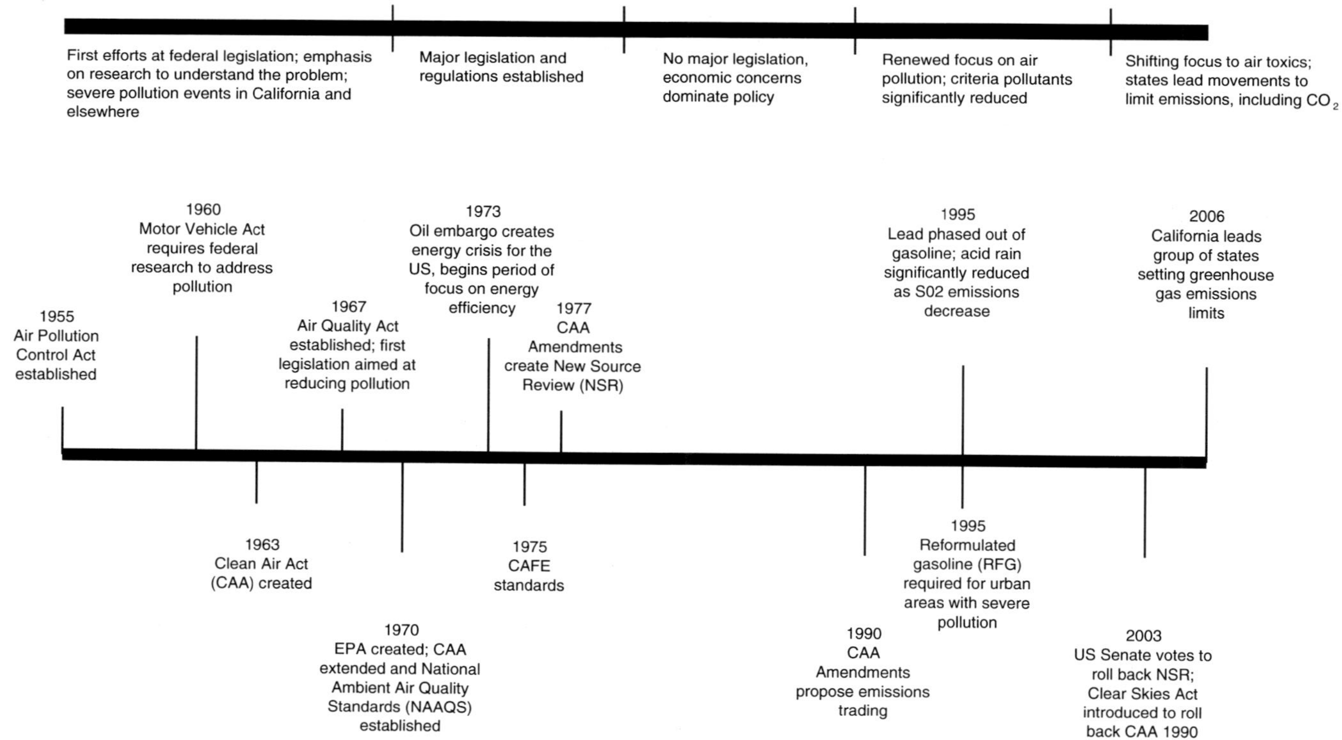

FIGURE 4-1 Timeline of air quality regulation trends in the United States.

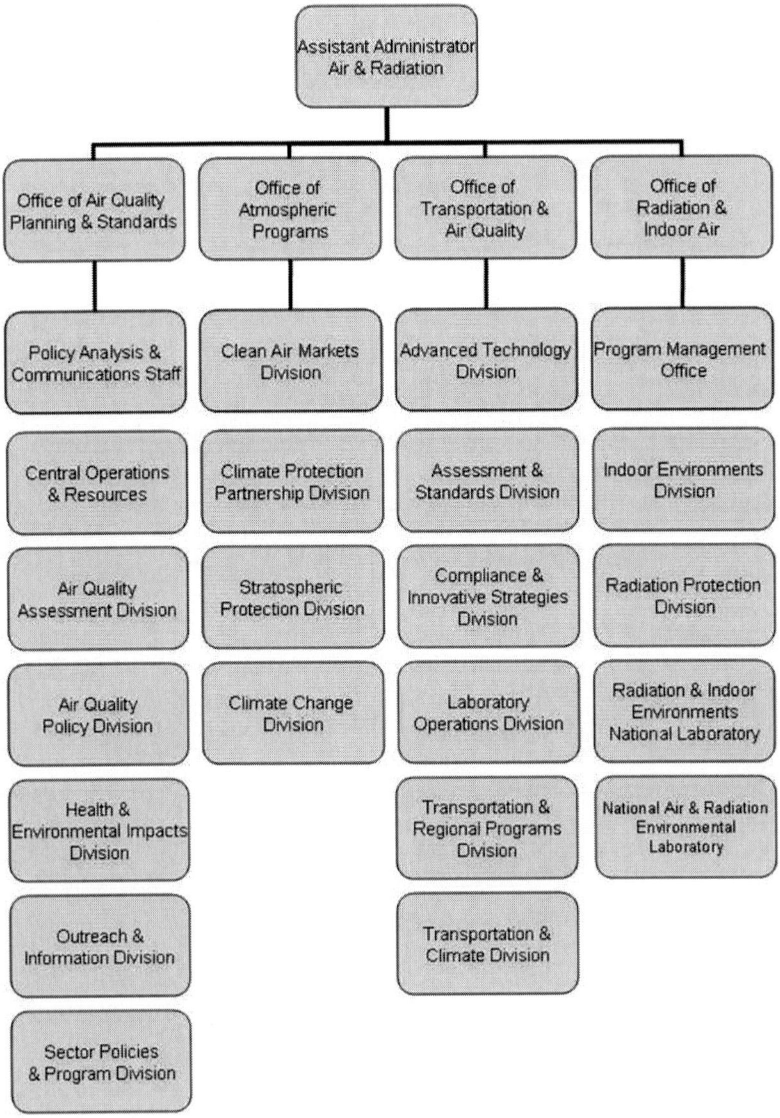

FIGURE 4-2 Organizational chart of EPA Office of Air and Radiation.

to attain ambient air quality standards that were set by the federal government. However, the Act does not leave implementation to the individual states alone; it also assigns enforcement powers to the federal government. These include both direct enforcement actions against sources that are violating provisions of the CAA, and sanctions against states that fail to attain the standards by established deadlines, or states that fail to comply with other statutory requirements, including requirements to develop and administer the construction and operating permit programs for stationary sources, enforce permit limits, monitor air quality, and track emissions.

State and local governments, environmental groups, and industry associations in the United States commonly use these provisions to seek the review of rules that EPA has developed, and to force the agency to undertake duties that Congress has assigned to it by statute. In addition to being incorporated explicitly into the statute, the CAA's citizen suit provisions are in keeping with core principles of administrative law in the United States, which in turn trace back to English common law principles of accountability of government officials to damage suits (Breyer and Stewart, 1979).

The citizen suit provisions of the CAA augment the federal Administrative Procedure Act (APA) of 1946, which established modern requirements for the interaction between federal courts, regulatory agencies, and the interested public. The APA, which largely codified preexisting judge-made principles of administrative law, provides opportunities for the public to participate in the formulation of regulations, requires agencies to establish a reasonable basis for their decisions in the public record, and establishes procedures for judicial review of agencies' reasoning and compliance with their authorizing statutes.

EPA's administration of the CAA is thus overseen by the President as head of the executive branch, by the courts in response to citizen suits, and by Congress, which exercises ultimate oversight through the power to amend or replace EPA's authorizing legislation. In fact, Congress has enacted two major sets of amendments to the CAA since 1970, in 1977 and again in 1990.

Other Agencies

The EPA is not the only U.S. government agency with responsibility for air quality. The U.S. Department of Transportation maintains an air quality unit that provides information related to air quality impacts of the nation's transportation systems, and supplies advice to transportation officials regarding air quality planning. While not all federal agencies have air quality-specific programs, several agencies target energy, largely because of energy consumption's impact on the environment, and in particular, on air quality. The U.S Department of Interior (DOI) runs a range of programs that are related to domestic energy production. As America's principal conservation agency, DOI must ensure that air pollution does not degrade natural systems and that energy resources are used wisely.

DOI oversees numerous smaller agencies that are tasked with managing specific resources such as the Bureau of Land Management, Fish and Wildlife Service, National Park Service, U.S. Geological Survey, and Bureau of Reclamation.

All federal agencies, including the ones mentioned above, are governed by the National Environmental Policy Act (NEPA). NEPA requires that federal agencies include environmental values in their decision making, by considering both the potential environmental impacts of proposed action and the impacts of reasonable alternatives to those actions. The manifestation of this requirement is that federal agencies must produce detailed reports called environmental impact statements (EISs), before taking significant action. NEPA regulations are overseen by the Council on Environmental Quality. Some actions are categorically excluded from needing a detailed assessment, if they meet certain requirements such as that type of action that has been previously evaluated by the agency as having no significant impact. The next level of complexity requires the agency to perform a preliminary environmental assessment. If the environmental assessment determines that no significant impacts are likely to occur, the agency issues a finding of no significant impact. If the environmental assessment suggests that more significant impacts may occur, the agency must continue with a more detailed EIS. EPA publishes EISs weekly in a "Notice of Availability" in the *Federal Register*. The public can attend public meetings on EISs, and submit comments directly to the federal agency. It is the agency's job to consider comments from the public during the comment period.

China

China's history of AQM is comparatively short. The 1972 UN Conference on the Human Environment in Stockholm marked the first time that China officially addressed environmental issues. The conference, and the nascent environmental movements of industrialized countries such as the United States, provided the impetus for China and other developing countries to establish national environmental policies. Previously, the Chinese Ministry of Health oversaw environmental matters, but did so with little authority or motivation to regulate. Yet in spite of somewhat weak enforcement, environmental laws and regulations have resulted in improved air quality in many Chinese cities over the past few decades (Aunan et al., 2006).

The real starting point for environmental protection in China is generally thought of as stemming from the first national conference on environmental protection held in 1973 (Editorial Board, 1994). The first organization concerned solely with environmental protection, the Environmental Protection Leading Group (Leading Group), was set up in 1974 under the State Council, which was formed to coordinate environmental protection at the national level. Since then, the institutions for environmental protection have developed and undergone changes several times (see timeline, Figure 4-3). In 1979, the provisional Environmental

Protection Law was promulgated, stipulating that "whosoever causes pollution is responsible for its elimination." For the next decade, China's environmental protection policies were focused primarily on pollution control (Qu, 1991; SCIO, 1996). Environmental protection departments not only formulated control plans, but also applied for investment in facilities designed for pollution abatement.

Environmental protection was proposed as one of China's national basic policies at the Second National Conference on Environmental Protection held in 1983. The government encouraged policies which combined pollution prevention and control measures (as opposed to the singular focus on control), and it also reinforced the principle of the "polluter pays." The basic spirit of these policies was to shift emphasis from control to prevention, but in a way that Chinese leaders considered to be suitable to their economic circumstance. In other words, pollution prevention was not sought at the expense of economic development, which maintained primary importance (Qu, 1997).

The Third National Conference on Environmental Protection in May 1989 was a landmark for environmental management in China. Key regulations and measures that were developed at that conference included:

- The Environmental Protection Objective Responsibility System;
- The Quantitative Examination System on Comprehensive Control of Urban Environment (QESCCUE);
- A pollution discharge licensing system; and
- Centralized control and deadlines for pollution elimination.

Earlier approaches based on point-source pollution controls were insufficient to improve regional environmental quality; thus Chinese leaders established the goal of moving toward integrated regional pollution control (Qu, 1991).

Encouraged by the outcomes of the Rio Summit in 1992, the Chinese government approved and promulgated *China's Agenda 21-White Paper on China's Population, Environment, and Development in the 21st Century,* in March 1994. This document put forward China's overall strategy for sustainable development and served as a guide to various departments and regional governments, as they developed their own plans of action. Some policies and regulations flowing out of this were the Total Emission Control policy (1996), Acid Rain and SO_2 Control Zones (1998), and the Cleaner Production Law (2002).

In March 1998, the Ninth National People's Congress radically reformed government administration. Amidst this massive effort to cut central government administration, the environmental protection administration emerged as a bureaucratic exception: after years of lobbying, it was finally upgraded from semi-ministerial status (as NEPA) to ministerial status (the current SEPA). In this reorganization, SEPA was the only agency which was elevated in terms of official rank. This unexpected promotion, during a time of strict administrative austerity,

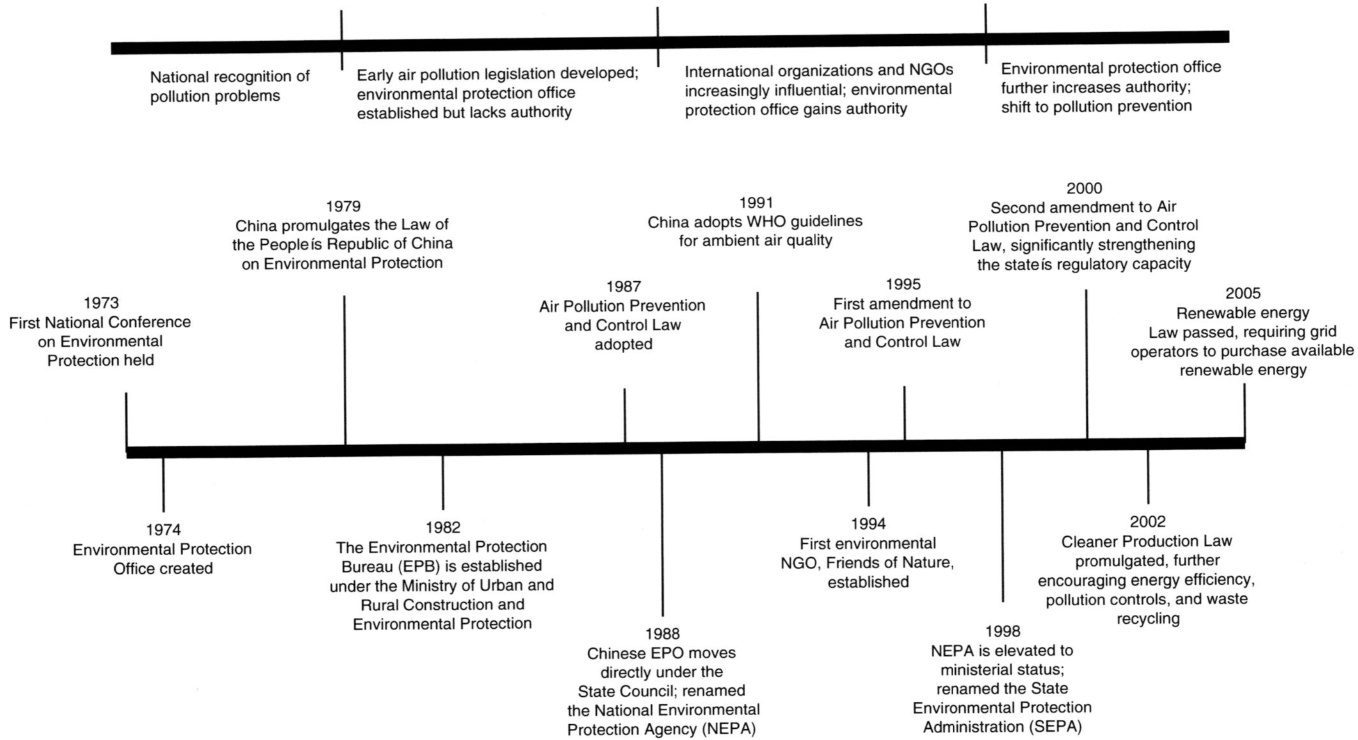

FIGURE 4-3 Timeline of air quality and energy regulation trends in China.

indicated that environmental problems were a serious central government concern in need of increased attention (Jahiel, 1998).

In China, the National People's Congress enacts the laws, governments at different levels take responsibility for their enforcement, the administrative departments in charge of environmental protection exercise overall supervision and administration—and the various departments concerned exercise supervision and administration according to the stipulations of the law. SEPA is the cognizant environmental protection administration agency under the State Council, whose task it is to exercise overall supervision and administration over the country's environmental protection work. The people's governments at the provincial, city, and county levels have also successively established environmental protection administration departments to carry out overall supervision and administration of the environmental protection work in their localities (Figure 4-4).

SEPA has a number of responsibilities related to air pollution monitoring and mitigation. In addition to helping formulate policies, laws, regulations, and administrative rules to be formalized by the National People's Congress, SEPA's other primary duties include:

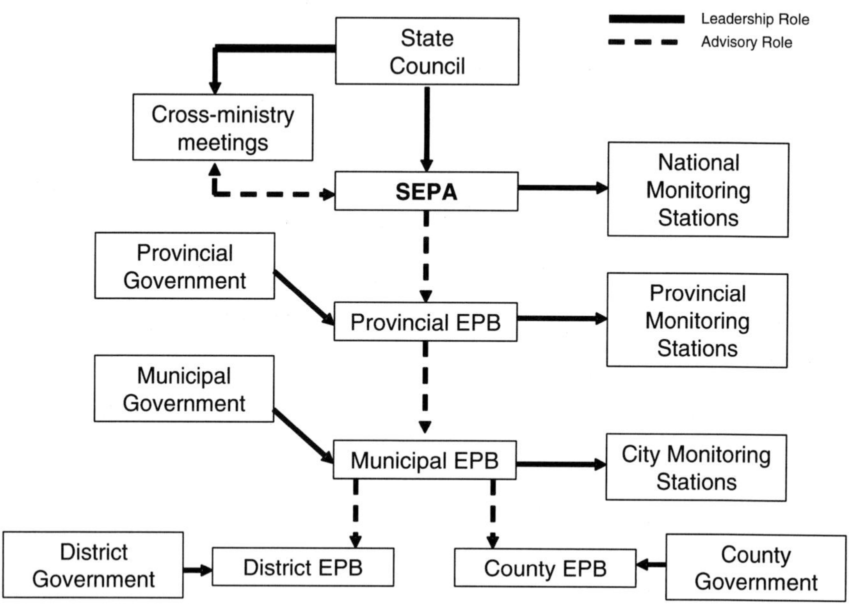

FIGURE 4-4 China's Environmental Protection Administrative System (adapted from Wang and Wu, 2004; Jahiel, 1998).

- Conducting environmental impact assessments for major economic and technical polices, programs, and development plans;
- Organizing the formulation and supervision of pollution prevention plans and ecological conservation plans in key regions identified by the Central Government;
- Investigating major environmental pollution accidents and ecological damage;
- Formulating national standards of environmental quality and pollutant discharge;
- Approving Energy Information Administration (EIA) reports on development and construction activities;
- Organizing research and development, and technical demonstration projects of environmental protection; and
- Environmental monitoring, data collection, and information dissemination.

Although SEPA's mandate is extensive and comprehensive, it is constrained by a limited staff (Fritz and Vollmer, 2006). Figure 4-5 illustrates the number of departments within SEPA with responsibilities for energy and air quality issues. SEPA employs approximately 220 people in its Beijing office, and while it has recently established five regional offices with approximately 30 staff in each, it still relies on provincial and lower-level Environmental Protection Bureaus (EPBs) to handle regional issues, both in monitoring and enforcement. This is the key divergence in responsibility between the central and local governments; enforcement is largely the job of provincial and local EPBs. Statistically there are nationwide more than 3,200 environmental protection administration departments with a total staff of 217,000 engaged in environmental administration, monitoring, inspection and control, statistics collection, scientific research, publicity, and education (SCIO, 2006). However, these numbers reflect *all* positions which could be construed as dealing with environmental protection; in reality, very few employees have responsibilities directly related to air quality monitoring and enforcement.

Local-level action has nonetheless met with some success. Cities have led efforts to adjust their urban energy structure, in order to deal with air pollution issues, advocating clean energy use and central heating—so as to reduce pollution caused by burning coal. Many cities, such as Dalian, have also undergone industrial relocation to the benefit of the urban core, though this strategy does not necessarily reduce total pollution (Bai, 2002). To reduce dust pollution caused by construction, cities are using ready-mixed concrete; concrete mixing is already prohibited in some larger cities.

The Chinese government regards industrial pollution prevention and control as the focal point of environmental protection. China's strategy in this regard is undergoing a major departure from the past, shifting from end-of-pipe solutions to comprehensive controls, from pollution concentration regulation to total pollutant

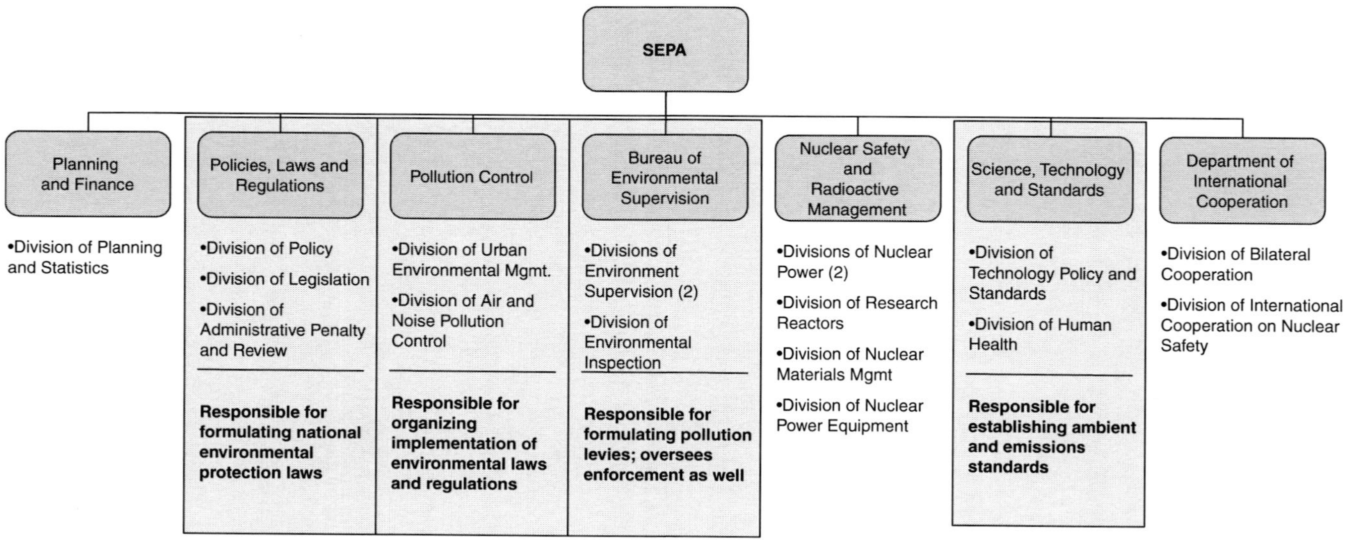

FIGURE 4-5 Departments within SEPA responsible for energy and air quality issues.

amount, from point-source control to comprehensive regional control, and from addressing pollution at the enterprise level to adjusting the industrial structure, promoting clean production, and developing a circular economy (see Chapter 5, Box 5-1). During the Ninth Five-Year Plan period (1996-2000), the State closed down 84,000 small enterprises that had caused both serious waste and pollution. As a result, the discharged amount of principal pollutants has kept declining, while the economic output of these sectors has increased year by year (SCIO, 2006).

AIR QUALITY REGULATORY MECHANISMS

United States

Standard Setting and Implementation

Under the CAA, efforts to reduce air pollution in the United States begin with the specification of National Ambient Air Quality Standards (NAAQS) for six common air pollutants: particulate matter (PM), carbon monoxide, sulfur dioxide, ozone, nitrogen dioxide, and lead. These six pollutants are known as "criteria" pollutants, because the standards that are required to be developed are based on "criteria" documents that review the current state of scientific knowledge of these pollutants' effects on health and welfare. The CAA requires primary standards to be set to protect human health with an adequate margin of safety, without regard to the cost of achieving these standards. The Act also allows for secondary standards to prevent harmful welfare effects (e.g., visibility degradation or harm to materials or ecosystems). In practice, however, EPA has often set secondary standards that are equal to the primary standards. EPA is required by statute to review the NAAQS and to update air quality criteria every 5 years; but because of insufficient resources and the complexity of the undertaking, this requirement has not always been met in practice.

The process of attaining the NAAQS begins by monitoring air pollution concentrations in communities across the country. EPA provides guidelines and a significant amount of financial support to the states to carry out this monitoring. The proposed FY 2007 EPA budget allocates \$185 million for funding state and local air quality grants. EPA then uses the monitoring data to classify communities according to whether or not they meet the standards. Those that do not, are designated as "non-attainment areas" and face statutory deadlines for reducing air pollution and for coming into compliance. Once non-attainment areas are designated, the states that contain them must develop State Implementation Plans (SIPs) to lay out their approach for timely compliance, and they must contain enforceable regulations for ensuring that emissions are reduced as necessary. The CAA provides for federal sanctions against communities that fail to achieve compliance with the NAAQS in the time allowed under the statute. However, these sanctions have been used only sparingly; more often deadlines for attainment have been extended when pollution problems have proven difficult to solve. Compliance with the NAAQS must ultimately be established, based on monitoring data that demonstrate that the standard is being met and maintained on an ongoing basis.

Along with relatively ubiquitous criteria pollutants, the CAA also addresses a long list of other air pollutants collectively known as hazardous air pollutants (HAPs) or air toxics. Many but not all of these pollutants are known or suspected carcinogens. The 1970 CAA amendments required EPA to develop a list of HAPs and then to issue national emissions standards for each of the pollutants that it had listed. However, EPA made little progress in responding to this requirement, due in large part to protracted debates over scientific uncertainty about risks posed by potential HAPs and the costs and benefits of prospective emissions standards. Between 1970 and 1990, EPA listed only eight compounds as HAPs and issued national emissions standards for only seven of those eight compounds. In response to this slow progress, Congress amended the CAA in 1990 to fundamentally revise the HAP provisions.

The 1990 amendments explicitly listed 189 hazardous air pollutants, and directed EPA to identify the categories of sources that contribute significantly to emissions of these pollutants. EPA is then required to set emissions standards based on the maximum reductions being achieved by sources within each category. Although control costs are a consideration in setting these technology-based standards, the new statutory scheme is designed to impose emissions reduction requirements relatively quickly, avoiding the contentious cost-benefit balancing exercises that had bogged down EPA's efforts to implement the earlier provisions. Under the new HAPs provisions, risk assessment was moved to a second stage of the process. In this stage, which EPA began only recently, the agency is required to assess the "residual" risk that remains after the imposition of the first round of technology-based standards, and then to undertake additional regulation, if the residual risk is unacceptably high.

Although states bear primary responsibility under the CAA for achieving compliance with the NAAQS, federal regulations have greatly assisted them in meeting this responsibility. In particular, the federal government has played a lead role in setting and in enforcing emissions standards for new motor vehicles and non-road engines, and in regulating the composition of motor vehicle fuels. The 1970 CAA amendments required automobile manufacturers to reduce light-duty-vehicle emissions of hydrocarbons and carbon monoxide emissions by 90 percent by 1975, and to reduce nitrogen oxides emissions by 90 percent by 1976. The 1977 CAA extended the deadlines for these reductions to be achieved, and reduced the stringency of the requirement for nitrogen oxides—but the standards were ultimately met and surpassed. Under its authority to regulate new mobile sources, EPA has subsequently set and enforced emissions standards for new heavy-duty on-road vehicles and for new non-road engines used in farm and construction equipment. The heavy-duty and non-road emissions standards are accompanied by limits on sulfur content of diesel fuel, which reduce the contribution of these sources to emissions of sulfur oxides, and which also enable the use of catalytic control devices that are susceptible to damage from sulfur in engine exhaust.

The CAA expressly states that federal regulation of new motor vehicles preempts regulation by the states, with one exception. In recognition of its leadership and technical capacity, California is allowed to set its own more stringent standards. Other states can adopt California's tailpipe standards in place of the federal standards, but cannot develop their own (NRC, 2004). States have greater responsibility and flexibility in devising and adopting measures to reduce emissions from in-use vehicles. In order to meet the NAAQS for carbon monoxide, ozone, and PM, the states have independently adopted in-use control measures such as fuel content specifications and vapor pressure limits, vehicle inspection and maintenance programs, and retrofit requirements for heavy-duty diesel trucks and buses.

Under the CAA, federal and state governments in the United States also play complementary roles in regulating air pollution from stationary sources. The CAA requires EPA to set emissions limits for new stationary sources, based on best demonstrated technology. For some types of sources, such as wood stoves, these "new source performance standards" (NSPSs) are enforced by the federal government through manufacturer certification rules, similar to those for new motor vehicles. For large stationary sources like power plants and industrial boilers, the limits are enforced through construction and operating permits, which are generally administered by the states. In the case of large stationary sources, the nationally applicable NSPSs are supplemented and usually surpassed in stringency by control requirements developed through case-by-case review of the control technology available at the time that a project developer applies for a construction permit. This latter analysis is known as new source review (NSR). In essence, the national NSPS limits set the floor for emissions reductions, but sources in many areas must go further to match the state of the art in control technology identified through NSR.

Cap-and-Trade Programs

The 1990 CAA amendments adopted a major innovation in the regulation of air pollution from stationary sources, allowing the trading of sulfur dioxide emissions credits, subject to a national cap. The primary objective of this program was to cost-effectively mitigate acid rain caused by sulfur dioxide emissions from power plants. By 2010 the program will cap SO_2 emissions from all power plants that are greater than 25 MW in capacity at 8.95 million tons, representing a 10 million ton reduction in emissions from 1980 levels. Allocations to individual power plants are made in the acid rain program based on historical rates of fuel consumption. To date, the acid rain program is credited with exceeding the targets for SO_2 reductions at costs that are far lower than predicted when the 1990 amendments were enacted. Improvements in atmospheric concentrations of

SO_2 and sulfate, wet sulfate deposition, and haze have been seen in the eastern United States. Acid neutralizing capacity is rising in many of the northeastern lakes, where acidification prompted the original action.[3] The acid rain program also included performance standards for nitrogen oxides emissions, but without emissions trading.

Since the 1990 CAA amendments were adopted, EPA has acted under its statutory authority to regulate interstate air pollution to adopt three additional cap-and-trade programs. In 1994, a group of states in the northeastern United States, working together as members of the Ozone Transport Region committed to a cap-and-trade program to limit nitrogen oxides emissions from large stationary sources during the summer ozone season. The program ran from 1999 through 2002. In October 1998, EPA adopted a federal rule that replaced and expanded this program to 22 eastern states and the District of Columbia, where interstate transport of pollution had been complicating state and local efforts to meet and maintain compliance with the NAAQS for ozone. This program capped emissions on a state-by-state basis, and allowed, but did not require, states to achieve compliance through emissions trading.[4] In March 2005, EPA finalized a cap-and-trade program for sulfur dioxide and nitrogen oxides, which is designed to further limit emissions of these two pollutants from power plants in the eastern United States, and thus is designed to address interstate transport contributions to fine PM and ozone pollution.[5] This program, known as the Clean Air Interstate Rule, is being challenged in court primarily by industry groups, cities, and state agencies, who object to the way emissions allocations were granted, or to the inclusion of certain areas in the program.

Also in 2005, EPA finalized a cap and trade program for mercury emissions from power plants. As proposed, the Clean Air Mercury Rule (CAMR) sets statewide caps on mercury that would reduce nationwide mercury emissions from coal-fired power plants from the current level of 48 tons per year down to 15 tons per year, by 2018.[6] States can, but are not required to, allow sources within their borders to meet the statewide caps through emissions trading. CAMR is being challenged in court primarily by states and environmental groups, who argue that the rule falls short of achieving the emissions reductions needed to protect human health and the environment, and that trading should not be allowed for toxic air pollutants such as mercury, because trading allows some sources to increase their emissions and to potentially increase local impacts, even when the overall cap is met.

Beyond these recent examples, which have been adopted using EPA's existing statutory authority, several bills have been introduced in the U.S. Congress

---

[3]Executive Office of the President of the United States, National Acid Precipitation Assessment Report to Congress: An Integrated Assessment, August 2005.

[4]63 Fed. Reg. 57356 (October 27, 1998).

[5]70 Fed. Reg. 25162 (May 12, 2005).

[6]70 Fed. Reg. 15994 (March 29, 2005).

in recent years that would expand the use of emissions trading programs for conventional pollutants such as sulfur dioxide and nitrogen oxides, and in some cases would also establish cap-and-trade programs for carbon dioxide and other greenhouse gases.

Existing emissions trading programs in the United States rely on continuous emissions monitors (CEMs) for nitrogen and sulfur oxides to track actual emissions and to compare them against allowances (see Box 4-1). The 1990 CAA amendments required the installation of CEMs for SO_2 and NO_x at power plants participating in the acid rain program. The EPA's Clean Air Markets Database posts emissions data on a publicly available website. The CAMR requires either continuous emissions monitoring, or a sorbent trap system that can collect an uninterrupted continuous sample.[7]

Emissions Fees and Taxes

Emissions fees or taxes present another economic means for controlling pollution emissions. Unlike trading scenarios, which permit emissions to a certain cap, emissions fees are assessed as a tax per ton of pollutant emitted. Polluting firms (or individuals) are able to decide whether and how to reduce costs. Emissions taxes are considered the most efficient way to internalize pollution costs, though that is not to say they are necessarily the most efficient way to reduce pollution. Setting appropriate fees is essential, and as China has experienced with its levy system, insufficient taxes will not bring about the desired effect.

Emissions fees are one way to decrease the likelihood of pollution "hotspots" developing in a given area. This is one critique of the more flexible permit trading scenario, although previous research has not shown any problems with hotspots under the SO_2 trading program (Burtraw and Mansur, 1999; Swift, 2000). Emissions fees could also be implemented regionally, as many states have explored. In this way, regional groups can better determine the costs of a particular pollutant (e.g., a metropolitan area on the U.S. east coast might consider a higher tax on NO_x emissions than would an area in the western United States with traditionally lower ambient levels).

A carbon tax is the often-cited example of an emissions fee, which could better incorporate the external costs of CO_2 emissions. Given the global impacts of CO_2 emissions, this option may be more efficient than implementing a global cap and trade system. As will be discussed in later chapters, U.S. cities and states have taken the initiative in curbing their CO_2 emissions, in some cases devising carbon taxes.

In the transportation sector, congestion fees represent another tax designed to reduce vehicular emissions. While not instituted broadly in the United States, they have met with success in metropolitan areas such as London. Some Chinese

[7]70 Fed. Reg. 15994 (March 29, 2005).

BOX 4-1
Components of an Emissions Trading Program

Market-based approaches to pollution control can be highly cost-effective because they help minimize the marginal costs of emission reductions among the regulated facilities, thereby reducing overall pollution control costs. Market-based approaches can also provide strong incentives for entities to implement innovative pollution-control technologies and best management practices, which may contribute to substantial cost savings and emissions reduction.

The most common market-based approach is a cap-and-trade program in which relatively small quantities of pollutant emissions are traded among sources. "Cap-and-trade" programs are conducted under an aggregate emissions cap that is based on a maximum amount of emissions that is possible while still protecting human health. These programs have been shown to be effective at achieving emission reductions at much less cost to the regulated facilities than traditional technology-based or performance-oriented standard.

In a cap-and-trade program, a regulatory agency sets a limit or cap on the amount of a given pollutant that can be emitted within its jurisdiction. All entities within that jurisdiction that emit the pollutant are given an allowance or credit towards the amount that they can emit. The credit represents the right to emit that amount, and the sum of all the credits among the regulated community should equal the cap for that pollutant in the jurisdiction. If a company emits more of the pollutant than its credit allows, it must buy credits from one or more other companies that do not exceed their allowances (the trade). Examples of successful trading programs in the United States include:

- The acid rain SO_2 emissions trading program—which has served as the paradigm for many United States cap-and-trade-programs
- NO_x emissions trading programs, including the South Coast Air Quality Management District program in Southern California
- The Northeast O_3 Transport Region NO_x Budget Trading Program
- The NO_x SIP Call Trading Program

A closer examination of the SO_2 trading program indicates that its success appears to be the result of a combination of factors, specifically:

- *Substantial emission reductions*—the mandated 50 percent reduction ensured that regional trading did not result in areas of emission increases
- *Simplicity*—a clear cap and few restrictions on implementation
- *Effective monitoring*—CEM systems aided compliance assurance, mitigating disputes between environmental advocates and the regulated industries
- *Transparency*—EPA's allowance tracking system promptly updated transactions and emissions data, lending transparency to the process
- *Certain penalties*—non-compliance led to financial penalty plus a loss of allowances for the subsequent year
- *Opportunity for banking allowances*—firms were able to insure against adverse conditions; this also led to 'buy-in' by regulated parties, because banking gave them an asset whose value depended on a stable and successful program.

cities are also considering implementing congestion fees, though most plans to date have been withheld over concerns of public backlash. Overall, this sort of tax presents equity issues, particularly for a developing country such as China, as it is viewed as unduly burdensome to poor vehicle owners.

Enforcement of Environmental Laws

In the United States, local, state, tribal, and federal agencies all have powers to enforce air pollution regulations adopted into law at their respective levels, and to impose penalties for non-compliance. When violations of air pollution regulations are initially found, government agencies generally have the option of working with the violator to help them achieve compliance, assessing administrative penalties, or bringing suit against the company and seeking either civil or criminal penalties. EPA's enforcement arm can levy civil penalties of up to $27,500 per violation per day for non-compliance with stationary source requirements. EPA also enforces fuel and motor vehicle emissions standards; it recently announced a $94 million settlement with DaimlerChrysler for defective catalytic converters and for on-board diagnostic systems on the company's1996 through 2001 model year vehicles.[8] Criminal penalties, including jail sentences, are available in case of serious environmental violations. Additionally, the CAA provides for citizen enforcement of the Act's provisions, through civil suits against the EPA administrator or against a state, for failure to meet CAA requirements, or against any facility that constructs or modifies a major source without a permit, or in violation of its permit conditions.[9]

Long-Term Planning and Policy Revision

The most significant driver for long-term planning and policy updates built into the CAA is the requirement for periodic reviews of air pollution effects on health and welfare and for corresponding adjustments to the NAAQS. Perhaps the most significant change these reviews have engendered is the shift in the size range of PM that is regulated in the United States—from total suspended particulate to particulate matter of less than 10 microns in diameter (PM_{10}), and particulate matter of less than 2.5 microns in diameter ($PM_{2.5}$). EPA is now considering a shift from using PM_{10} as the indicator for coarse PM, to the use of the size fraction between 2.5 and 10 microns. Significant shifts in regulatory approaches have also arisen from advances in the understanding of atmospheric chemistry and transport, especially the increased appreciation of the role played by interstate transport in ozone and $PM_{2.5}$ pollution, and an understanding of the contribution of nitrogen oxides to regional-scale ozone pollution.

---

[8]U.S. EPA, Compliance and Enforcement, Cases and Settlements, http://Cfpub.epa.gov/compliance/cases (accessed June 12, 2006).

[9]42 USC 7604.

As discussed above, while EPA has revised air pollution regulations within existing statutory frameworks, Congress has also adopted major amendments to the CAA, in response to concerns about the cost and efficacy of air pollution control efforts, and recognition of new air pollution problems and impacts. For example, the 1990 CAA amendments extended deadlines for states to come into compliance with the NAAQS for PM and ground-level ozone, and also addressed two previously underappreciated problems—acid deposition and depletion of the stratospheric ozone layer. No major amendments have been passed since 1990, but several bills have been introduced in recent years that would constitute significant changes if enacted. The common thrust of these bills has been moving toward increased reliance on market-based cap-and-trade programs, especially for power plant emissions. A more controversial issue has been whether to adopt mandatory limits on emissions of carbon dioxide and of other greenhouse gases.

China

Responsibility for the development and implementation of environmental policy in China is split between the national and local governments. In general, the central government provides policy direction and the general framework, while local governments are responsible for implementation and enforcement, and they have some flexibility in choosing appropriate measures. This division of responsibility is similar in some ways to that in the United States, where primary and secondary standards for criteria pollutants are established at the national level, and states are expected to come into compliance with these standards through State Implementation Plans. A key difference, however, and one that complicates the enforcement of most environmental regulations in China, is that the local governments, specifically the local EPBs, are responsible for enforcing the laws.

China's legal system is divided into seven types and three levels of law. The types are the constitution and constitution-related laws, civil and commercial law, administrative law, economic law, social law, criminal law, and the law on lawsuit and non-lawsuit procedures. The levels are state laws, administrative regulations, and local statutes. Additionally, canonical documents can be issued by the State Council and by responsible administrative departments. China pays great attention to environmental legislative work, and has now established an environmental statutory framework that takes the Constitution of the PRC as the foundation, and the Environmental Protection Law of the PRC as the main body of its environmental legislation.

The pollution levy system (PLS) is one of the earliest environmental policies established and implemented in China. Building on the "polluter pays" principle, the Environmental Protection Leading Group of the State Council put forward the PLS in late 1978, and in 1979 the first pilot pollution levy program was implemented in Suzhou city, Jiangsu province. Following some additional pilot programs, the PLS was made effective in 1982. The original levy system made

21 airborne pollutants subject to discharge fees. When more than one type of pollutant was discharged at a single point, fees were collected only for the pollutant which was subject to the highest charges, and fees were calculated on the basis of how much the discharge exceeded the standard. In most cases, it was cheaper for a company to pay the pollution discharge fees than to invest in abatement measures. Therefore the levy system has been widely criticized because the fees prescribed under it are too low to induce companies to adopt cleaner technologies. Implementation of the PLS has also been subject to criticism; originally the levies were assessed on large enterprises, and the levies were recycled back (80 percent after the local EPB took 20 percent for its own operating costs) to the polluting firms, ostensibly to invest in pollution abatement. However, there were no provisions to regulate the use of the discharge fees, and fees were often negotiated, mishandled, or misappropriated.

In an effort to make the PLS more meaningful and to exert a greater impact on the behavior of companies, China revamped the system several years ago. The State Council issued the *Regulations for the Administration of the Levy and Use of Pollutant Discharge Fees* on January 2, 2003, and the Ministry of Finance, SEPA, the National Development and Reform Commission (NDRC), and the State Economic and Trade Commission (SETC), issued the *Measures for the Administration of Rates for the Collection of Pollution Discharge Fees* on February 28, 2003. Companies make a filing each year before December 15 declaring the types, quantity, and density of pollutants that they will discharge during the next year under normal operating conditions, based both on their actual pollution discharge experience in the current year and their production plans for the next year. Local EPBs must then examine and approve these declarations before January 15 of the next year.

The pollutant discharge fee must be included in the EPB's budget, managed as a special fund for environmental protection, and mainly used for allocating grants and subsidies for the pollution prevention and control projects of including regional control, technology development and demonstrations, application of new processes, and various control projects set forth by the State Council. Any amendment to the standard for levy of pollutant discharge fee shall be made after an advance notice. However, deficiencies in emission measurement often lead to negotiated payments only roughly—if at all—related to actual emissions. Negotiated payments are an effective and perhaps initially unavoidable way to raise revenue; however, for the pollution levy to significantly affect abatement behavior, the incidence of the levy must be closely correlated with actual emissions (Ellerman, 2002). The SO_2 emissions levy has been a useful indicator of China's approach to environmental policy (see Box 4-2). First applied in 1982 and set at 0.04 RMB/kg, the levy has continually been increased, often drastically, based on experiences with pilot programs. Most recently, the government has decided to set the rate at 1.26 RMB/kg by 2010, doubling the prevailing rate, which had been tripled only 4 years earlier. This proposed rate is based on expe-

BOX 4-2
Emissions Trading in China

In China, emissions trading pilot programs are structured to use a cap-and-trade system similar to the system used in the United States. In April 2002 the Hong Kong SAR Government and the Guangdong Provincial Government reached a consensus to reduce regional emissions of SO_2, NO_x, inhalable suspended particulates (RSPs) and volatile organic compounds (VOCs) by 40 percent, 20 percent, 55 percent, and 55 percent, respectively, by 2010, using 1997 as the base year. To achieve these emission reduction targets, the two governments are working on a regional AQM plan that will set out the improvement measures that Hong Kong and Guangdong will take. The Hong Kong Environmental Protection Department announced in December 2003 that it is considering the feasibility of an emissions trading scheme in the Pearl River Delta region as one way of trying to achieve desired objectives. In 2006, high levels of air pollution and smog in Hong Kong focused much public attention on the need to finalize a plan for the Pearl River Delta region.

The State Council issued a document at the end of 2005 calling for a strengthening of environmental protection, using the cap-and-trade system. But controversy has arisen over whether the plants should have to pay the State for their allowance quotas. In September 2006, SEPA and the Ministry of Finance (MOF) began working together to start a cap-and-trade system to curb China's world-leading emissions of SO_2.

riences in Beijing, as well as projected abatement costs, though these will range depending on sources and local conditions.

Local Statutes

To implement the State's environmental protection laws and regulations, the local people's congresses (or standing committees) at the provincial level or "the comparatively larger cities"[10] may enact and promulgate local laws on environmental protection, based on local conditions. For example, in 2000 the southern industrial province of Guangdong enacted two regulations addressing regional air pollution issues: the Guangdong Province Environmental Protection Regulation and the Guangdong Province Vehicular Emissions and Pollution Prevention Regulation.

SEPA establishes the national standards for environment quality. The provinces, autonomous regions, and municipalities directly under the central

[10]A "comparatively larger city" refers to a city where a provincial or autonomous regional people's government is located or where a special economic zone is located, or a city approved as such by the State Council. There are 19 such cities.

government may establish their own local standards for environment quality for items not specified in the national standards for environment quality, and shall report them to SEPA for the record. SEPA also establishes the national standards for the discharge of pollutants, in accordance with the national standards for environment quality and with the country's economic and technological conditions. Units that discharge pollutants in areas where the local standards for the discharge of pollutants have been established shall observe such local standards.

Responsibility and Examination Systems

According to the Environmental Responsibility System (ERS), each governor, mayor, and county magistrate shall sign a responsibility contract that specifies the environmental objectives and tasks that he or she should meet at the end of the term. ERS aims to assign responsibility for overall environmental quality to officials in local government. Thus the provincial governor is held responsible for the overall environmental quality in a province, and the mayor of a municipality or magistrate of a county is responsible in their respective jurisdictions (Sinkule and Ortolano, 1995; Qu, 1991). This regulation was introduced in the early 1980s, due to the fact that responsibility for environmental quality lay with the EPB, but it had no power over the economic and social development decisions that affected environmental quality (Qu, 1991).

The ERS is implemented through written contracts or agreements, each of which stipulates the environmental targets for a particular time period. There are two types of ERS contracts: contracts among heads of different levels of government and contracts between governmental officials and factory directors (Sinkule and Ortolano, 1995; Qu, 1991). ERS has its legal basis in the Constitution of PRC (Article 26), and in the Environmental Protection Law (Article 16; 24).

The QESCCUE examination system mentioned earlier adopts quantitative methods to assess the performance of city governments in comprehensive urban environmental control (CUEC). CUEC refers to the work aiming at protecting and improving the urban environment through legal, economic, administrative, and technical means, under the unified leadership of city governments. The main components of CUEC (or treatment) include three aspects: prevention and control of urban industrial pollution, construction of urban infrastructure, and urban environmental management.

QESCCUE is managed at different levels. Key cities first carry out self-examination, and the results are submitted to the provincial EPB for review and examination, before final submission to SEPA. After examination, SEPA reviews the results and then reports to the public. SEPA takes charge of the examination on 113 national key cities for environmental protection. Apart from publicizing the evaluation results through major media, SEPA has also published the Annual Report on Urban Environmental Management and Comprehensive Urban Environmental Control of Key Cities for Environmental Protection. In 2004, SEPA reported for the first time to the public the quantitative examination results conducted by each province, autonomous region, and municipality directly under the Central Government. QESCCUE has been introduced in over 500 Chinese cities (more than 70 percent of all cities).

Another program, prompted by a directive in the Ninth Five-Year Plan to establish "model cities," is the National Model City of Environmental Protection (NMCEP) program. NMCEPs provide examples for other cities in terms of improving their urban environment and implementing environmentally sustainable urban development. This campaign meets the increasing requirements of the public for good living environment and brings benefits to the people. Therefore, it has received widespread support and participation by the public. The model cities have experienced an improvement in environmental quality. NMCEPs, like other cities in China, use the Air Pollution Index (API) to evaluate ambient air quality (Table 4-1).

The number of days with an API under 100 increased by 5 percent, and the green space in urban areas increased nearly 12 percent. In 2003, the urban air quality of the NMCEPs was Grade II or better 92 percent of the time. The annual average concentration of SO_2 in the air of those cities was 11 $\mu g/m^3$ lower than the national average level, and the concentration of inhalable particles was 55 $\mu g/m^3$ lower than that of the national average level (SEPA, 2005). By the end of 2005, there were 53 cities and 3 urban districts achieving the distinction of NMCEP, with more than 100 cities (or districts) in the process of applying or under review (SEPA, 2006b). These examination indicators aim to reflect such contents as the ability and competitive power of sustained urban development, the socioeconomic development level, and the degree of harmony between economic growth and the environment.

TABLE 4-1 China's Air Pollution Index (API)

	Pollutants and Concentrations (mg/m^3)							
API	SO_2 Daily Average	NO_2 Daily Average	PM_{10} Daily Average	CO Hourly Average	O_3 Hourly Average	API Ranges	Air Quality Level	Assessment
50	0.05	0.08	0.05	5	0.12	0–50	I	Excellent
100	0.15	0.12	0.15	10	0.2	51–100	II	Good
200	0.8	0.28	0.35	60	0.4	101–200	III	Lightly polluted
300	1.6	0.565	0.42	90	0.8	201–300	IV	Polluted
400	2.1	0.75	0.5	120	1	301–500	V	Heavily polluted
500	2.62	0.94	0.6	150	1.2			

RESEARCH, MONITORING, AND EDUCATION

United States

Energy and Air Pollution Research

Energy and air pollution research receives funding from both the private sector and from government sources. Federal agencies providing funding include EPA, the National Science Foundation (NSF), and the U.S. Department of Agriculture. EPA's funding is coordinated through the National Center for Environmental Research and much of it is carried out "in-house," through the ORD, for example, the Mercury Research Strategy. The NSF runs numerous funding programs under the general category of environmental engineering research, several of which address air quality and energy issues. The USDA provides state-level funding through the Cooperative State Research, Education, and Extension Service. Energy and air pollution research programs are also prominent at numerous universities throughout the United States.

The Department of Energy (DOE), with primary responsibility for the nation's energy research, is increasingly giving importance to improving air quality through cleaner alternatives. As one example, the Fossil Energy Research & Development program is developing pollution-control technologies and generation processes to virtually eliminate criteria pollutant emissions from fossil-fuel power plants. Clean Cities, another program, was developed to help local decision makers adopt practices which reduce petroleum consumption. DOE runs or oversees 21 national laboratories and specialized offices, including Argonne National Laboratory, the National Renewable Energy Laboratory, and the National Energy Technology Laboratory—the only U.S. laboratory devoted to fossil fuel technologies. More than 30,000 scientists and engineers perform research in these laboratories; and while their research may not always directly address air quality, changes to the nation's energy system that result from the work of DOE have a major influence on the ability of various regions to meet the NAAQS.

Air Pollution Monitoring Systems

EPA runs several air pollution monitoring programs. The National Air Toxics Program includes maximum achievable control technology standards, residual risk standards, area source standards, mobile source rules, the utility mercury reductions rule, local-scale projects, and the Great Waters project. The PM Supersites program is a research program that monitors ambient air and provides valuable data for the benefit of human health, exposure research, and the atmospheric sciences community.

Additional EPA programs include the Photochemical Assessment Monitoring Stations program, the Interagency Monitoring of Protected Visual Environments program, and the Environmental Technology Verification Program. The EPA's overall strategy for air pollution monitoring is outlined in the National Ambient Air Quality Monitoring Strategy. The EPA also maintains a database of emission inventories, emission factors, emissions modeling, and emissions monitoring knowledge, and runs an Emission Measurement Center. The EPA mandated a network of approximately 4,000 monitoring stations, referred to as the NAMS/SLAMS network (National/State and Local Air Monitoring Stations), to assess NAAQS attainment status. Most of these are state and local monitoring stations, and thus size and distribution are up to the state. However, a subset (~1,000 stations) are NAMS, and are located in areas with the highest expected pollutant concentrations and population exposures (NRC, 2004).

Air Pollution Modeling

EPA is the primary federal agency responsible for air pollution modeling. The EPA's Air Quality Monitoring Group operates the Support Center for Regulatory Atmospheric Modeling, to provide direction in the full range of air quality models and to conduct modeling analyses to support policy and regulatory decisions. The site contains links to detailed information on air quality models, modeling applications and tools, modeling guidance and support, and meteorological data and processors. The three most common types of air pollution modeling are dispersion modeling, photochemical modeling, and receptor modeling. Preferred dispersion models include the AERMOD Modeling System and the CALPUFF Modeling System. The website also lists several recommended modeling systems for photochemical and receptor models with links to more information on each model.

Air Pollution Control Technology

A wide range of air pollution control equipment is available in the United States and can be selected based on the requirements of the project. The New and Emerging Environmental Technologies Clean Air Technologies Database is an online source for useful information on new and emerging technologies designed to prevent, destroy, remove, monitor, sample, or model air pollution emissions from stationary or mobile sources. Listings include both technologies that are commercially available and those that are still in the development stage. Additionally, the Air Pollution Control Equipment Selection Guide has a primer on air pollution control, followed by listings for technologies such as biofilters, electrostatic precipitators, and fluidized bed scrubbers. Other useful information on control technologies can be found at the air pollution control equipment retailer page and the Air Pollution Control Equipment Selection Guide.

China

Research and Education

In China, on-the-job training has been an essential aspect of environmental management capacity building. In 1981 the Environmental Administrative Personnel Training College was established for the purpose of offering on-the-job training, continuing education, and academic-level education to administrative personnel in the environmental protection departments throughout the country. Theses trainees are playing an important role in promoting the nation's environmental protection work.

Research into environmental science and technology in China began in the 1970s. For some major environmental research subjects the Chinese government has formulated corresponding research programs and plans for environmental protection, while organizing forces to tackle key scientific and technological problems. Additionally, China has expanded its research into comprehensive prevention and control of regional environmental pollution, environmental background values and environmental capacity, pollution control technology, and global environmental problems. As a result, the country has made substantial scientific and technological achievements in some research areas, such as the capacity of the atmospheric environment, and acid deposition's impact and control strategies.

Public Education

The Chinese government regards it a strategic task to popularize environmental protection knowledge among the people and to raise their consciousness about environmental protection and gradually to cultivate fine environmental ethics and codes of conduct. Virtually all newspapers, radio, and TV stations offer environmental protection programs, and, in particular, the media pays special attention to severely polluted areas and units. Meanwhile, the various provinces and cities have also developed such activities.

In 1983 China established the first national-level professional newspaper on environmental protection in the world—the China Environment News, with an annual circulation of nearly 300,000 copies. In 1980 the China Environmental Science Press was established. Since 1990 the China Environment Yearbook has been published, and its English version has been published since 1994. Besides, there are more than 30 local environmental newspapers and several hundred professional periodicals.

In April 1992, China set up the China Council for International Cooperation on Environment and Development composed of more than 40 leading specialists and well-known public figures from China and other countries. The Council has put forward valuable concrete proposals on energy and the environment, resources accounting and the pricing system, public participation, and the implementation of the environment laws and regulations, which have aroused the attention and response of the Chinese government.

Air Pollution Monitoring Systems

SEPA is responsible for national ambient air quality monitoring in China. In addition to establishing a monitoring system, SEPA formulates the monitoring norm and, in conjunction with relevant departments, organizes the monitoring network. The China National Environmental Monitoring Centre (CNEMC), established in 1980, is an institution directly affiliated to SEPA. It provides technical support supervision and technical service for the environmental supervision management of SEPA, plays a role as network center, technical center, information center and training center of national environmental monitoring, and provides professional management and guidance for the national monitoring system. It is responsible for collecting, verifying, and managing environmental monitoring information and environmental data.

There are 2,389 environmental monitoring stations directly under the government, including one state-level station, 41 province-level stations, 401 city-level stations, 1,914 county-level stations, and 32 radiation monitoring stations. The national network includes more than 1,000 air quality monitoring sites now. SEPA and upper-level EPBs regularly issue reports on environmental situations. Local EPBs at or above the county level, in conjunction with relevant departments, investigate and assess the environmental situation within areas under their jurisdiction and draw up plans for environmental protection which shall, subject to overall balancing by the department of planning, be submitted to the people's government at the same level for approval before implementation. Local EPBs at or above the county level or other departments invested by law with power to conduct environmental supervision and management shall be empowered to make on-site inspections of units under their jurisdiction that discharge pollutants.

As of 2003, there were 631 urban automatic air quality monitoring sites and by the end of 2004, more than 45,000 people were employed nationwide in environmental monitoring stations (CNEMC, 2006). An ambient air quality daily report is issued according to China's API (see Table 4-1), along with a forecast using the API range. Reported pollutants are SO_2, NO_2 and PM_{10}. Each pollutant reported has an index ranging from 0 to 500, with 50 corresponding approximately to Class I of the CNAAQS, 100 corresponding to Class II, 200 corresponding to Class III, and 500 corresponding to significant harmful effects. Cities are expected to meet Class II standards.

Air Quality Reporting

China's urban air quality reporting system has played an active role in raising pubic environmental awareness and in intensifying environmental monitoring.

The main practice of this system is to report the API that reflects the extent of air pollution. In 1997, 13 key cities[11] including Beijing, Shanghai, Chongqing, Dalian, and Xiamen started issuing urban air quality weekly reports. Forty-two additional environmental protection key cities began issuing ambient air quality daily reports in June 2000, including 32 provincial capitals, municipalities directly under the Central Government and municipal capitals, ten coastal cities, and key tourist destinations. In June 2001, they were joined by 47 more key cities. At present, there are 180 cities including all 113 key cities implementing the air quality reporting system. The ambient air quality daily report and ambient air quality forecast of 47 environmental protection key cities are sent to CNEMC every day; the ambient air quality daily report is sent to CCTV, CETV (China Education TV), People's Daily, China Daily, Xinhua news agency, and related websites (most importantly, SEPA's).

Recently, some Chinese cities have cooperated to develop regional reports on air quality. In southern China, the Guangdong Provincial EPB and Hong Kong's Environmental Protection Department (EPD) issued a regional air quality report in 2006, the first such effort in China. The report presents results from the Pearl River Delta Regional Air Quality Monitoring Network's Regional Air Quality Index and includes data on ozone, available on both the EPB and EPD websites.[12]

NON-GOVERNMENTAL ORGANIZATIONS

United States

Hundreds of NGOs in the United States are involved in shaping and in promoting the enforcement of air pollution laws. These groups include environmental and public health groups that generally advocate for more protective regulations and enforcement policies, regulated companies, and industry associations that often seek to ensure that regulatory burdens are minimized, but may also support science and engineering research of mutual interest to their members, and professional societies of engineers, scientists, lawyers, or other professionals, who work in the field of environmental management and join together primarily for purposes of continuing education and information exchange.

Environmental and public health advocacy organizations in the United States range from groups with a few members that focus on issues of local interest, to national organizations with hundreds of thousands of members and broad portfolios. National organizations that actively work on air pollution issues include the Sierra Club, the National Resources Defense Council, Environmental Defense, and the American Lung Association. These groups have professional staffs that number in the hundreds, and budgets of millions

[11]Key cities are nationally designated "key cities for environmental protection."

[12]Guangdong EPB: http://www.gdepb.gov.cn; Hong Kong EPD: http://www.epd.gov.hk.

of dollars per year, funded by donations from private individuals, foundations, corporations, and government grants. Their history of involvement in environmental issues in the United States dates back more than a century; for example, the Sierra Club was founded in 1892 and the American Lung Association in 1904. These groups work on all stages of development and implementation of air pollution regulation at local, state, and national levels. Their primary activities related to environmental regulation include commenting on regulatory proposals as local, state, and federal agencies develop them, and challenging the regulations in court when they deem them to be unsatisfactory. These organizations may also seek to initiate regulatory action, for example, by suing to force EPA to act on statutory duties that it has failed to undertake. Non-profit environmental and public health organizations are generally organized as charitable "501(c)(3)"organizations for tax purposes, meaning supporters can deduct their contributions, but the organizations are limited in the amount of legislative lobbying that they can undertake.

Industry stakeholders that participate in public processes to develop and implement air pollution regulations include both individual companies and associations of companies that are impacted by these regulations. Trade associations like the National Mining Association and the Edison Electric Institute (an association of shareholder-owned electric companies) are most often funded by dues assessed to their members. Alongside environmental and public health organizations, these industry groups are entitled to have their views heard as regulations are developed, and to utilize the citizen suit provisions of the Administrative Procedures Act and the CAA to gain judicial review of EPA's compliance with their provisions. Industry associations also lobby state and federal legislators, seeking to advance their interests through this arena. Non-profit industry associations ("501(c)(6)" organizations for tax purposes) can undertake unlimited lobbying, but cannot offer charitable deductions to their contributors. While industry associations and environmental groups are often viewed as adversaries, this is not always the case. These groups sometimes work together to advance their joint interests, for example in seeking federal funding to support research and development of cleaner engine or power plant technology.

A third category of NGOs that contribute to air quality management efforts in the United States involves those professional societies whose mission is to foster the exchange of technical information, while remaining neutral on most direct policy debates. The U.S.-based Air and Waste Management Association (AWMA) is a prime example. Founded in the early 1900s (under a different name), the AMWA's main activities are the publication of a peer-reviewed journal covering engineering, science, and management-focused air pollution research, and the organization of an annual conference on air pollution and waste management that is attended by thousands of environmental professionals from countries around the world. AWMA has nearly 10,000 individual members from 65 countries, including academics, consultants, industry, and government employees, working on air pollution and waste management issues.[13]

NGOs are also actively involved with the development of energy policy. For example, the NGO energy caucus is an international assembly of NGOs that are working on issues related to sustainable development and, in particular, on issues or projects directly affected by energy production and consumption. The caucus works to develop consensus policies and to present these policies to the United Nations. The caucus contains several U.S.-based NGOs engaged in energy-related work, including the Center for Energy and Environmental Policy, Citizens' Network for Sustainable Development, Communities United for Responsible Energy, Rocky Mountain Institute, and the World Resources Institute. An example of how NGOs attempt to influence energy policy can be seen through the Union of Concerned Scientists (UCS). The UCS is actively engaged in the debate surrounding the development of the Energy Policy Act of 2005, by deploying and organizing advocacy, analytical, and media tools in an attempt to influence the direction of the bill. NGOs provide a crucial pathway for the public to influence the policy-setting process in the United States with regard to air quality and energy policies.

China

NGOs are starting to play an important role in environmental protection in China. The mass media, too, has been involved by exposing cases of violation of environmental laws and regulations, providing environmental data and information to the public, and reporting on pollution episodes and accidents. This helped to mobilize the public to exert pressure on business behavior and governmental decision making. The 1996 State Council Decision Concerning Certain Environmental Issues signaled a turning point, strongly encouraging both the media and citizens to expose illegal actions that caused environmental damage. By the late 1990s, the media and environmental NGOs had become increasingly influential. NGOs worked with the media to cover environmental affairs, to publicize NGO activities, and to gain public support. The campaign for the selection of Beijing as a venue for the 2008 Olympic Games contained a highly visible environmental aspect due to public pressure. NGOs respond to political conditions as well as to opportunities offered by the media, the Internet, and international NGOs (Yang, 2005).

All Chinese NGOs must be registered and approved by the government. Indeed, many are established to meet the objectives of government authorities. Some administrations at the subnational level still limit freedom of speech, or the right to associate—effectively making it impossible to form a voluntary group. The laws regulating the registration of civic organizations change frequently.

[13]http://www.awma.org.

BOX 4-4
A Brief Introduction to *Friends of Nature (FON)*

The first Chinese environmental NGO, formally registered in March 1994 as the Academy for Green Culture, is an affiliate to the non-governmental Academy for Chinese Culture. FON is the first membership-based, non-profit, public welfare NGO in China, and is funded wholly by membership fees and public support.

Mission:
To promote environmental protection and sustainable development in China by raising environmental awareness and providing a vehicle by which ordinary citizens can express their concerns about China's deteriorating environmental situation.

Objectives:
To develop informal environmental education, such as field trips and vacation camps for students, lectures, seminars, and training classes for teachers, and publications and activities for the general public.

Anyone who agrees with FON's mission and volunteers to help in its activities may become a member upon application. Foreigners may also be admitted as "associate members." Since its founding, FON's membership has grown quickly and steadily to include over 8,000 members—3,000 of which are active—and 30 group members.

China's 1998 Registration Regulations for Social Organizations imposes a number of requirements to establish NGOs. These include the need to have a sponsoring institution, more than 50 members, and a minimum level of financial resources. The regulation also disallows the existence of two organizations in the same field or sector, and in the same jurisdiction. Those organizations that choose to avoid these restrictions and remain unregistered, are unable to enter into contractual relations, such as obtaining telephone lines or leasing office space. Nor can they offer personnel benefits like pensions and medical insurance, or have their own bank account—making it harder to attract staff and funding.

Environmental NGOs in China can be divided into four types. The first type is government-sponsored NGOs, including research institutions, training centers, and other academic organizations such as the All-China Environment Federation (ACEF), but also includes regional environmental science institutes, environmental protection industrial associations, and some wildlife conservation societies. The second type consists of the independently established organizations, which are registered in civil departments. The original one of these, and still the best known, is Friends of Nature (see Box 4-4). The third type of NGO is made up of university student associations and their unions. Finally, some international environmental NGOs maintain regional offices in China, most of these in Beijing.

By the end of 2005, there were 2,768 environmental NGOs in China. Among these, 1,382 are government-sponsored environmental NGOs, accounting for 50 percent; 1,116 are university student associations and its union, accounting for 40 percent; 202 are independently established environmental NGOs, accounting for 7 percent; and 68 are international environmental NGOs with a regional office in China, accounting for slightly less than 3 percent (ACEF, 2006).

The Chinese government has endeavored to boost public participation in environmental protection. The Environmental Impact Assessment Law requires public participation in the work, and demands that appraisal meetings or hearings be held or that other means be pursued to collect the opinions of the relevant authorities, experts, and the public on the EIA report for any plan or construction project that may cause an unfavorable impact on the environment. In February 2006, the environmental authorities released the Provisional Measures for Public Participation in Environmental Impact Assessment, which clearly stipulates the scope, procedure, and form of organization regarding public participation.

CONFLICT AND CONSENSUS

United States

In the United States, members of the public influence the development and implementation of environmental and energy law and policy through their votes for elected officials at local, state, and federal levels. However, the ability of the general public to influence policy through the ballot box is diluted by the number of voters who participate in elections at any level of government, by the number of issues that may weigh into an election, and by the extent to which media attention and campaign financing reduce the accountability of elected officials to the public at large. Particularly at the national level, special interest groups that can afford to focus time and money on environmental and energy issues have much more than representative influence in shaping policy. Those interest groups in the United States that are not precluded from lobbying (because of tax-exempt status as charitable organizations) may seek to gain influence with public officials through campaign contributions and by brandishing their ability to influence how their members vote in public elections. Additionally, much of the information that members of Congress receive on policy issues comes via interest groups, who may filter the information through their particular perspectives.

Once a statute is enacted and the scene shifts to the development of implementing rules, the general public and interest groups in the United States are guaranteed an opportunity to participate in rulemaking processes. EPA and other regulatory agencies must give notice of pending rules, publish proposed language, and accept and respond to public comments submitted on the proposal. Agencies may also hold public hearings to present proposals and to take oral comment. Environmental groups, industry groups, and state and local regulatory agencies actively comment on EPA's regulatory proposals. The number of comments received by the agency on important or controversial rules may number in the tens of thousands. State law generally provides for similar opportunities for public participation in state-level rulemaking.

Once a final rule is published, interested parties (including individual citizens) may seek judicial review of whether the rule comports with the authorizing statute and whether the rationale for the agency's decision is sound. Again, environmental groups, industry groups, and state and local governments commonly seek judicial review to challenge environmental regulations that they dislike. The CAA provides explicitly for citizen suits to challenge final actions by the EPA. However, access to the courts for judicial review is not unfettered. A party wishing to challenge a regulation is required to establish standing to sue, meaning that the individual or group has a particular stake in the outcome of the rulemaking. The doctrine of standing is designed to implement the constitutional principle in the United States that the role of the courts is to protect the rights of individuals, while the elected legislative and executive branches have the prerogative of deciding how best to advance the public interest. Consequently, petitioners in court must demonstrate that they have a particularized stake in the outcome of litigation, not just an interest shared by the general public.

Under the CAA, citizens and public interest groups play a role in the enforcement of regulations, as well as in their development. As mentioned above, the CAA authorizes citizen suits to directly enforce the statute's requirements. Additionally, EPA and state, local, and tribal agencies look to the public to report environmental violations. EPA encourages the filing of tips and complaints from the public through its enforcement website.[14]

China

There is a well-established policy framework and institutional structure for environmental management in China. However, there are some major factors affecting their effectiveness, for example, whether the policies have been designed to reflect and address existing local and regional conditions, whether enough monitoring instruments are in place, whether an enforcement scheme has been set up for non-compliance, or whether all the target groups have been fully informed. Additionally, at an institutional level, SEPA must coordinate with numerous other agencies, some of which exert considerably more authority over energy, environmental, and general development issues (Figure 4-6). As a result, SEPA has tended to focus primarily on issues not requiring coordination (Wang and Wu, 2004). Environmental protection is often faced with pressure from local governments

[14]http://www.epa.gov/compliance/complaints/index.html (accessed June 12, 2006).

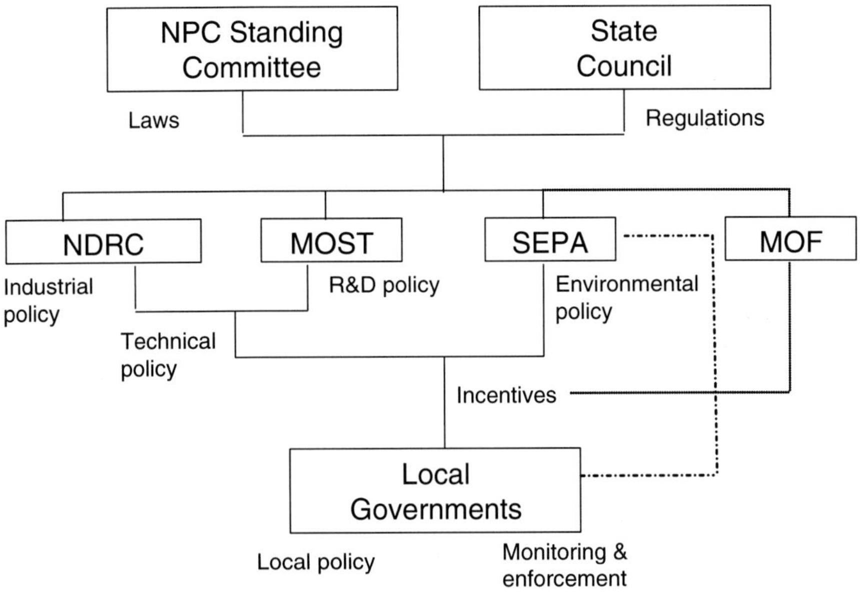

FIGURE 4-6 Main national institutions impacting China's environmental management. NOTES: NDRC is National Development and Reform Commission, MOST is Ministry of Science and Technology, MOF is Ministry of Finance.

to ease environmental regulations for the sake of economic development. This further compromises enforcement efforts; local EPBs, though guided by national standards, ultimately answer to the local government, and economic growth still tends to trump local environmental concerns. Moreover, slight adjustments to smokestack heights, monitoring station locations, and industrial siting may satisfy the local requirements without fully addressing the air pollution problems.

Enforcement

The transfer of polluting industries and wastes from one region in China to another is a subject that the Chinese are hesitant to discuss. Nevertheless, there is evidence that over the past decade, polluting firms—particularly when banished from prosperous coastal cities—have relocated to less-developed inland locales. In many of these areas, local government officials have courted any kind of investment, regardless of environmental impact, so long as it appears profitable, and opposition by environmental officials has been thwarted. Several areas in China, particularly poorer, rural ones, have now become the repository for wastes from wealthier urban areas where consumerism has taken hold. Though this transfer often involves China's richer coastal regions, it is not restricted to these areas.

In documents such as the 11th Five-Year Plan, China recognizes that it lacks adequate technology and the research funding required to properly implement a clean energy policy. Aside from setting general emissions-reduction targets, the PRC has begun experimenting with emissions trading systems in pilot locations. The pilot programs have mainly focused on SO_2 emissions. In 2003 and 2004, the Chinese government implemented emission trading programs in four provinces (Henan, Shandong, Shaanxi, Jiangsu) and three cities (Shanghai, Tianjin, Liuzhou). A separate pilot project was developed in Taiyuan, in cooperation with the Asian Development Bank, SEPA's research wing (CRAES), and international NGOs, to implement a cap-and-trade system (Morgenstern et al., 2004). While these pilot programs have been in operation for the past few years, little data are publicly available to determine progress toward their goals.

Most pollution data in China are still considered confidential, and this above all else fundamentally limits progress in developing an air quality management regime. Data are classified and distributed only to high-level government officials.[15] Although many cities now report their daily API, the data used to calculate this API, as well as more specific information on sampling sites, are withheld. This information, however, is crucial to understanding the true characteristics of the local air quality, in order to make informed decisions. Moreover, NGOs have limited ability to obtain information. Notwithstanding the limited access to information and other restrictions faced by NGOs, they have undertaken campaigns to stop polluting activities and have conducted studies of environmental issues aimed at influencing national leaders. Some NGOs are also carrying out research that explores new approaches to environmental planning and decision making (ACEF, 2006).

ENERGY POLICY AND THE RELATION TO AIR QUALITY

Although energy and air quality are inextricably linked, energy policy and air quality policy are not consistently developed together. However, both countries have shown progress in developing policies which can further both energy and air quality goals.

United States

The energy supply chain in the United States is occupied by both private firms and federal, state, and local government agencies, but is dominated by private industry. Mining companies and power generators and distributors, although highly regulated, are completely privatized. Natural resources and the lands from which they are extracted are generally privately owned, but the federal government

[15] This refers to two regulations in particular, issued by SEPA: *Ordinance on the Administration of Nationwide Environmental Monitoring*, and *Report System on Environmental Monitoring*.

does own approximately 30 percent of the country's land; it leases a portion of it to private firms for mineral exploration and mining. Energy distribution and transmission lines in the United States are primarily owned by private entities.

The DOE is the cabinet-level agency responsible for ensuring the supply and delivery of energy. It was created in 1977 in response to the energy crisis of 1973. The Federal Energy Regulatory Commission (FERC) is an independent regulatory agency within DOE. FERC reviews and approves proposals for siting oil and natural gas pipelines, hydroelectric projects, and energy storage facilities. EPA also works with DOE on issues related to energy policy. On the regulatory side, EPA is responsible for ensuring compliance with the new Renewable Fuels Standard mandated by the Energy Policy Act of 2005. On the technology side, EPA and DOE jointly administer the Energy Star program to promote energy-efficient appliances and practices.

U.S. Government Incentives

The U.S. federal government has historically had a de facto energy policy in that it encouraged, promoted, and supported the development of domestic U.S. energy resources in many diverse ways. It is estimated that federal financial incentives totaled $644 billion through 2003 (in 2003 dollars), classified within the six generic categories defined below, and listed in Table 4-2, which illustrates the distribution of incentives among the different policy options and support mechanisms. These incentives sometimes favor the development of energy sources which support improved air quality:

- *Research and Development*—federal R&D funding.
- *Regulation*—federal regulations and mandates.
- *Taxation*—special exemptions, allowances, deductions, credits, etc. related to the federal tax code.
- *Disbursements*—direct financial subsidies such as grants.
- *Government Services*—assistance provided by the federal government without direct charge.
- *Market Activity*—direct federal involvement in the marketplace.

Research and Development Of the $644 billion in total federal incentives, R&D funding comprised about 18.7 percent—$120.7 billion. These R&D funds were not distributed evenly among technologies, and from Table 4-2 it is clear that three energy technologies—nuclear energy, coal, and renewable energy—have received 86 percent of all R&D support. Between 1976 and 2003 the federal government spent six times as much on coal R&D as it had the previous quarter century, and more than ten times as much on renewables R&D. Most recently, major new energy R&D initiatives have been implemented and proposed that are

TABLE 4-2 The Total Cost of Federal Incentives for Energy Development Through 2003 (billions of 2003 dollars)

	Nuclear	Hydro	Coal	Oil	Natural Gas	Renewables	Geothermal	TOTAL	Percent
Research and Development	60.6	1.2	27.3	6.7	5.6	16.4	2.9	120.7	18.7
Regulation	9.9	4.1	6.2	106.1	2.9	0	0	129.2	20.1
Taxation	0	10.5	26.7	155.4	75.6	11.7	1.4	281.3	43.7
Disbursements	-8.3	1.4	6.4	2.1	0	1.5	0	3.1	0.5
Government Services	1.2	1.3	12.6	27.2	1.3	1.7	0	45.3	7.0
Market Activity	0	54.1	1.7	4.5	1.7	1.3	1.4	64.7	10.0
TOTAL	63.4	72.6	80.9	302.0	87.1	32.6	5.7	644.3	
Percent	9.8	11.3	12.6	46.9	13.5	5.1	0.9		100

SOURCE: Management Information Services, Inc., 2006.

related to climate change, fuel cells, and hydrogen. These have been primarily targeted toward renewables and coal.

Regulation Federal mandates and regulatory actions have been an important part of energy policy, accounting for $129.2 billion (20.1 percent) of energy incentives. There are essentially two types of regulatory actions that the Federal government can undertake to promote energy development: (1) exemption from federal regulations; (2) payment by the federal government of the costs of regulating the technology. An example of the latter type of regulatory incentive relates to nuclear energy, and through 2003 the federal government expended $9.9 billion on regulating the nuclear energy industry. These expenditures include the cost of administering the Nuclear Regulatory Commission and are net of the regulatory user fees paid by utilities. Federal payments for regulating the nuclear energy industry were phased out during the 1980s, and since 1991 the industry has been paying for the costs of regulation.

Taxation Tax policy has been, by far, the most widely used incentive mechanism, accounting for $281 billion (43.7 percent) of all federal incentives. One example of this policy relates to the oil and gas industries, which have utilized the percentage depletion and intangible drilling provisions of the federal tax code as an incentive for exploration and development. Federal tax credits and deductions have also been utilized to encourage the use of renewable energy.

Disbursements Direct federal grants and subsidies have played only a small role in energy policy, accounting for only $3.1 billion (0.5 percent) of incentive costs. An example of federal disbursement subsidies has been, for the oil industry, subsidies for the construction and operating costs of oil tankers.

Government Services This category refers to all services traditionally and historically provided by the federal government without direct charge, and totaled $45.3 billion through 2003, representing 7 percent of total incentives. Relevant examples pertain to the oil industry; the policy of the U.S. government is to provide ports and inland waterways as free public highways, and in ports that handle relatively large ships, the oil tankers represent the reason for deepening channels. They are usually the deepest draft vessels that use the port and a larger-than-proportional amount of total dredging costs are allocable to them.

Market Activity Federal energy incentives consisting of direct federal government involvement in marketplace activities totaled $64.7 billion through 2003—10 percent of all energy incentives. Most of this effort was expended on behalf of hydroelectric power, and, to a much lesser extent, on behalf of the oil industry. Market intervention incentives for hydroelectric energy include the prorated costs of federal construction and operation of dams and transmission facilities.

BOX 4-5
Natural Gas Regulation in the United States

Historically, the use of natural gas has been favored in the United States because it is cleaner burning fuel. However, the net results of a network of legislation, regulations, and financial incentives to use a fuel are not always internally consistent and can lead to outcomes which distort supply, demand, and market dynamics. When governments artificially lower energy prices by regulation or subsidies to make them more affordable for consumers, demand and, consequently, the resulting environmental emissions can increase, particularly in countries without modern pollution control equipment and regulations. From the 1930s until the 1980s, most of the interstate natural gas industry was highly regulated in the United States. Many of these regulations were in conflict, and low, regulated prices constrained supply growth while demand grew rapidly. During the 1970s these policies resulted in gas shortages. Additional regulations in the late 1970s attempted to allocate and curtail gas deliveries to some customers, such as industrial consumers and electric generators. These regulations exacted a significant cost on U.S. industry and consumers, and ultimately on the U.S. economy. Price controls on natural gas were effectively removed during the 1980s and competitive markets emerged.

At present, many U.S. regulations and policies affecting natural gas are still in conflict. Public policies promote the use of natural gas as an efficient and environmentally attractive fuel, leading to restrictions on fuels other than natural gas for the siting of power generation and industrial facilities, restrictions on fuel switching, and fuel choice limitations. Other laws and regulations have been enacted that limit access to gas-prone areas, and there are outright bans on drilling in certain regions. There are laws and regulations that restrict pipeline and infrastructure siting or interfere with the functionality of the market in ways that lead to inefficiencies. Overall, these conflicting policies, encouraging use, but not addressing supply, have contributed to the current tight natural gas supply/demand balance in the United States, with higher and more volatile prices.

EIA data show that U.S. natural gas prices were relatively stable in constant dollars from 1987 through 1998.[a] However, as shown in Figure Box 4-5 beginning in 2000, prices began to escalate—they were roughly 50 percent higher in 2000 compared to 1998. Prices in late 2003 and early 2004 further increased 25 percent over 2000, and by early 2006 prices were three times higher than in 1998.

These higher prices have had a particularly adverse impact on the U.S. manufacturing sector, which is highly dependent on natural gas. The U.S. Department of Commerce estimated that from 2000 to 2004, higher natural gas prices caused an average of 489,000 civilian jobs to be lost each year, 79,000 of these in the manufacturing sector. Employment in industries such as chemicals, foundries, glass, paper, and fertilizer has been significantly reduced or, in some cases, virtually eliminated as facilities have closed or moved overseas, in some measure to areas with secure, long-term natural gas supplies. Natural gas–once a strength of the U.S. energy portfolio—is now characterized by highly volatile and increasing prices.

In 2005 alone, natural gas prices for industrial consumers ranged from \$6.84 to \$11.92. This volatile situation makes it very difficult for many manufacturing firms to effectively plan energy costs, and undercuts their cost competitiveness in world markets.

These problems will become even more serious as domestic supplies continue to decline and demand increases. Supply and demand will balance at a higher range of prices than historical levels. That price range will be primarily driven by demand response moderated by efficiency, conservation and fuel flexibility, the ability to increase conventional and non-conventional supply from North America including the Arctic, and increasing access to world resources through LNG.

---

[a]Natural Gas Markets and EIA's Information Program, March 2000.

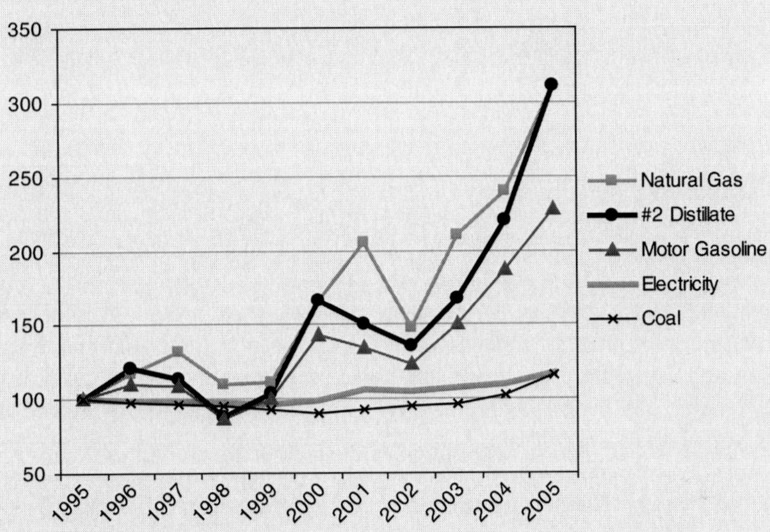

FIGURE Box 4-5 Indices of selected energy price levels in the United States, 1995-2005 (1995 = 100).
SOURCE: U.S. Energy Information Administration and Management Information Services, Inc., 2006.

These costs are prorated because, beginning in the 1930s, federal dams and water resource projects have been multipurpose. The results of these investments include flood control, navigation, recreation, regional development, and other benefits in addition to hydroelectric power. It is thus necessary to estimate that portion of the net investment in the construction and operation of dams that is allocated to power development and to the relevant transmission facilities.

Relevant Energy Legislation

The Energy Policy Act of 2005 (Public Law 109-58, 42 USC 15801) The Energy Policy Act of 2005 established energy research and development programs covering energy efficiency, renewable energy, oil and gas, coal, Indian tribal energy, nuclear matters and security, vehicles and motor fuels (including ethanol), hydrogen, electricity, energy tax incentives, hydropower and geothermal energy, and climate change technology. The Act extended tax credits until 2008 for renewable energy facilities (i.e., wind, closed- and open-loop biomass, geothermal energy, small irrigation power, landfill gas, and trash combustion facilities). It also provided tax credits for investment in bonds funding clean renewable energy and investment in advanced coal projects and certified gasification projects. The Act also repealed the Public Utility Holding Act of 1935.

Public Utility Regulatory Policies Act (Public Law 95-617) Also known as PURPA, this Act was promulgated to promote the use of renewable energy. It created a market for non-utility electric power producers, such as small power production facilities and cogeneration plants, by requiring large electric utilities to purchase power from these “qualifying facilities” at avoided cost (the utilities’ incremental cost to produce the power), in an attempt to provide equitable rates for consumers. Congress encouraged the entry of qualifying facilities into the market by exempting them from rate and accounting regulations enforced by FERC and by the Securities Exchange Commission. PURPA was amended by the Energy Policy Act of 2005.

National Energy Conservation Policy Act (Public Law 95-619) This Act sought to encourage conservation of non-renewable energy resources and to reduce growth in demand for energy without limiting economic growth.

Power Plant and Industrial Fuel Use Act (Public Law 95-620) This Act promoted the greater use of coal and other alternate fuels over oil and natural gas as primary energy resources, and encouraged the use of coal as the primary energy source for existing and new electric power plants. The goals of the Act were to increase the nation’s energy self-sufficiency by promoting the use of domestic energy resources in lieu of petroleum imports.

China

The NDRC, formerly known as the State Development Planning Commission, was established in 2003 under the State Council of China, to develop national macroeconomic policy and strategies and to implement the Energy Conservation Law. The SETC was originally responsible for implementing the Law, but was discontinued in March 2003 when its responsibilities were assigned to the NDRC. Among its responsibilities, the NDRC is tasked with developing a strategy for sustainable development, including cleaner power production and pollution prevention.

The State Bureau of Quality and Technical Standards (SBQTS) developed energy-efficiency standards for consumer appliances such as refrigerators, air conditioners, and fluorescent lamps. SBQTS also created the Energy-Conserving Products Certification Commission to award an energy conservation label to products that meet its standards, similar to the EPA's Energy Star label. SBQTS is now a part of the General Administration of Quality Supervision, Inspection and Quarantine of the People's Republic of China, which oversees industry standards.

The State Electricity Regulatory Commission (SERC) was established in 2003 as a law enforcement agency under the State Council of China. Its primary task is to regulate the electric power sector and to monitor market competition in the electricity industry. It issues licenses to electric power generators, monitors operations and pricing, and will develop an electricity trading market. SERC has authority to enforce environmental laws and regulations as well as safety and technical standards. In practice, though, SERC has had to compete with the powerful NDRC. The original reform plan which established SERC dictated that SERC and NDRC share responsibility in areas such as electricity price setting. As a result, NDRC has typically exercised most control over the power sector (IEA, 2006).

Relevant Energy Legislation

Coal Law of the People's Republic of China China's Coal Law, promulgated in August 1996 and enforced in December of that year, promotes the use and protection of the nation's coal resources and the development of its coal industry and provides for the standardization of coal production and management. It establishes that coal resources are owned by the state.

Electricity Law of the People's Republic of China China's Electricity Law, promulgated in December 1995 and enforced in April 1996, regulates the construction, production, supply, and utilization of electric power and specifies that utilities should protect the environment, adopt new technologies, minimize discharge of poisonous waste, and prevent pollution and other public hazards. The State encourages and supports electricity generation by using renewable and clean energy resources. And the planning for electric power development shall reflect the principles of rational utilization of energy, coordinated development of the power sources and power networks, increasing economic benefits, and being conducive to environmental protection.

Energy Conservation Law China's National Peoples' Congress (NPC) created the Energy Conservation Law (ECL) in 1997 to address energy conservation management, rational energy utilization, technology progress, and legal liabilities. It applies to all energy sources in China, including coal, oil, and oil by-products, natural gas, electricity, coke, heat, liquid petroleum gas, and biomass. The ECL regards energy conservation as managing energy use in a manner that limits energy loss and waste in the production and consumption chain. Article 4 emphasizes that conserving energy is a "long-term strategy" for developing the nation's economy. In response to the ECL, some provinces wrote energy conservation regulations that granted authority to provincial agencies to enforce energy conservation and these provincial regulations are more stringent than the ECL.

The NPC Standing Committee is currently researching the effectiveness and enforcement of the ECL, with the goal of strengthening its legal framework. The government is also working to incorporate approaches taken by other countries, such as the United States in its Energy Policy Act of 2005.

Cleaner Production Promotion Law The Standing Committee of the NPC created the Cleaner Production Promotion Law in June 2002. The comprehensive law was enacted to promote cleaner production of energy, increase the efficient use of raw materials in generating energy, minimize pollution, protect the environment and human health, and promote sustainable development of the economy. It is administered by the NDRC.

The law requires the State Council to develop fiscal and tax policies that will provide an incentive to implementing cleaner production of energy. It also requires the State Council to establish a system to periodically identify obsolete production technologies, processes, and equipment that are hazardous to the environment and that utilize resources inefficiently. When technology is upgraded, the law requires that enterprises select processes and equipment that have high resource utilization rates and reduced generation of pollutants.

Renewable Energy Law The Renewable Energy Law was adopted by the NPC on February 28, 2005, and came into effect on January 1, 2006. The Renewable Energy Law defines "renewable energy" to include the following: wind energy, solar energy, hydropower, biomass power, geothermal energy, ocean energy, and other types of non-fossil energy. It sets forth a framework under which the NDRC is designated as the authority in charge of the regulation and formulation of policies for the development, use, and pricing of renewable energy. Under the Renewable Energy Law, scientific research associated with the development of renewable energy is to be included in technology and industry development plans, and hence will be supported by designated funding. All reports and plans are to be shared with the public.

Significantly, the Renewable Energy Law imposes a mandatory obligation on energy distributors to purchase renewable energy. For instance, power grid operations are obliged to enter into grid-connection agreements with renewable energy generation enterprises, to purchase all of the on-grid electricity generated from renewable energy within the coverage of their grids, and to feed such electricity to the grid. Gas and heat network operators as well as petroleum distributors are required to feed gas, heat, and/or fuel generated from biomass resources into their network, as long as the technical standards for grid connection are met. The Law outlines penalties for failures to comply with the requirements set forth within it.

Vehicle Pollution Controls

Under Chinese law, as in most countries of the world, controlling emissions from new vehicles is the responsibility of the national government, specifically SEPA. Local governments can petition the State Council to allow earlier introduction of the national standards, but only after SEPA has adopted them. Therefore, SEPA has a very important ongoing role in controlling one of the major pollution sources in Chinese cities.

In response to the growing air pollution problems related to motor vehicles during the 1990s, China initiated a serious motor vehicle pollution control effort. It moved aggressively in the late 1990s to eliminate the use of leaded gasoline and quickly followed this with the introduction of EURO I standards for new cars and trucks.[16] Since prohibition of leaded gasoline, lead emission has been reduced by 1,500 tons each year.

Most recently, it phased in the EURO II standards in 2003-2004 for cars, trucks, and buses, and lowered the sulfur content of both gasoline and diesel fuel to 500 ppm maximum.[17] Nevertheless, the emissions requirements for new vehicles across China lag behind those of the industrialized world by almost a decade. In an effort to narrow this gap, SEPA decided to introduce Euro III, Euro IV, and (for heavy-duty trucks only), Euro V standards in 2007, 2010, and 2012, respectively,[18] These emission standards will require advanced vehicle

---

[16]This is the same set of standards introduced in Europe in 1992.

[17]Beijing introduced Euro II standards one year earlier than the rest of the country in 2003.

[18]In December, the State Council approved the implementation of State Phase III and IV (similar to Euro III and IV) vehicle emission standards in Beijing. From December 30, 2005, Beijing applies Phase III requirements for light-duty gasoline and gaseous fuels vehicles, and for heavy-duty diesel and gas-fueled engines. From January 1, 2007, Beijing will apply Phase IV emission requirements for light-duty diesel vehicles. The State Council requires Beijing to prepare qualified diesel fuel in advance, and before this date, Beijing is allowed to take appropriate measures to discourage light-duty diesel vehicles.

emission control technologies, which, in turn, require much better fuel quality for gasoline and diesel. Without additional improvements in fuel quality, greater tightening of new vehicle standards will not achieve their full benefits and some advanced pollution-control technologies will be precluded.

Beyond new vehicle controls, there has been recognition of the need to introduce improved vehicle inspection and maintenance programs to maximize the air quality benefits of stringent new vehicle standards. This is the responsibility of the local governments. Beijing, for example, has mandated the requirement that all in-use vehicles undergo an annual loaded mode (ASM) emissions inspection, and is considering a further tightening to a VMASS based system. However, most if not all other I/M programs in the country remain quite weak with only limited improvements in a few cities such as Shanghai. Some cities have also started clean vehicle campaigns, actively promoting the use of low-pollution vehicles fueled by natural gas and liquefied petroleum gas.

REFERENCES

ACEF (All-China Environment Federation). 2006. Report on the Development Situation of Environmental NGOs in China.

Aunan, K., J.H. Fang, T. Hu, H.M. Seip, and H. Vennemo. 2006. Climate change and air quality—Measures with co-benefits in China. Environmental Science & Technology 40(16) 4822-4829.

Bai, X. 2002. Industrial relocation in Asia: A sound environmental strategy? Environment 44(5):8-21.

Beijing EPB (Beijing Municipal Environmental Protection Bureau). 2005. Report on the State of the Environment in Beijing 2004.

Breyer, S.G. and R.B. Stewart. 1979. Administrative Law and Regulatory Policy. Little Brown and Co.

Burtraw, D. and E. Mansur. 1999. The Environmental Effects of SO_2 Trading and Banking. Environmental Science and Technology. 33(20):3489-3494.

CNEMC (China National Environmental Monitoring Center). 2006. Official website, http://www.cnemc.cn.

Davis, D. 2002. When Smoke Ran Like Water: Tales of Environmental Deception and the Battle Against Pollution. New York: Basic Books.

Editorial Board. 1994. China's Environmental Protection Administration Over 20 Years. Beijing: China Environmental Science Press.

Ellerman, A.D. 2002. Designing a tradable permit system to control SO_2 emissions in China: Principles and practice. Energy Journal 23(2).

Fritz, J.J. and D. Vollmer. 2006. To what extent can technology compensate for institutional failure in an urban environmental setting: The case of China. Technology in Society 28(1/2):95-104.

He, K., H. Huo, and Q. Zhang. 2002. Urban air pollution in China: Current status, characteristics, and progress. Annual Review of Energy and the Environment 27(1):397-431.

IEA (International Energy Agency). 2006. China's Power Sector Reforms: Where to Next? Paris: IEA/OECD.

Jahiel, A.R. 1998. The Organization of Environmental Protection in China, The China Quarterly 156 (Special Issue: China's Environment). 757-787.

Lewis, J. 1985. The Birth of EPA. EPA Journal, November. Washington, D.C.: U.S. Environmental Protection Agency.

Meng, F., F.H. Cai, J.X. Yang, and Y.F. Pu. 1999. Management and Monitoring of SO_2 Emissions Sources in China. Paper presented at the First SEPA-EPA Workshop on SO_2 Emissions Trading, Beijing, China, November 1999.

Morgenstern, R.D., P. Abeygunawardena, R. Anderson, R. Greenspan Bell, A. Krupnick, J. Schreifels, C. Dong, W. Jinan, W. Jitian, and S. Larsen. 2004. Emissions Trading to Improve Air Quality in an Industrial City in the People's Republic of China, Resources for the Future, Discussion Paper 04-16.

NRC (National Research Council). 2004. Air Quality Management in the United States. Washington, D.C.: The National Academies Press.

Powell, F.M. 1992. Trespass, nuisance, and the evolution of common law in modern pollution cases. 21 Real Estate Law Journal 182.

Qu, G. 1991. Environmental Management in China. Beijing: China Environmental Science Press.

Qu, G. 1997. We Need a Transformation. Changchun: Jilin People's Publishing House.

SCIO (The State Council Information Office). 1996. Environmental Protection in China.

SCIO. 2006. Environmental Protection in China (1996-2005).

SEPA (State Environmental Protection Administration). 2005. China Urban Environment Management. http://www.zhb.gov.cn/eic/650501886859804672/20050602/8240.shtml.

SEPA. 2006a. State Environmental Protection Website. Retrieved Oct. 1, 2006 from http://www.zhb.gov.cn.

SEPA. 2006b. Annual Report of Quantitative Examination System for Comprehensive Urban Environment Control 2005.

Sinkule, B.J. and L. Ortolano. 1995. Implementing Environmental Policy in China. Westport, Conn.: Praeger.

Stewart, R.B. and J.E. Krier. 1978. Environmental Law and Policy, 2nd Ed. Bobbs-Merrill.

Swift, B. 2000. Allowance Trading and SO_2 Hot Spots—Good News from the Acid Rain Program. Environment Reporter. 31(19):954-959.

Wang, H. and C.H. Wu. 2004. Environmental Institutions in China. In Urbanization, Energy, and Air Pollution in China: The Challenges Ahead. National Research Council, National Academy of Engineering, Chinese Academy of Sciences, Chinese Academy of Engineering. Washington, D.C.: The National Academies Press.

Wang, M., Y. Jiang, D. He, and H. Yang. 2005. Toward a Sustainable Future: Energy, Environment and Transportation in China, edited by W. Zhou and J. Szyliowicz. Beijing: China Communications Press.

Yang, G.B. 2005. Environmental NGOs and institutional dynamics in China. China Quarterly (181):46-66.

Zhang, S.Q. 2001. Environmental regulatory and policy framework in China: An overview. Journal of Environmental Sciences-China 13(1):122-128.

5

Energy Intensity and Energy Efficiency

Energy intensity is a measure of the energy required per unit of output or activity. At a national level, the ratio of the energy consumption to GDP is often used as a measure of energy intensity. For the economy as a whole, this is a broadly useful measure to compare countries and is frequently used as a proxy for energy efficiency, but it does aggregate numerous underlying factors and thus can obscure the lessons learned on specific efficiency improvements.

Energy efficiency improves when a given level of service is provided with reduced energy inputs, or when services are enhanced for a given amount of energy input (EERE, 2007a). Efficiency improvement is generally the lowest cost method to reduce emissions, including CO_2, and it also has a significant impact on consumption, thereby decreasing demand, saving money, and improving energy security. Energy efficiency can be predicted reliably and, in that respect, can be viewed as another energy resource, since it is technology-based (as opposed to conservation measures which are dependent upon behavioral changes[1]). Efficiency improvements also provide a shorter time frame to meet energy needs and to reduce emissions relative to other approaches, such as increased usage of renewable energy technologies. This is not meant, however, to downplay the importance of investments in renewables in the medium to long term.

The State of California has demonstrated remarkable decreases in energy demand relative to GDP growth over the last 35 years, due to improvements in energy efficiency. The Los Angeles region alone saves an estimated $700 million annually through energy-efficiency measures (Rosenfeld, 2007). Extrapolations

[1]Efficiency refers to improving productivity per unit energy versus conserving a given quantity of energy.

suggest that, if similar measures were employed nationwide, annual energy savings of $20 billion would be realized along with more than $250 billion in net societal benefits—though this would necessitate a four-fold increase in energy efficiency investments, which currently amount to less than $2 billion annually (NAPEE, 2006). Numerous recent reviews have likewise affirmed the centrality of improved energy efficiency in China's drive towards sustainable development (Sinton et al., 2005; CASS, 2006). Despite its potential, energy efficiency remains underutilized as a way to modify energy demand in the United States (NAPEE, 2006).

This chapter looks at energy intensity and energy efficiency in the United States and China broadly, both on the supply side, particularly in the power sector, and on the demand side. It will provide a more detailed look at some of the most successful energy efficiency efforts.

ENERGY INTENSITY

Figure 5-1 shows the energy demand and GDP per capita for a variety of countries. To a rough level of approximation there is a "universal" relationship between energy and GDP, with the United States as a significant outlier, in that it has a much higher energy consumption per capita. The good news is that, as U.S. GDP has increased over the past 10 years, the energy consumption per capita has remained relatively constant. China, like most of the developing world, is experiencing a growth in energy demand per capita as living standards increase.

Figure 5-2 indicates that energy intensity in the United States has declined since 1985, suggesting that the United States economy as a whole has improved its energy efficiency. However, this does not capture some fundamental shifts, such as the structural change from a manufacturing economy (energy intensive) towards a services economy (less energy intensive), which is not related to energy efficiency per se.[2] A newer measure, the intensity index, shown in the chart, was developed in order to account for some of these non-efficiency changes. From 1985 to 2004, it declined from 1 to 0.9, somewhat less rapidly than E/GDP, indicating an underlying improvement of efficiency of 10 percent.

In China's case, the economy has been shifting from agricultural to industrial, marked by strong GDP growth (Figure 5-3). China's energy intensity declined markedly from 1980 to 2000, but energy consumption has outpaced GDP growth since 2000. Broadly speaking, China's energy intensity is presently higher than that of many developed countries, albeit at a relatively low absolute value of energy consumption per capita. Both are rising rapidly, however, which will have implications both domestically and internationally.

[2]The traditional measure of energy intensity (E/GDP) is captured by the line with the most negative slope. The Intensity Index, which attempts to account for structural, behavioral, and weather changes unrelated to efficiency, will be used throughout this chapter since it is a better approximation of changes in energy efficiency. An explanation of this methodology is available at http://intensityindicators.pnl.gov/methodology.stm.

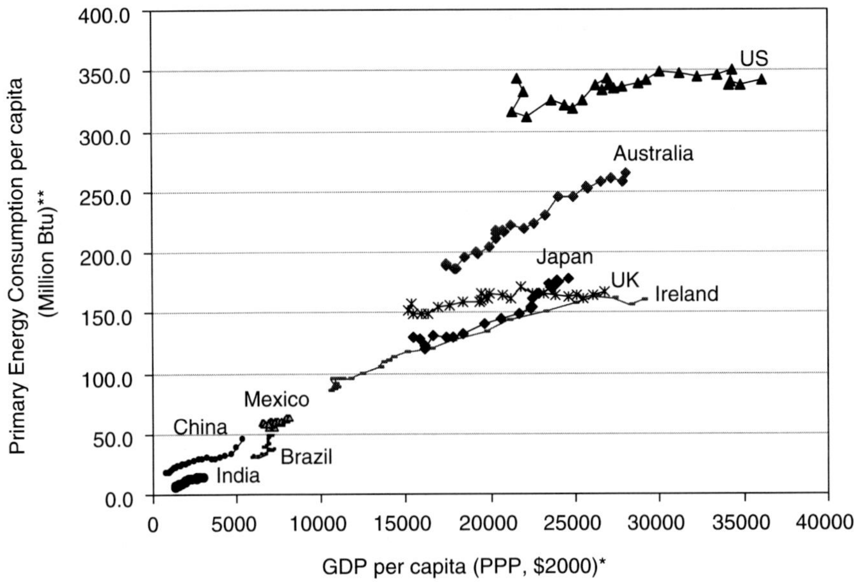

FIGURE 5-1 Energy demand and GDP per capita for select countries, 1980-2004.
*Heston, A., et al., 2006.
**EIA, 2006. International Energy Annual 2004. U.S. Department of Energy.

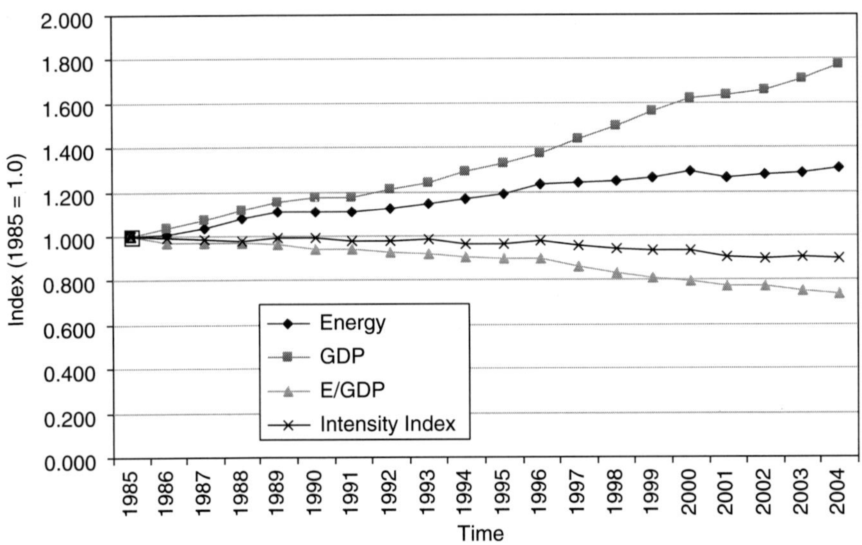

FIGURE 5-2 Energy intensity for the U.S. economy, 1985-2004.
SOURCE: EERE, 2007a.

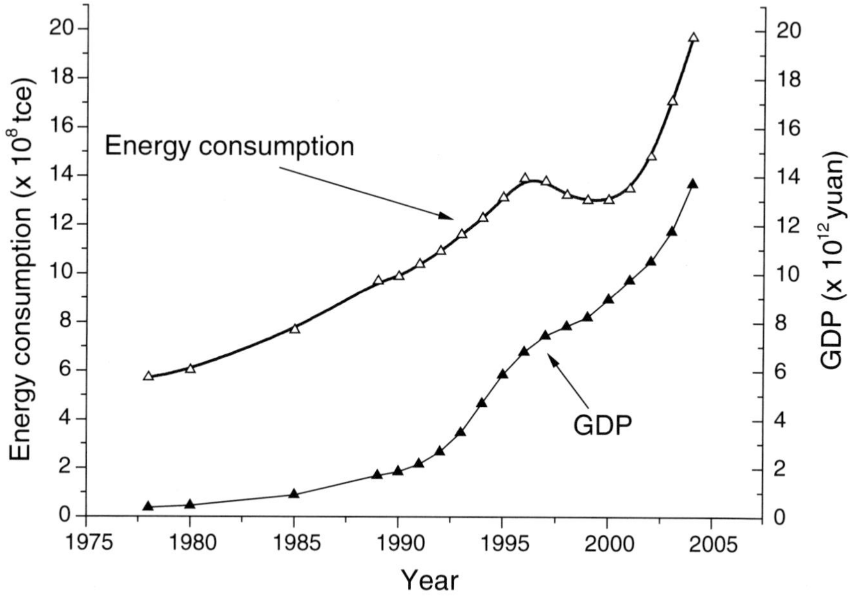

FIGURE 5-3 Energy consumption and GDP in China, 1978-2004.
SOURCE: NBS, 2005,

SUPPLY-SIDE ENERGY EFFICIENCIES

Coal Combustion Efficiency in Electricity Generation

The efficiency of electricity generation can be increased by improving coal-fired plant thermal efficiency; this simultaneously reduces both CO_2 emissions and conventional emissions such as SO_2, NO_x, particulates, and heavy metals. In the United States, the thermal efficiency of new supercritical pulverized coal (PC) plants and the projected efficiency of integrated gasification combined cycle (IGCC) plants (discussed further below) under commercialization are both in the range 39-43 percent on higher rank coals. This represents an 8-10 percent efficiency improvement over new subcritical PCs and over the two operating IGCC units, and more than a 20 percent improvement over the efficiency of the bulk of older operating coal generation plants.[3] Figure 5-4 illustrates the relative effi-

[3]Ultrasupercritical PC and advanced next generation IGCC will increase efficiency to the 43-48 percent range, with CO_2 reductions of an additional 10-15 percent. Efficiency improvements can be achieved for combustion technologies by operating at higher temperature and pressure steam conditions, utilizing advanced materials. The efficiency of IGCC can be increased by incorporating advanced heat recovery and improved plant component designs, including gas turbines.

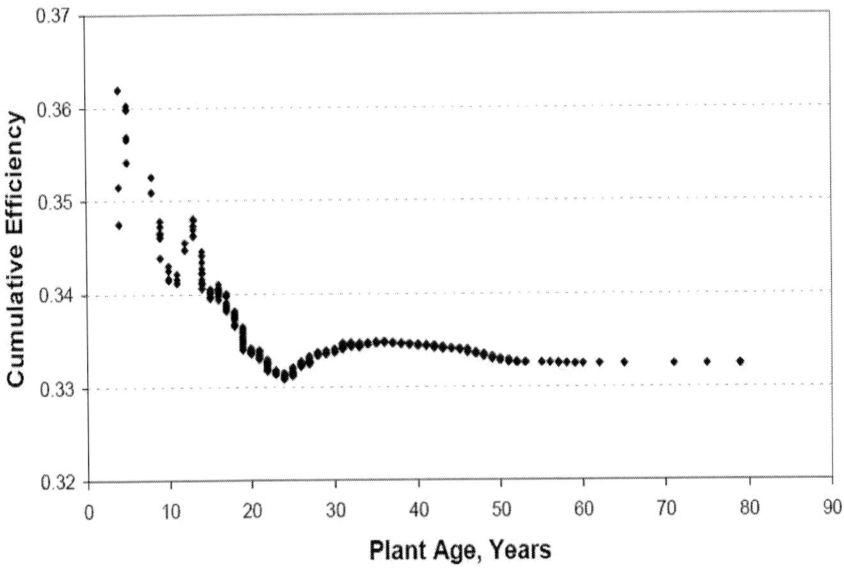

FIGURE 5-4 Efficiency of U.S. coal-fired power plants as of 2002.
SOURCE: NETL, 2002.

ciencies for U.S. coal-fired utilities (also see Chapter 6, Box 6-1, for an extended discussion of the future direction of efficient coal-based power).

There is considerable potential for improving plant thermal efficiency in China. Compared with the international advanced level of power generation, Chinese coal power plants are inefficient, averaging about 30 percent efficiency. This challenge is compounded by the rate at which such plants are being constructed; it is estimated that a new plant comes online every 7-10 days. In 2004, coal consumption for power generation was about 376 g/kWh, which was 55 g/kWh higher than that in advanced countries (Chinese Electric Power Yearbook, 2005). This was due mainly to the use of many small, inefficient generation sets; less than 60 percent of power generation sets had a capacity exceeding 200 MW, and 27 percent had a capacity less than 100 MW. Thus, small-scale power generation sets have impeded China's drive towards greater energy efficiency, although recent efforts to close smaller inefficient plants have moved forward, and plans call for additional closures in 2007 and beyond.

While further improvements can be achieved, China's power plants have made efficiency gains over the past 25 years. The standard coal consumption rate of electricity generation decreased from 398 gce/kWh in 1985 to 343 gce/kWh in 2005, an average annual decrease of 2.8 gce/kWh. The standard coal consumption rate for total power supply also fell from 431 gce/kWh in 1985 to 370 gce/kWh in 2005, a 3.1 gce/kWh annual reduction (China Energy Research Society, 2004,2006).

A challenge for both countries is the large number of large-capacity but older, less efficient coal-fired plants. According to DOE's National Energy Technology Laboratory, many existing U.S. plants are reaching their expected lifespans and face decisions on whether to modernize or to retire, if more stringent pollution standards cannot be met (NETL, 2002). These older plants lack pollution controls due to the fact that many were "grandfathered" under the Clean Air Act of 1970. Similarly, China's newer power plants exhibit improved but not necessarily state-of-the-art efficiencies and, with an average lifespan of 50 years, these plants are essentially locked in for decades.

Electricity Transmission and Distribution

As mentioned in Chapter 2, electricity transmission and distribution continue to present challenges in terms of system losses. Transmission refers to electricity moving from the power generation station to a substation. In order to reduce losses, transmission occurs at high voltages (110 kV or above), typically via overhead power lines. Distribution refers to electricity moving from the substation to consumers at much lower voltage. Efficiency within these existing systems can be improved primarily by one of three ways: increasing the transmission voltage, decreasing transmission distances, or improving transformer efficiencies.

High-temperature superconductivity (HTS) presents the most significant opportunity to improve efficiency in this sector. HTS cables can carry 3 to 9 times the AC power of conventional copper cables and can be either retrofitted for overhead lines or buried underground without significant losses (OETD, 2005). Conventional transmission lines are seldom buried underground due to higher costs and drastic power loss. HTS transformers also exhibit improved efficiency at approximately half the electric loss of conventional transformers. Commercial versions of these technologies are under development and could be available by 2010.

Distributed Energy Systems

Distributed energy (DE) is a strategy which makes use of small, modular generating systems located close to points of use, thereby improving efficiency in the transmission and distribution of electricity. Moreover, DE systems, particularly because they are located in or near populated areas, are characterized by cleaner technologies. Wind turbines and small-scale gas turbines are examples of technologies utilized in DE systems. In addition to improved efficiency, these systems offer other advantages such as reduced peak demand charges and increased system reliability, since they can be tied into the power grid.

Most DE systems in the United States and China currently rely on natural gas turbines, sometimes in combination with renewable sources. In the South Coast Air Basin (which includes Los Angeles), it was estimated that a realistic scenario of extensive distributed energy use would require 50 percent of its power from gas turbines, while photovoltaics and fuel cells might contribute 5 and 10 percent, respectively (Brouwer et al., 2006). In China, there are also significant opportunities to establish DE systems based on renewable sources, particularly in remote areas currently lacking access to an electrical grid.

Integrated Energy Systems

In terms of energy use, there are a number of opportunities to combine currently available technologies into more efficient systems. These follow the principles of the cascade utilization of energy, where different technologies are arranged in a cascade way, according to their preferred energy quality (Wu, 1988; Jin et al., 2005, 2007a). Combined-cycle systems (typically gas and steam) are widely employed in the United States (and increasingly in China) and represent just one of many opportunities to use energy more efficiently.

Combined Heat and Power (CHP) and Combined Cooling, Heating, and Power (CCHP)

China is already making extensive use of CHP, particularly in its urban areas. These systems generate electricity and convert waste heat into steam for central heating. As of 2004, there were over 2,300 CHP units of at least 6 MW capacity, totaling 48 GW of installed capacity, or more than 12 percent of China's total installed capacity, according to the China Electricity Council. CHP systems provide energy and heat more efficiently than do separate steam turbines and small-scale boilers and, as a result, can reduce coal consumption and its attendant emissions. CHP systems provide 82 percent of steam for heating and almost 27 percent of hot water nationwide, and the National Development and Reform Commission's (NRDC's) Energy Bureau has established plans to double CHP's share of total installed electrical capacity by 2020.

CCHP technology has also developed rapidly in recent years. These systems combine distributed electricity generation with high-efficiency utilization of thermal energy, with the energy saving ratio as high as 20-30 percent. These systems can take many forms based on their fuel flexibility. Using fossil fuel (natural gas) as the major source, a CCHP system can integrate use of renewable energy to reduce the consumption of fossil fuel. A small gas turbine can be used as its energy supply and the system may incorporate solar energy, geothermal energy, a heat pump, and energy storage technologies (fuel cells). The gas turbine is used to generate electric power, and recovery of the exhausted heat is applied to produce refrigeration or heat. Refrigeration can be generated by the combination of absorption refrigeration and compression refrigeration. An absorption heat pump and compression heat pump are integrated to ensure that the heat supply system operates reliably even when the ambient temperature is very low.

In North China, solar energy can provide domestic hot water in summer and can be used as a heat source for an absorption heat pump in winter. A geothermal energy system, adopted with an absorption heat engine, acts as a heat sink in summer and heat source in winter. Presently, there are few CCHP systems using only renewable energy due to the expensive initial investment, low profitability, and immature technology. However, natural gas-powered CCHP systems also face challenges in China, due to high natural gas prices and the inability to sell power back to the electrical grid, a feature which can help offset investment and operating costs. CCHP systems may also provide an opportunity to be combined with desalination technologies. Desalination is generally energy intensive and requires heat at a temperature comparable to waste heat given off from CCHP systems. Integrating these two components into a system in coastal areas might reduce the cost of desalinating sea water for domestic and industrial use.

Coal Gasification and Polygeneration Systems

IGCC is an integration of the technologies of coal gasification, gas purification, gas turbines, heat recovery steam generation, and steam turbines. It includes an air separation unit if the system uses the pure oxygen gasification process. Coal gasification is widely used in the chemical industry and most of the technologies being adopted for power generation are mature ones. The gas turbine and steam turbine subsystems adopt the existing technologies of oil- or natural gas-based combined cycle, with the attendant advantages of mass production. The air separation unit can employ the technologies of chemical engineering and metallurgy as well. At present, IGCC is mainly based on the gas turbine combined cycle and the power supply efficiency has reached 43-45 percent; it is expected to achieve 50-52 percent. However, China's primary interest in IGCC technology is integrating it into polygeneration systems, producing power as well as important by-products, including liquid fuel (F-T fuels, methanol, and DME). Cities such as Huainan are considering gasification plants which will provide a 50-50 split between power generation and chemical production.

Polygeneration systems are able to produce synthetic liquid fuels from coal, natural gas, biomass, heavy oil, and coke. The amount of coke produced in China is about 180 million tons per year, which is about half of the total coke production in the whole world. Coke oven gas (COG) by-produced in the coke-making process is about 36 billion cubic meters per year (0.65 EJ),[4] of which about half is currently utilized. The rest of the COG is directly burned and exhausted to the atmosphere, which results in energy waste and pollution.

[4]Based on an energy conversion of 18 MJ per cubic meter of COG.

In the coal gasification-based methanol production process, composition adjustment is required for the feed gas, because the H_2/CO ratio of raw syngas is much lower than the standard value required for methanol synthesis reaction. On the other hand, COG is higher in H_2 content (about 60 percent in volume) than the standard value required for methanol synthesis reaction. Thus, coal syngas and COG can be mixed to provide the proper H_2/CO ratio for methanol or DME production. Chinese engineers expect that a polygeneration system for methanol and power production based on both COG and synthesis gas will have economic benefits for both fuel consumption and initial investment (Jin, 2007b).

DEMAND-SIDE ENERGY INTENSITY AND EFFICIENCIES

Industry and Manufacturing

As mentioned above, the U.S. economy has been transitioning from secondary industries (e.g., manufacturing) to services (largely captured under "commercial" activity in terms of energy use). However, the industrial sector has made improvements in efficiency, as measured by its energy consumption divided by its contribution to GDP. Figure 5-5 shows the total energy consumption in the industrial sector and indicates that the sector's energy intensity has declined by 19 percent since 1985, most of this occurring after 1993 (EERE, 2007a).

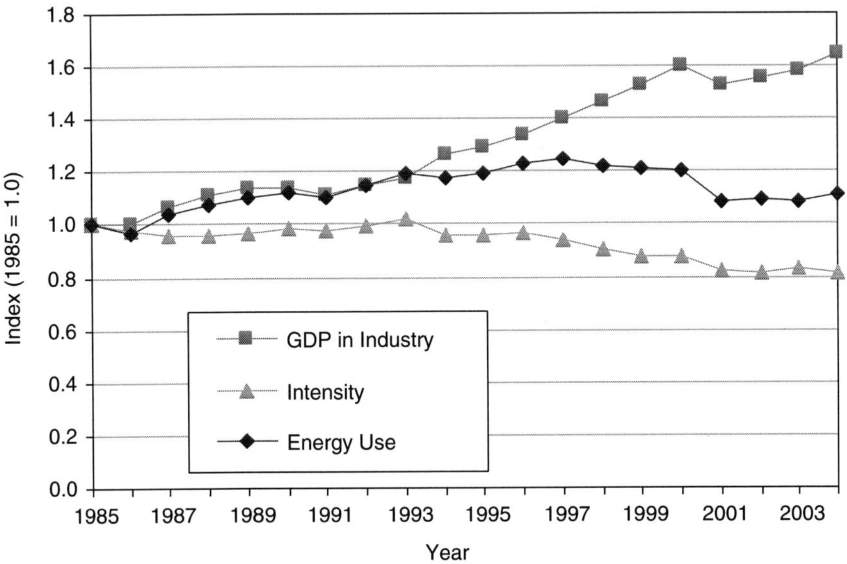

FIGURE 5-5 Energy intensity in the U.S. industrial sector, 1985-2004.
SOURCE: EERE, 2007a.

As an example, Table 5-1 provides a closer look at recent gains in energy efficiency in the U.S. iron and steel sectors. Gains made between 1998 and 2002 can be largely attributed to a decrease in coal consumption (and natural gas to a lesser extent), offsetting a slight increase in electricity consumption. Efforts aimed at reducing energy intensity in the U.S. industrial sector have focused on seven energy-intensive industries (aluminum, chemicals, forest products, glass, metal casting, mining, and steel)—though there are several additional industries (notably, petroleum refining which accounts for 17 percent of industrial energy consumption), which could benefit from energy-saving technologies (NRC, 2005).

In China, 70 percent of energy is consumed by industries (compared to 32 percent in the United States). In recent years, industrial production has accounted for about 50 percent of GDP. Energy consumption of major industries remains high, and per unit energy consumption for major industries (e.g., iron and steel), was on average 40 percent higher than the international advanced level.

China has been transitioning to a heavy industrial economy since the early 1990s. The proportion of heavy industry in gross value of industrial output increased from 50.6 percent in 1990 to 66.5 percent in 2004, and the energy consumption for heavy industry was about three times more than that for light industry. Between 1980 and 2000, the annual average increase ratio of energy consumption for industry was 4.2 percent. The building materials, steel production, and chemical engineering industries were the main energy consumers, accounting for 54 percent of energy consumption for all industries in 2000 (National Energy Strategy and Policy Report, 2004). At present, another interesting transition is taking place. Some heavy industry is being shifted from China to countries such as Vietnam that have lower labor costs, while China is gradually shifting to more sophisticated, higher value-added industries. Thus, China may be beginning to experience the same changes experienced by the United States and other developed nations over the past century.

While energy per unit of production for the main energy-intensive products exceeds international norms, it has decreased annually. As shown in Table 5-2,

TABLE 5-1 Consumption of Energy for All Purposes per Value of Production, 1998 and 2002

	Survey Years	
Iron and Steel Mills	1998	2002
Total	31.4	26.6
Net Electricity	3.1	3.7
Natural Gas	9.8	8.5
Coal	13.5	10.1

NOTE: 1000 Btu per constant 2000 dollar.
SOURCE: EIA, 2006.

TABLE 5-2 Energy Consumption for Main Energy-Intensive Products

	1980	1990	1995	2000	2002	2005
Energy Consumption for Steel Production (kgce/ton)	1201	997	976	898	823	741
Total Energy Consumption for Cement (kgce/ton)	218.8	201	199.2	172	162	149
Total Energy Consumption for Ethylene (kgce/ton)	2013	1580	1277	1097	1028	986

NOTE: Energy Policy Research, No. 6, 2006 (Data of 2000-2005); 1 kgoe = 10,000 kcal, 1 kgce = 7,000 kcal.
SOURCE: China Energy Statistical Yearbook 2000-2002 (data before 2000).

between 1980 and 2005, the energy consumption per ton of steel production decreased 38 percent; energy per ton of ethylene production was reduced by 51 percent, and, for cement, by about 32 percent. The gap of energy intensity for highly energy intensive products between China and other developed countries decreased gradually in the same period.

China's building material industry has the highest energy consumption rate of all industries. The amounts of coal consumption can exceed 200 Mt (6 EJ) of standard coal per year. Among building materials, cement production is the most highly consumptive. In 2000, energy consumption for cement accounted for 70.5 percent of total energy consumption for building materials. From 1990 to 2000, cement production rose from 210 Mt to 597 Mt, and plate glass production increased from 80,700 kilo weight cases[5] to 184,000 kilo weight cases. The production of architectural ceramics and sanitary ceramics increased several times, though the average consumption increased just 5.9 percent, because the energy intensity for producing the main building materials decreased. For example, from 1990 to 2005, the coal consumption for cement per ton decreased from 201 kgce to 149 kgce, and the consumption for plate glass per weight case decreased also from 30 kgce to 22 kgce between 2000 and 2005 (China Energy Research Society, 2006; National Energy Strategy and Policy Report, 2004).

A major challenge for improved industrial efficiency in China is the prevalence of coal-fired boilers. At present, there are about 530,000 small- and medium-sized industrial boilers in China, which consume about 25 percent of the total coal production to heat water or generate steam for industrial and residential heating. However, their average energy efficiency levels only lie in the range of 60-65 percent, or 10-15 percent lower than that of the international advanced level. Pollution from industrial boilers nationwide is second only to that of power plants,

[5]A weight case refers to the total weight of plate glass with thickness of 2 mm and area of 10 m^2, ~50 kg.

and it even exceeds power plant emissions in some cities. Currently, the industrial boilers in China use raw coal and scattered coal as fuels. The coal particle size does not meet the requirements of the combustion equipment; about 45~65 percent of the raw coal particles are smaller than 3 mm—so the thermal loss of mechanical incomplete combustion ranges from 10-27 percent. In addition, there are no effective dust-capturing or desulfurization equipments used to treat the boiler flue gas, so that the emissions of SO_2 and dust typically exceed the standards.

Both countries set energy efficiency targets as part of their broader energy strategies. Improving energy efficiency contributes to a variety of different goals, from pollution reduction to economic savings to energy security. In China, the NDRC currently sets these targets as part of the Five-Year Plan (FYP). Table 5-3 illustrates China's key energy efficiency targets through 2020.

The NDRC's energy efficiency targets for the 11th FYP (2006-2010), announced in March 2006, indicated that the government would slash energy consumption, both per capita and per unit GDP, by 20 percent in 5 years. This equates to a 4 percent reduction per year. Given the experience from 2000 to date in which GDP growth has been outpaced by energy consumption, this seems like a very ambitious goal, one which is not typically observed in developing countries. But, if China can recapture the experience over the 20 years from 1980-2000, where GDP increased 9.7 percent annually and energy consumption increased just 4.6 percent annually, this goal would be reachable. However, in January 2007, China announced that it had failed to reach its first-year target. Specifics were not provided, but only six cities met their individual targets (Xinhua, 2007). This failure was attributed to strong economic and population growth, which, in effect, outpaced gains in energy efficiency. Nonetheless, Chinese leaders will continue to pursue their energy efficiency targets, particularly as part of a strategy to develop a "circular economy" (Box 5-1).

TABLE 5-3 China's Specific Energy Consumption Reduction Targets

	2000	2010	2020
Coal Consumption for Power Supply (thermal power plant) (gce/kWh)	392	360	320
Energy Consumption for Steel Production (kgce/ton)	898	685	640
Total Energy Consumption for 10 Non-ferrous Metal (tce/ton)	4.81	4.60	4.45
Total Energy Consumption for Synthetic Ammonia (kgce/ton)	1372	1140	1000
Total Energy Consumption for Cement (kgce/ton)	172	148	129
Total Energy Consumption for Ethylene (kgce/ton)	1097	930	860

NOTE: Data for 2000 are actual values.
SOURCES: NDRC, 2005; China Energy Research Society, 2006.

BOX 5-1
China's Circular Economy

In recognition of the need to reconcile its rapid economic development with mounting environmental degradation, China has challenged its cities to develop a "circular economy." This concept has quickly become a key theme of the 11th Five-Year Plan (2006-2010). Put simply, the circular economy borrows on principles such as the three R's (reduce, reuse, recycle) and puts them into the context of a sustainably developing country. China's experiments with developing such an economy build upon the efforts of Germany and Japan in the 1990s. The circular economy combines three major objectives:

- Increased efficiency in the use of raw materials (including energy and water);
- Improved management of wastes, including enhanced reuse and recycling; and
- Conservation and enhanced ecological sustainability through improved spatial planning and economic coordination.

In order to develop a circular economy, Chinese cities also draw on the principles of industrial ecology, as embodied in the building of Eco-Industry parks. In 1999, China began developing such parks, which create networks or chains of firms within a region which can better utilize or share resources and utilize cleaner production methods. There are at present over 100 such parks spread across 20 provinces, all supported by the central government. Far more parks have been developed and supported by provincial and local governments. Early successes include the Guitang Sugarcane Eco-Industrial Park (also the first-national level park), a cluster of companies including an alcohol plant, a pulp and paper plant, a sanitary paper plant, a cement plant, a calcium carbonate plant, and a power plant, among others, all utilizing in some way one another's by-products.

Rather than simply improve efficiency and reduce waste, city leaders are encouraging the development of new industries to recycle or even reuse wastes. In this way, resource efficiency is being viewed as a new economic opportunity. As of 2005, 10 provinces and cities have developed pilot projects to develop circular economies and, through 2010, hundreds more will be embarking on similar experiments.

Residential and Commercial

In the United States, these sectors are traditionally considered separately, although they measure the same energy consumption factors: electricity use (for heating, cooling, lighting, appliances) and any additional energy consumption (typically gas) for heating and cooling. Commercial buildings tend to be less susceptible to weather fluctuations, but have more demand for appliances (including lighting). In China, data on energy efficiency are compiled for buildings, which essentially measure the same factors. However, data on residential energy consumption are less precise and therefore total consumption may not be sufficiently accounted for.

Figure 5-6 shows the decrease in energy intensity for the U.S. residential sector from 1985 to 2004. It is important to note here that this index is based on energy use *per unit area*, and thus gains made in efficiency, which are reflected here, are somewhat mitigated by the steeper increase in both the number and size of housing units.

The U.S. commercial buildings sector is the one sector which has experienced an increase in energy intensity between 1985 and 2004 (Figure 5-7). Owing to a number of confounding factors such as building type, an explanation for this observed trend is not currently available (EERE, 2007a).

Presently, the rate of new construction in China far outpaces that of developed countries. By 2020, building floorspace in China will be approximately twice what it was in 2000. This might be viewed as an opportunity to build highly efficient buildings with currently available or emerging technologies. However, as of 2004, 95 percent of the newly built buildings and 99 percent of existing buildings were categorized as high-energy-consumption buildings (Lin, 2006). The peak loads for air conditioning can reach 45,000 MW, equal to 2.5 times the output of the

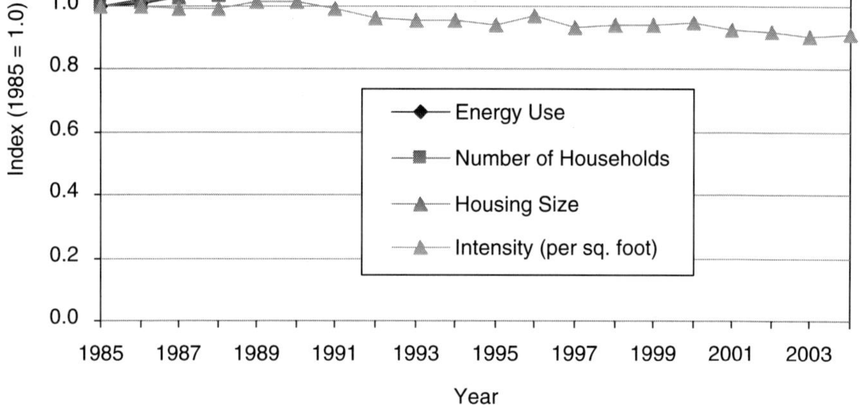

FIGURE 5-6 Energy intensity in the U.S. residential sector.
SOURCE: EERE, 2007a.

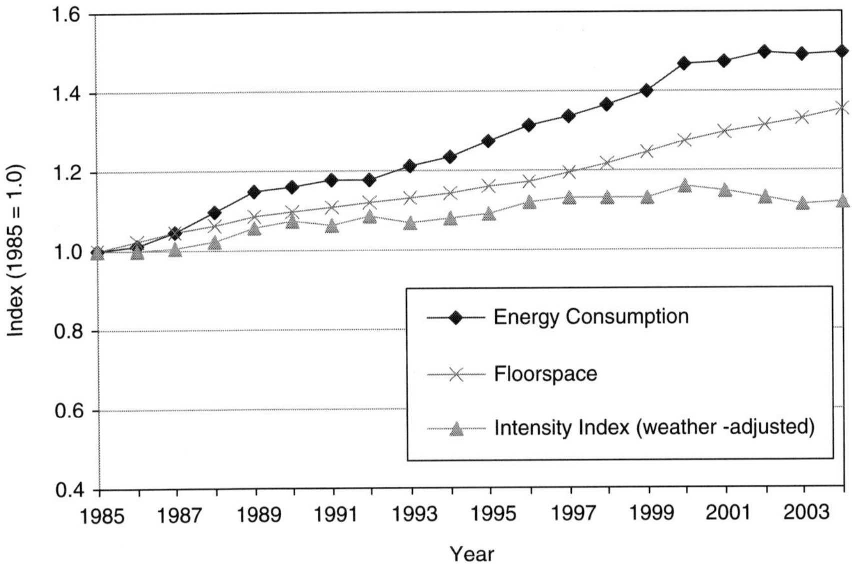

FIGURE 5-7 Energy intensity in the U.S. commercial sector.
SOURCE: EERE, 2007a.

Three Gorges hydropower station (located in central China). Presently the total energy consumption for buildings approaches 30 percent of the total energy consumption in China. According to the present trends, the energy consumption for buildings in 2020 will reach 1.1 billion tons (33.3 EJ) of standard coal, three times that in 2000; and the peak loads of air conditioning will correspond to 10 times the output of the Three Gorges hydropower station at full capacity. Moreover, the energy consumption for space heating in China's buildings is about two to three times as great as that of developed countries with similar climate conditions (NDRC, 2005).

Transportation

Both countries' transportation sectors are dominated by petroleum-based fuels and thus gains made in energy efficiency result either from improvements in fuel economy or shifts from one mode of transportation to another. This sector combines passenger and freight transportation for all modes (highway, air, rail, and water). Figure 5-8 shows the changes in energy intensity for the U.S. transportation sector between 1985 and 2004.

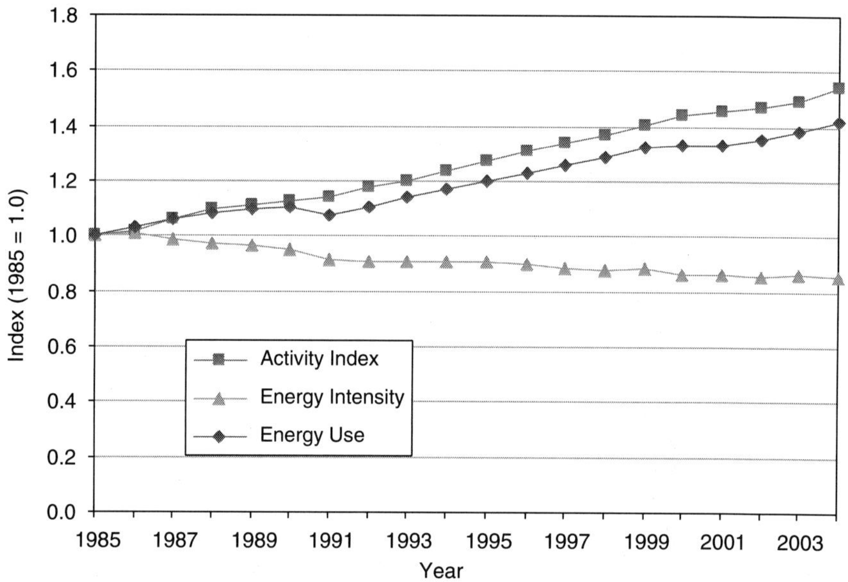

FIGURE 5-8 Energy intensity in the U.S. transportation sector.
SOURCE: EERE, 2007a.

Similar to the case for residential energy consumption and efficiency, the net effects of improvements in transportation efficiency are somewhat masked by the increase in total consumption. For example, on an energy per passenger-mile basis, air travel is less energy intensive than is highway travel (EERE, 2007a). Therefore a modal shift from highway travel to air travel can increase total mileage, while displaying less of an increase in energy consumed, resulting in a decrease in energy intensity. This same logic can be applied to public transportation systems, which offer even greater energy savings on a per passenger-mile basis. These changes are considered to come under the category of "other explanatory factors," and are not reflected as efficiency gains or losses (which are measured within subsectors, e.g., air travel), though their impact on total energy consumption is noticeable.

At present, China does not have aggregate data on energy intensity in the transportation sector. However, it is estimated that the fuel efficiency gap between vehicles in China and in developed countries is large (National Energy Strategy and Policy Report, 2004). China's automobiles are thought to consume 20 percent more fuel per mile, and light- and middle-duty trucks consume 25 percent and 10 percent more, respectively. One possible reason for this is the shortage of research and development of automobile technologies, which are 10-20 years behind advanced countries (Lin, 2006). Older automobiles, many of which consume 5-30 percent more fuel than new automobiles, account for 25 percent of all vehicles. Ninety percent of the trucks used for freight traffic are uncovered vehicles, which decreases their fuel economy. The number of diesel trucks is small and the diesel fuel used is often of poor quality. Low fuel prices may also serve as a disincentive to improving vehicle fuel efficiency. However, China's recently enacted fuel economy standards (discussed later), if enforced, in combination with the fact that new vehicles will continue to supplement China's rapidly growing fleet, will certainly improve the transportation sector's overall efficiency in coming years.

DEMAND-SIDE EFFICIENCY INITIATIVES

Both China and the United States have made progress in demand-side energy efficiency initiatives. Broadly, this refers to energy management, building and appliance standards, transportation energy efficiency policies, and other government programs and financial incentives. As climate change and air pollution and energy security increasingly influence energy policy, so too will the importance of demand-side energy efficiency increase. As one review of select U.S. programs (appliance standards, financial incentives, informational and voluntary programs, and government energy use) reveals, demand-side improvements in efficiency can save up to four quads of energy per year (4.2 EJ), and reduce carbon emissions by as much as 63 million metric tons (Gillingham et al., 2004). Electric utilities have also utilized demand-side management programs, though this peaked in the mid-1990s, after which some utilities were deregulated and consequently cut back or terminated such programs, which had previously been mandated (EIA, 2005). This final section looks in detail at some of the key programs in the United States and China.

Energy Service Companies

In the United States, many energy utilities have abandoned the business of power station operation and now focus their efforts on transmission and distribution, labeling themselves energy service companies/providers. This has paved the way for these service providers to offer ratepayers additional options, such as purchasing renewable or green power, typically at a higher rate. Recently, some cities have broadened the energy management concept to treat energy efficiency improvements as a resource in and of themselves. Rather than build new plants to expand capacity, they seek efficiency improvements through demand-side management, which creates "virtual power plants" that obviate new construction.

Austin, Texas, can claim, perhaps, the nation's first such virtual power plant. The local utility made use of enforced energy efficiency building codes, rebates for more efficient appliances, and other programs and policies intended to significantly reduce local demand for energy. Over a period of 12 years, estimated savings totaled 550 MW, allowing the city utility to remove a coal-fired power plant from its planning books.[6] The advantages of such plants are obvious: they are cheaper than new construction, are emissions-free, and create local jobs (EERE, 2007b).

In China, a prototype may exist in some World Bank-sponsored pilot projects. The World Bank's Energy Conservation Project has established three pilot Energy Management Centers in Shandong, Beijing, and in Liaoning; these centers have supported and promoted several small-scale local energy-efficiency projects. The Bank has also funded 11 global environmental facilities in China that focus on either renewable energy or energy conservation and efficiency.

Appliance Technologies and Standards

Appliance and equipment efficiency standards have contributed substantially to energy savings in the U.S. residential and commercial sectors. Interestingly, because many of these standards have been implemented without controversy, their effectiveness is not fully known or appreciated (Dernbach, 2007). Refrigerators are typically the largest energy consumer in U.S. households, and DOE-supported research led to a reduction of more than two-thirds in the average electricity consumption of refrigerators since 1974—even as average unit sizes increased, performance improved, and ozone-depleting substances were removed (NRC, 2001).

Energy Star is one of the most successful current U.S. programs focused on energy efficiency. EPA established its Energy Star program in 1992. Originally created to promote energy-efficient computers, the program has expanded to include more than 35 product categories for homes and businesses. Since its beginning, American consumers have purchased more than one billion Energy Star products, including 100,000 new homes that meet Energy Star standards. Thousands of buildings have been upgraded through energy-efficient improvement projects. EPA estimates that the Energy Star program has contributed to savings of 100 billion kWh of electricity, prevented discharge of more than 20 million metric tons of carbon elements, and saved more than $7 billion.

China's State Economic Trade Commission (now part of the NDRC) established the China Green Lights Program in 1996 to promote energy-efficient lighting technologies. The program has had success in increasing the production and use of efficient lighting technologies, but has also been challenged by the high initial cost of more efficient technologies and the limited quality of efficient technology produced by China. The Efficient Lighting Institute (ELI) is an international branding system for high-quality energy-efficient lighting products. The

[6]More information on Austin's innovative virtual power plant is available at http://www.austinchamber.org/DoBusiness/TheAustinAdvantage/Energy.html.

original ELI program tested the quality certification and labeling concept and focused on seven countries during the period 2000 through 2003. In 2005, the China Standard Certification Center (CSC) was commissioned by the International Finance Corporation, with funding from the Global Environment Facility, to develop and expand the ELI certification and branding system globally. The expanded ELI program is operated by a new institute, the ELI Quality Certification Institute, which is led by CSC with assistance from a team of international experts from Asia, North America, and Latin America.

Finally, the CFC-Free Energy-Efficient Refrigerator Project, established by the EPA and China's State Environmental Protection Agency (SEPA), began in 1989 to promote the manufacture and sale of CFC-free energy-efficient refrigerators and, secondarily, to provide sustainable economic and environmental benefits to refrigerator manufacturers and owners (Phillips, 2004). The project focused on CFC substitutes research, energy-efficient design, developing prototypes, and testing in the field. Participants included Chinese refrigerator and compressor manufacturers, the China Ministry of Finance, the NDRC, the China State General Administration for Quality Supervision, Inspection, and Quarantine, the University of Maryland, and several Chinese industry trade groups. The project focused on "technology push" and "market pull" approaches to overcoming barriers to the adoption of energy-efficient technologies, such as lack of awareness of benefits of energy-efficient refrigerators, lack of expertise in energy-efficient design, and dealer reluctance to sell energy-efficient products. The key products of the project were a technical training program, a standards and labeling program, an incentive program for refrigerator and compressor manufacturers, and programs for retailers and customers.

The project achieved the following results:

- An increase in the production and sale of energy-efficient refrigerators (consuming less than 55 percent of the current energy use standard) from less than 400,000 units in 1999 to almost 5 million units in 2003;
- A majority of refrigerators produced by a number of manufacturers are now energy-efficient products; and
- It exceeded its goals of 20 million energy-efficient units sold, a lifetime product emissions reduction of 100 million tons of CO_2 and energy savings of 66 billion kWh by a factor of 2 or more.

Building Technologies and Standards

New residential and commercial buildings in the United States are subject to energy-efficiency standards. These standards are primarily set by individual states through residential and commercial building codes, but updates to the codes do not apply to existing buildings. However, there is still great opportunity in the form of renovations and upgrades to existing structures (Dernbach, 2007). Moreover, a comprehensive evaluation indicates that the net realized economic benefits associated with DOE's energy-efficiency programs for the building sector were approximately $30 billion (1999 dollars)—substantially exceeding the roughly $7 billion (1999 dollars) in costs from 1978-2000 (NRC, 2001).

The Energy Star label can be applied to buildings, and is available for new homes, renovation projects, and businesses. Buildings rating in the top 25 percent of energy-efficient buildings are eligible, calculated through a free online Portfolio Manager.[7] Over 3,200 buildings in the United States have the Energy Star label, consuming on average 35 percent less energy, with some exceeding 50 percent in energy savings (Energy Star, 2007). Installing low-emissivity or selective film windows can be a cost-effective renovation to an existing structure, which can cut energy consumption in half. Adding reflective roofs (white or another specific pigment to reflect near-infrared radiation) can also significantly reduce building cooling costs and lessen the urban heat island effect. This latter technology has applications for automobiles as well.

Green building is another movement which has taken hold in both countries, aided in part by the U.S. Green Building Council's Leadership in Energy and Environmental Design (LEED) rating system. Although green building encompasses more than just energy resources, energy efficiency is one of its five key areas, and among these, it provides the most economic return (Energy Star, 2006). The LEED system has been the preferred rating system for green builders locally, nationally, and even internationally. Installing solar panels and purchasing electricity from renewable sources will improve a building's rating under most systems. However, there are a series of more conventional elements, from HVAC systems to passive heating and lighting, which dollar for dollar can have even larger impacts on energy performance.

Vehicle Fuel Efficiency

The largest efficiency gains in the transportation sector will come from improved fuel economy. In the United States, the need for improved fuel efficiency arose in the wake of the 1973-1974 oil embargo. In 1975 the Energy Policy and Conservation Act was adopted, mandating the U.S. Department of Transportation to govern increased fuel efficiency for automobiles. The result was the still intact Corporate Average Fuel Economy (CAFE) standards. The passenger vehicle fleet, in general, is regulated by the CAFE standards. CAFE refers to the sales weighted average fuel economy (miles per gallon) of a manufacturer's cars and light trucks with gross vehicle weight ratings of less than 8,500 lbs. Fuel economy values are evaluated using protocols developed by the EPA. Congress requires that CAFE standards be set at the maximum feasible level, considering technological feasibility, economic practicality, and effect of other standards on fuel

[7]Available at http://www.energystar.gov/index.cfm?c=evaluate_performance.bus_portfoliomanager.

economy and on the need of the nation to conserve energy. In the United States, vehicle specifications are categorized within a two-class system: cars and light duty vehicles. A significant loophole in the standards is that light-duty vehicles weighing more than 8,500 lbs (including many pickup trucks and sports utility vehicles) are exempted from the standards and tend to have significantly lower fuel economy ratings that smaller vehicles.

Tightening the CAFE standards is frequently proposed as a means of combating vehicle pollution and rising fuel use in the United States. In 2002, the U.S. National Academies examined the effectiveness and impact of the CAFE standards and concluded they had reduced oil consumption by about 2.8 million barrels per day (6.27 EJ per year), or about 14 percent, and contributed to reduced emissions (NRC, 2002). Further analysis building on this study has indicated that enhanced standards could reduce oil consumption and automobile emissions, save drivers money (in fuel costs), but also increase GDP and create job growth (Bezdek and Wendling, 2005). It is noted in the 2002 report, however, that other approaches, such as higher fuel taxes, tradable credits for fuel economy improvements, taxes on light-duty vehicles that fall below CAFE standards combined with rebates for vehicles exceeding the standards, and/or standards based on vehicle attributes, such as weight, size, or payload, might be more successful at improving fuel economy.

In 2004, the Chinese government proposed a set of vehicle fuel efficiency standards in an attempt to reduce the country's rising dependency on oil. Designed to regulate China's rapidly growing automotive industry, these standards have the power to change the way manufacturers behave by altering vehicle production. The Chinese standards are separated into two implementation phases: Phase 1, which began in 2005/2006, and Phase 2, which will begin in 2008. Unlike U.S. standards, Chinese-sold vehicles need to meet fuel-efficiency standards according to their weight class. The Chinese standards require that each vehicle within one of sixteen designated weight categories meet specific mpg (miles per gallon) standards. For example, according to the 2005 standards, the heaviest vehicles must reach 19 mpg and the lightest vehicles must achieve 38 mpg (Sauer and Wellington, 2004). If the standards are enforced correctly, China should see an increase in the amount of fuel-efficient and technologically advanced vehicles on the road. The demand for small cars is continuing to grow, due to increasing gas prices and strict vehicle fuel-efficiency standards. In the future, the demand for smaller cars is expected to rise in China's domestic auto industry, as the government continues to implement the fuel efficiency standards (Li, 2006).

Figure 5-9 displays comparative fuel economies for passenger vehicles in a number of countries. Though California is currently not permitted to enact fuel economy standards higher than the national standard, it is estimated that the more stringent California greenhouse gas emission standards would save up to $150 billion each year in fuel costs if adopted nationwide (Rosenfeld, 2007).

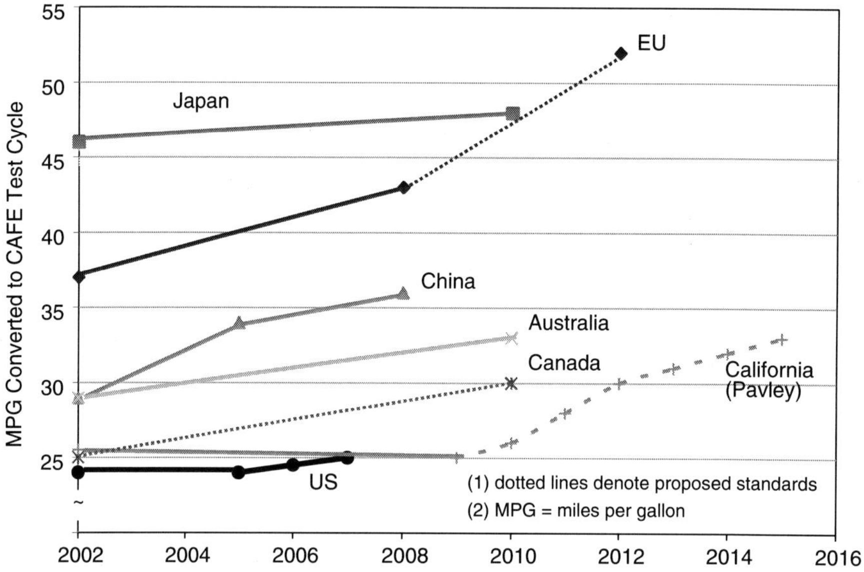

FIGURE 5-9 Comparison of fuel economy for passenger vehicles.
NOTE: California's standards pertain to greenhouse gas emissions and not fuel economy.

Hybrids

Hybrid electric vehicles have been commercially available since 1999 (1997 in Japan) in most markets. Their combination of an internal combustion engine and electric motor result in significant energy-efficiency gains, generally two to three times more efficient than conventional automobiles (EERE, 2007c). The most common hybrids do not need to be plugged in, as their electric battery is recharged using regenerative braking or by an on-board generator. They are also fuel flexible; hybrids have been developed to run on gasoline, methanol, compressed natural gas, hydrogen, or other alternative fuels.[8] A deterrent to the use of such vehicles in the United States is that the increment of initial higher purchase cost will not be returned in the cost of fuel saved at present prices in the lifetime of the vehicle

A new type of hybrid, the plug-in hybrid, is in the demonstration stage, and the U.S. National Renewable Energy Laboratory is leading efforts to develop such a vehicle, which would allow the driver to drive much longer on electric battery

[8]This is not to be confused with Flex-Fuel Vehicles which are designed to run on gasoline or an ethanol blend (E85). Rather, this is a reference to the fact that hybrids do not strictly have to be designed to run on conventional gasoline.

power, which is cleaner and far less petroleum-consumptive. These plug-ins would literally be plugged into the electrical grid to recharge the batteries; the fuel tank is retained and easily filled for long trips beyond the charge of the batteries. The primary challenge to overcome is increased battery weight and cost.

Transit Oriented Development

Transit-oriented development (TOD) is another means for increasing efficiency in the transportation sector. Congestion has steadily grown in urban areas of the United States over the past two decades, and the problem has been perhaps more acute in Chinese cities. The response in both countries has largely been to build more roads to accommodate the burgeoning vehicle use, but in neither case has new construction been able to keep pace with demand. As a result, cities have developed laterally, increasing commute times while decreasing fuel efficiency (as a result of lower velocities), and creating challenges for more efficient public transportation systems.

Perhaps no area demonstrates this conundrum more so than the Los Angeles metropolitan area. Yet, even Los Angeles is incorporating TOD into its urban planning. It announced in early 2007, plans to build a large mixed-use facility at an existing rail station, in order to reduce congestion and personal vehicle travel (LACMTA, 2007). Many TOD projects in the United States are developing around existing rail stations, although they are not limited to rail transportation systems, and indeed, in other countries, notably Latin America, similar developments are taking place in conjunction with bus rapid transit systems.

In general, TOD is characterized by dense settlements which encourage the use of public transit. These developments have mixed uses, all within walking distance of public transit (TRB, 2004). By encouraging public transportation, efficiency in the transportation sector improves as personal vehicle trips and congestion both decrease.

REFERENCES

Bezdek, R.H. and R.M. Wendling. 2005. Potential long-term impacts of changes in US vehicle fuel efficiency standards. Energy Policy 33:407-419.

Brouwer, J., D. Dabdub, G.S. Samuelsen, M. Carreras, and S. Vutukuru. 2006. Urban Air Quality Impacts of Distributed Generation in the South Coast Air Basin and San Joaquin Valley. Presented to committee on April 5, 2006, University of California, Irvine.

CASS (Chinese Academy of Social Sciences). 2006. Understanding China's Energy Policy: Economic Growth and Energy Use, Fuel Diversity, Energy/Carbon Intensity, and International Cooperation. Background paper prepared for Stern Review on the Economics of Climate Change.

Chinese Electric Power Yearbook. 2005. China Electric Power Press.

China Energy Research Society. 2004. Energy Policy Research. No. 6.

China Energy Research Society. 2006. Energy Policy Research. No. 6.

Dernbach, J. 2007. Stabilizing and Then Reducing U.S. Energy Consumption: Legal and Policy Tools for Efficiency and Conservation. ELR News & Analysis. Washington, D.C.: Environmental Law Institute.

EERE (U.S. Department of Energy, Office of Energy Efficiency and Renewable Energy). 2007a. Indicators of Energy Intensity in the United States. http://intensityindicators.pnl.gov/total_highlights.stm (accessed February 23, 2007).

EERE. 2007b. State Energy Alternatives—Clean Energy Virtual Power Plant. http://www.eere.energy.gov/states/alternatives/virtual_power_plant.cfm.

EERE. 2007c. Clean Cities—Hybrid Electric Vehicles. http://www.eere.energy.gov/cleancities/hev.

EIA (Energy Information Administration). 2005. Annual Energy Review 2005. U.S. Department of Energy.

EIA. 2006. Manufacturing Energy Consumption Survey, 2002. U.S. Department of Energy, February 2006.

Energy Star. 2006. Green Buildings and Energy Efficiency: Diligence Pays. Off the Charts. Washington, D.C.: U.S. Environmental Protection Agency.

Energy Star. 2007. Energy Star website, http://www.energystar.gov (accessed February 23, 2007).

Gillingham, K., R.G. Newell, and K. Palmer. 2004. Retrospective Examination of Demand-Side Energy Efficiency Policies—Discussion Paper. Washington, D.C.: Resources for the Future.

Heston, A., et al. 2006. Penn World Table Version 6.2, Center for International Comparisons of Production, Income and Prices at the University of Pennsylvania, September 2006.

Jin, H.G., H. Hong, et al. 2005. A new principle of synthetic cascade utilization of chemical energy and physical energy. Science in China (Ser. E Engineering & Materials Science) 48(2):163-179.

Jin, H.G., B.Y. Li, et al. 2007a. Integrated energy systems based on cascade utilization of energy. Frontiers of Energy and Power Engineering in China 1(1):16-31.

Jin, H.G., S. Sun, et al. 2007b. Integration and characteristic of novel polygeneration system for producing coke and power. Journal of Engineering Thermophysics (in press).

LACMTA (Los Angeles Country Metropolitan Transportation Authority). 2007. Developers Break Ground on Major Joint Development at Hollywood & Vine Metro Red Line Station.

Li, Z. 2006. Government Encouraging Smaller Cars, Improved Fuel Efficiency. World Watch Institute.

Lin, B. 2006. China Energy Outlook 2006. Beijing: China Metrology Publishing House.

NAPEE (National Action Plan for Energy Efficiency). 2006. National Action Plan for Energy Efficiency Report. Washington, D.C.: U.S. Department of Energy and U.S. Environmental Protection Agency.

National Energy Strategy and Policy Report. 2004. Development Research Center of China's State Council.

NBS (National Bureau of Statistics). 2005. China Energy Statistical Yearbook. Beijing: China Statistics Press.

NDRC (National Development and Reform Commission). 2005. China Medium and Long Term Energy Conservation Plan. People's Republic of China, January 2005.

NETL (National Energy and Technology Laboratory). 2002. NETL Coal Power Database: Update for 2000 Operations. U.S. Department of Energy.

NRC (National Research Council). 2001. Energy Research at DOE: Was It Worth It? Energy Efficiency and Fossil Energy Research 1978 to 2000. Washington, D.C.: National Academy Press.

NRC. 2002. Effectiveness and Impact of Corporate Average Fuel Economy (CAFÉ) Standards. Washington, D.C.: The National Academies Press.

NRC. 2005. Decreasing Energy Intensity in Manufacturing: Assessing the Strategies and Future Directions of the Industrial Technologies Program. Washington, D.C.: The National Academies Press.

OETD (Office of Electric Transmission and Distribution). 2005. Superconductivity for Electric Systems: Program Plan FY 2005-2009. Washington, D.C.: U.S. Department of Energy.

Phillips, R. 2004. China CFC-Free Energy Efficient Refrigerator Project. Presented at the IEA-India Workshop on Energy Efficiency Standards and Labelling, Bangalore. October 13-14, 2004. United Nations Development Programme. http://www.iea.org/textbase/work/2004/bangalore/philips.pdf.

Rosenfeld, A.H. 2007. Energy Efficiency in the U.S. and California. Presentation to the committee, February 9, 2007. http://www.energy.ca.gov/commission/commissioners/rosenfeld.html.

Sauer, A. and F. Wellington. 2004. Taking the High (Fuel Economy) Road. What Do the New Chinese Fuel Economy Standards Mean for Foreign Automakers? World Resources Institute.

Sinton, J.E., R.E. Stern, N.T. Aden, and M.D. Levine; with T.J. Dillavou, D.G. Fridley, J. Huang, J.I. Lewis, J. Lin, A.T. McKane, L.K. Price, R.H. Wiser, N. Zhou, and J. Ku. 2005. Evaluation of China's Energy Strategy Options. Lawrence Berkeley National Laboratory, LBNL-56609.

TRB (Transportation Research Board). 2004. Transit-Oriented Development in the United States: Experiences, Challenges, and Prospects. Washington, D.C.: The National Academies Press.

Wu, Z. 1988. Cascade Utilization of Energy and Total Energy System of Gas Turbine. Beijing: Mechanical Industry Publishers.

Xinhua News Agency. 2007. China to fail energy reduction target. Xinhua Economic News Service, January 25, 2007.

6

Coal Combustion and Pollution Control

Coal accounts for a major fraction of the combustion material used worldwide to produce electricity, and coal generation plants account for a major fraction of the stationary sources of air pollution worldwide. As such, the technologies for coal combustion and control of its air pollution are so significant that they warrant an extended discussion.

There are a variety of combustion processes in both countries that have variable impacts on the environment. Traditional processes tend to dominate the energy scenario, although both countries are paying increasing attention to energy efficiency and to cleaner technologies, in order to address rising fuel costs and concerns over emissions. This section describes in detail some of the main processes, their purposes, and their relative contribution to emissions.

COMBUSTION PROCESSES

Pulverized Coal

Traditional pulverized coal (PC) plants account for 99 percent of all coal-fired plants in the United States and over 90 percent worldwide. The technology is well developed and suitable for a variety of coals, although it is not always appropriate for high-ash coals. High combustion temperatures lead to NO_x formation, which enters the flue gas along with SO_2. CO_2 is released in the flue gas at atmospheric pressure and at a relatively low volume (10-15 percent), which presents difficulties in sequestering the carbon, as is discussed later in the chapter.

The main distinction among boiler types is the pressure at which they operate. Subcritical boilers are a first-generation technology, with thousands in use throughout the world. Supercritical plants are widely used in the United States, Europe, Russia, and Japan, with a limited number in operation in South Africa and China. Europe and Japan have also constructed ultra-supercritical plants, which demonstrate even higher efficiencies than do their supercritical counterparts. However, without successful commercial application in the United States, the ultra-supercritical technology is still considered to be unproven and a potential technical and economic risk (EPA, 2006). Supercritical power generation technologies have been considered to be the standard in the electric power industry. Though larger demonstration plants are under development, the average capacity of supercritical plants is between 300 and 600 MW.

Increasing thermal efficiency has been a goal of many PC-fired plants, although low energy prices often serve as a disincentive to implement more efficient methods. Nevertheless, this represents a potentially cost-effective way to reduce CO_2 and other emissions and to decrease fuel consumption. Methods such as reducing the excess air ratio, reducing stack gas exit temperature (while recovering the heat), and increasing steam pressure and temperature have all been utilized at various times over the past several decades. Fuel type has an impact on efficiency, as does the type of plant. Older subcritical plants using poor-quality coal can have thermal efficiencies as low as 30 percent, while modern subcritical plants tend to operate between 35 and 36 percent. Meanwhile, modern supercritical plants typically operate in the 43-45 percent range (IEA, 2006a). As a reference, a 1 percent increase in efficiency can reduce specific emissions such as CO_2, NO_x, and SO_2 by 2 percent (World Bank, 2006b). Moreover, installation costs are only 2 percent more for supercritical plants as opposed to subcritical, while operation costs are comparable and fuel costs, due to higher efficiency, are lower for supercritical systems.

China has recently been focused on adapting supercritical technologies from abroad. The Henan Qinbei power plant, a 600 MW demonstration plant, is a supercritical coal-fired plant utilizing domestically designed technologies, and has been online since late 2004. Another domestic supercritical demonstration plant with a capacity of 1,000 MW—Zhejiang Yuhuan—is under construction. Table 6-1 illustrates the growth of China's power generation industry since 2000, and highlights the dominant role of thermal power, 98.7 percent of which comes from coal combustion.

Fluidized Bed Combustion

Fluidized bed combustion (FBC) is considered to be a clean coal technology and can be particularly useful for high-ash coals. It evolved as a result of efforts to develop a combustion process able to control pollutant emissions without implementing external controls, such as scrubbers. Coal particles are fed into a combustion chamber, suspended on jets of forced air, and combusted at 800-900°C, yielding less NO_x formation in comparison to PC combustion. In the process, the

TABLE 6-1 Development of China's Installed Power Capacity and Power Generation in Recent Years

		2000	2001	2002	2003	2004*
Installed Capacity (MW)	Thermal capacity	237540	253012	265547	289771	324900
	Proportion of total (percent)	74.4	74.8	74.5	74.0	73.7
	Annual growth (percent)	6.3	6.5	5.0	9.1	12.1
	Hydro capacity	79352	83006	86074	94896	108260
	Proportion of total (percent)	24.9	24.5	24.1	24.2	24.6
	Annual growth (percent)	8.8	4.6	3.7	10.3	14.1
	Nuclear capacity (MW)	2268	2268	4586	6364	7014
	Proportion of total (percent)	0.71	0.67	1.3	1.6	1.6
	Total	**319321**	**338487**	**356571**	**391408**	**440700**
	Annual growth (percent)	**6.9**	**6.0**	**5.3**	**9.8**	**12.6**
Power Production (billion kWh)	Thermal power	1107.9	1204.5	1352.2	1579.0	1807.3
	Proportion of total (percent)	81.0	81.2	81.7	82.9	82.6
	Hydro power	243.1	261.1	274.6	281.4	328.0
	Proportion of total (percent)	17.8	17.6	16.6	14.8	15.0
	Nuclear power	16.7	17.5	26.5	43.9	50.1
	Proportion of total (percent)	1.2	1.2	1.6	2.3	2.3
	Total	**1368.5**	**1483.9**	**1654.2**	**1905.2**	**2187.0**
	Annual growth (percent)	**11.0**	**8.4**	**11.5**	**15.2**	**14.8**

SOURCE: CEC, 2004.

flue gas is brought into contact with a sulfur-absorbing material such as limestone, resulting in 95 percent of the sulfur being captured inside the boiler without need for external controls. Although this represents less than 2 percent of the world total of coal-fired power, it has grown significantly between 1985 and 1995, and is utilized in hundreds of small units in China (IEA, 2006a).

FBC's popularity is ascribed to its fuel flexibility as well as to its ability to control SO_2 and NO_x emissions independent of costly add-on controls. In terms of fuel flexibility, although these units can be designed for co-firing and can accommodate low-grade fuels including municipal waste, they operate most efficiently when utilized with the originally intended design fuel. A number of atmospheric FBC boilers of sizes from 250 to 300 MW are in use commercially, but are more common at smaller sizes for process heat and on-site power supply. The thermal efficiency of these units is typically lower (3-4 percent) than are similar size PC combustion boilers, as heat loss is considerable (IEA, 2006a).

Pressurized FBC builds on the earlier atmospheric pressure technologies. While still in the demonstration phase, it already shows advantages over atmospheric FBC and traditional PC combustion. Its compact design is suitable to modular construction and makes for easier retrofits compared to conventional FBC units. It has lower capital costs than IGCC plants or PC plants outfitted with pollution controls, and has a higher potential (45 percent); it has demonstrated (40-42 percent) thermal efficiency in contrast to many PC plants. While initial capital costs are typically higher than the cost range for a PC plant, other factors, specifically add-on pollution controls, make FBC plants cost-competitive (World Bank, 2006a). Size seems to be the limiting factor, since the demonstration plants are all 70 MW plants. However, Japan is constructing a 350 MW demonstration plant.

Pressurized FBC is categorized into two technologies: bubbling-bed and circulating. Bubbling-bed technology is the first-generation pressurized FBC technology and is in use as a demonstration at the Tidd Plant in Ohio, the result of a joint project between the U.S. Department of Energy (DOE) and the American Electric Power Corporation. A second generation pressurized fluidized bed combustor uses circulating fluidized-bed (CFB) technology and a number of efficiency enhancement measures. CFB technology has the potential to improve operational characteristics, by using higher air flows to entrain and move the bed material, and by recirculating nearly all of the bed material with adjacent high-volume, hot cyclone separators. The relatively clean flue gas goes on to the heat exchanger. This approach theoretically simplifies feed design, extends the contact between sorbent and flue gas, reduces the likelihood of heat exchanger tube erosion, and improves SO_2 capture and combustion efficiency.

In China and potentially in other coal-rich developing countries, CFB may be one of the most important power generation technologies in the future. This technology is mainly adopted in low-quality coal-burning power plants, and is also used in cogeneration (heat and power) plants. CFB boilers were first introduced in China in the 1980s at low capacities. A 300 MW CFB boiler demonstration project is under construction at the Neijiang Baima power plant in Sichuan province. Domestic CFB boilers with capacities less than 200 MW are already in use at the industrial scale; 300 MW CFB boilers with independent intellectual property rights are also currently at the research and production stage.

Integrated Gasification Combined Cycle

Integrated gasification combined cycle technology, or IGCC, represents perhaps the most promising technology for utilizing coal, while at the same time decreasing environmental impacts. In the United States, the Environmental Protection Agency (EPA) and DOE have been cooperating to advance and commercialize the technology. In the coming decades, it will likely be the competing technology, along with PC combustion, in coal-based power generation. Its potential for CO_2 capture is particularly promising. Overall, these plants have less of an environmental impact than do traditional PC plants; emissions are lower, they consume significantly less water, and they generate less solid waste (EPA, 2006).

The United States currently has 20 coal gasification plants, two of which produce power. Many of these plants were installed at petroleum and chemical plants. Additional plants are in various stages of development, with plans to use coal to generate electricity along with co-products, including hydrogen, ammonia, chemicals, fertilizers, and Fischer Tropsch liquids (NETL, 2005). These plants have been made possible by government subsidies and have experienced technical and commercial problems, as is common for many new technologies (EPA, 2006). These early plants can play an important role in demonstrating and perfecting the technology to make IGCC-based power plants suitable for wider commercial deployment (NRC, 2003). Twenty-four additional coal-fired plants using IGCC have been proposed as of June 2006, thanks in part to a 20 percent investment tax credit for IGCC, as part of the U.S. Energy Policy Act of 2005. Little research has been done on applying low-rank coals to IGCC technology; current plants use bituminous coals and, therefore, comparisons to sub-bituminous or lignite coals are difficult (NRC, 2003). According to the recent EPA study, when assessing commercial applications within the United States, IGCC has better thermal performance than do subcritical and supercritical PC plants (see Table 6-2).

China has been an industry leader in exploring IGCC technologies. In Yantai, Shandong province, an IGCC power plant of capacity 300-400 MW has been in development for years. However, due to economic constraints, it has yet to be constructed. In 2005, the Chinese Huaneng Group Corporation brought forward the coal based polygeneration system, nicknamed "Green Coal Power." In October of the same year, the Huaneng Group joined the U.S. DOE's "Future Gen" Enterprise Alliance; and in December, Huaneng formed Green Coal Power Ltd with the Chinese Shenhua Group Corporation, the Chinese Coal Energy Group Corporation, and with five power corporations as member companies (see Box 6-1). The primary aim of this company is to research and demonstrate coal gasification-based energy systems producing hydrogen as a by-product and power source, and outfitted with CO_2 separation technologies. At present, they are working on changing gas and steam combined-cycle power generation sets with a capacity of 100 MW into an independent intellectual property rights IGCC system, with a capacity of 120 MW.

New coal power plants are long-term construction projects, requiring 3-4 years to place into service after groundbreaking. Adding site evaluation, permitting, financing, and other upfront project planning time means that even plants currently under initial development will not provide power until 2010-2012 at the earliest.

TABLE 6-2 Generation Performance Comparison

Performance[a]	Bituminous Coal				Subbituminous Coal				Lignite Coal			
	IGCC Slurry Feed	Sub-critical PC	Super-critical PC	Ultra-Super-critical PC	IGCC Slurry Feed	Sub-critical PC	Super-critical PC	Ultra-Super-critical PC	IGCC Solid Feed	Sub-critical PC	Super-critical PC	Ultra-Super-critical PC
Net Thermal Efficiency, percent	41.8	35.9	38.3	42.7	40.0	34.8	37.9	41.9	39.2	33.1	35.9	37.6
Net Heat Rate, Btu/kWh	8167	9500	8900	8000	8520	9800	9000	8146	8707	10,300	9500	9065
Gross Power, MW	564	540	540	543	575	541	541	543	580	544	544	546
Internal Power. MW	64	40	40	43	75	41	41	43	80	44	44	46

[a]Based on a net 500 MW plant.
SOURCE: EPA, 2006.

BOX 6-1
United States-China Cooperation on FutureGen

Led by the FutureGen Industrial Alliance, Inc., and with support from the U.S. Department of Energy, FutureGen is a government-industry partnership created in response to President George W. Bush's call for an increase in hydrogen power research to reduce the negative impact of current energy emissions on global climate change. The FutureGen Alliance is a non-profit organization, created in 2005, that represents major coal users and producers. On December 15, 2006, China entered into an agreement with the United States to join the Government Steering Committee. In addition to government participation, key energy companies from both countries have joined the partnership. The China Huaneng Group, a member since 2005, is one of the top ten energy producers in world and China's largest coal-fired power producer.

By recognizing the need for clean energy produced from coal, the FutureGen Alliance believes that its operation can facilitate continuing economic growth within the United States, while serving as a model for future hydrogen-based energy plants around the world.[a] Instead of releasing harmful emissions during energy production, the FutureGen plant will store carbon dioxide in deep saline formations and will emit hydrogen and other particles that can be used by other industries in their production processes. By exploring current technologies on coal gasification, electricity generation, emissions control, carbon dioxide capture and storage, and hydrogen production the FutureGen plant will be a pioneer in the field of hydrogen power research by testing all of these technologies at one facility. By converting the carbon within coal into a gas, the FutureGen plant will produce a gas that is primarily made up of hydrogen and carbon monoxide, thus producing an environmentally clean by-product for use in powering turbines to produce electricity. Other uses for the hydrogen by-products include fuel cells, combustion turbines, and other hydrogen-based technologies.

Another important factor in hydrogen power-related research is the facility's location. FutureGen has selected four candidate locations for their test facility: Mattoon, IL, Tuscola, IL, Heart of Brazos near Jewett, TX, and Odessa, TX. These sites were selected based on the environmental and financial considerations of using each site for FutureGen's operation. In the fall of 2007, a decision will be made by the FutureGen Alliance as to which site will be home to the FutureGen facility.

There are multiple benefits to choosing this process. First, FutureGen's technologies could have a major impact on national energy security. FutureGen's process of energy production also has a very low impact on climate change by storing the carbon dioxide removed from coal during the gasification process. Thus, carbon dioxide emissions, which negatively impact the earth's climate, will drastically be reduced during this process. Through FutureGen's process of using coal to produce a cleaner, cost-efficient energy source, countries around the world could rely on their domestic supply of coal to provide energy to growing populations. When the FutureGen plant is operational, it will be the "environmentally cleanest fossil fuel-fired power plant in the world."[b] In 2012, FutureGen anticipates to be in operation and will be the first plant producing hydrogen from coal and electricity at the same time.

---

[a]FutureGen Alliance. 2006. http://www.futuregenalliance.org/.

[b]U.S. Department of Energy. 2006. "FutureGen—Tomorrow's Pollution-Free Power Plant." http://www.fossil.energy.gov/programs/powersystems/futuregen/.

For the first round of new capacity, most generators in the United States have chosen a subcritical or supercritical PC or CFB, as well as a higher temperature European/Japanese ultra-supercritical PC (USC/PC). All of these plants are being designed to significantly exceed new source performance standards. Within the next 5 years, the industry will have to introduce the next generation of advanced IGCC and U.S-designed USC/PC, bringing continued advancements in efficiency and environmental performance, including the capability for CO_2 capture. Box 6-2 details some of these clean coal technologies. Federal, state, and local support may be required to coordinate permitting and approvals, and regulatory and permitting agencies need to support the introduction of each generation of new technology.

BOX 6-2
Comparing Clean Coal Technologies

Several clean coal technologies are currently available, categorized as advanced high-efficiency combustion-based technologies and gasification-based technologies, and technologies are being developed that will allow capture of carbon dioxide in both advanced combustion and gasification plants. Advanced combustion technologies combust coal in the presence of air or oxygen. Gasification technologies use a partial combustion of coal, in the presence of either air or oxygen, to produce a synthetic fuel gas (NCC, 2006).

Emerging technologies at initial commercialization include:

- IGCC with air or oxygen blown gasification to produce syngas for use in combustion turbines, with plant efficiencies of 39-43 percent
- PC/Ultra-supercritical steam plants of European and Japanese design, providing efficiencies of 40-42 percent

Developing technologies include:

- Advanced IGCC with hydrogen production and CO_2 capture (FutureGen)
- PC/USC steam plants, providing efficiencies of 48 percent
- Advanced USC PC/CFB, with efficiency goals of 50 percent prior to CO_2 capture
- Innovative post-combustion capture technologies with reduced cost and power usage for PC/CFB technologies
- Advanced PC/CFB with oxygen combustion to facilitate capture of CO_2 emissions

Future technologies include:

- Hybrid cycles (IGCC with fuel cell)
- Chemical looping combustion and gasification
- Next-generation PC/CFB oxyfuel plants

POLLUTION CONTROLS

Installing pollution controls is another means to reduce specific emissions. Most of these controls are designed to reduce a particular pollutant (e.g. SO_2), although there are some co-benefits to installed controls, such as mercury emissions reductions, as will be discussed. Since these pollution controls have developed in response to concerns over air pollution, or in many cases, regulations regarding emissions, they are often installed as a retrofit onto an existing facility. Retrofits have become an essential pollution-control strategy for U.S. energy producers, in order to come into compliance with mandated emissions reductions. Retrofits necessarily entail additional costs, because the new technology must be incorporated into an existing structure (IEA, 2006b). As an example, the original cost of the Plant Bowen coal-fired power plant in Georgia was $400 million in the early 1970s (roughly $1.4 billion in 2005$). Adding NO_x control devices in the 1990s cost about $400 million more. The current addition of sulfur dioxide scrubbers is costing another $900 million (Marr, 2006). Altogether, Georgia Power (which operates Plant Bowen and nearby Plant Hammond) plans to spend $1.3 billion on pollution-control devices by 2010, in order to come into compliance with federal regulations.

China has taken an important first step by requiring new coal-fired power plants to install desulfurization equipment. This has been a direct result of concerns over China's world-leading SO_2 emissions. However, a similar requirement for NO_x controls does not currently exist, though the government began pilot projects in 2004 to implement a NO_x levy of 0.6 RMB/kg. While some plants in China are voluntarily installing low-NO_x burners on new facilities, fewer are investing in additional NO_x controls. The importance of building in pollution controls cannot be overstated. This of course does not diminish the importance of retrofitting existing facilities with pollution controls, but as both countries consider expanding power generation capacity, much of this coal-fired, there is a significant opportunity to plan for these pollution controls at the outset. While this entails higher initial costs, as experience has shown, costs and installation times increase significantly for retrofits. Finally, in light of the possibility that future regulations may limit CO_2 emissions, some power companies are taking steps to prepare for installing further equipment to capture and sequester CO_2, should it become a regulated emission.

Coal-fired power plants are also a source of dioxins and furans, 2 of the 12 persistent organic pollutants (POPs) that China, the United States, and numerous other countries are beginning to regulate. China ratified the Stockholm Convention on POPs in 2004 and is making progress in eliminating the manufacture and use of certain POPs (mostly pesticides) (NRC, 2007). Reductions in dioxin and furan emissions, which are unintentional by-products of combustion processes and chemical manufacturing, can be partially achieved by improving pollution controls on combustion sources. Installing particulate controls has the co-benefit of reducing dioxin emissions.

Coal Preparation

Coal preparation, in a broad sense, is any treatment of mined coal to remove waste, and includes crushing, screening, and washing. Approximately 70 percent of raw coal in China is currently burned without crushing and screening it; this has an important impact on emissions of sulfur dioxide and particulate matter, accounting for 70-80 percent of China's total emissions (NDRC, 2005). However, coal preparation can increase heating values, decrease transportation costs, and reduce pollutants before they are released into the air, notably SO_2 and Hg. Coal washing in China could reduce sulfur emissions by 20 percent, but requires investments at the mines, which nationwide would total at least 16 billion RMB (NRC, 2004). Additionally, coal briquettes with limestone additive (to reduce SO_2) are used in domestic heating and cooking, but are also appropriate for small industrial boilers: an investment of 2 billion RMB could yield 113 million tons of briquettes, which would help reduce ground-level pollution from the millions of small industrial boilers (NRC, 2004). Though burning low-sulfur coal is the most economical way to reduce SO_2 emissions, wide-scale substitution would require major disruptions and changes to the mining and transportation networks.

Particulate Controls

Electrostatic Precipitators

Electrostatic precipitators (ESPs) are particulate collection devices which use an induced electrical charge to remove particles from flue gas. ESP has been the preferred technology for use at coal-fired power plants. They are highly efficient at particulate removal (typically 99.0-99.5 percent) and have minimal impact on air flow through the device. Since they do not require a large pressure drop, they have less of an impact on plant efficiency, compared to fabric filters. One primary challenge to an ESP's efficiency is electrical resistance, which can result from combustion of low-sulfur coal. Though this is typically a dry process, it is possible to spray incoming air with moisture, which can improve the capture of fine particles, as well as reduce the electrical resistance of the incoming particles. Dry ESP waste is adsorbed onto metal plates, then rapped to remove the particulate matter for disposal or potentially reuse (e.g., fly ash used in cement). Wet ESP waste is flushed with water for treatment or disposal. At present more than 96 percent of the coal power plants in China have ESP.

Fabric Filters

Fabric filters, alternately referred to as baghouses, have been employed more widely than ESP since the 1970s, largely at the industrial scale (IEA, 2006c). China has seen a similar increase in the use of baghouses, not only for industrial purposes but also for use at power plants. The choice between ESP and fabric filters depends on coal type, plant size, and boiler type and configuration; additionally, if regulations require removal efficiency above 99.5 percent, fabric filters may be more cost-effective (World Bank, 2007). However, fabric filters require a decrease in pressure and thus a decrease in plant efficiency.

Flue Gas Desulfurization

Wet flue-gas desulfurization (FGD), commonly referred to as wet scrubbers, is a well-established process for significantly reducing SO_2 emissions. Design efficiencies range from 80 to 95 percent SO_2 removal, and with additives, this can be improved by 5-10 percent. FGD continues to be the preferred technology for both retrofits and new construction of power plants, particularly in Europe, Japan, and the United States (World Bank, 2006d). Although its effectiveness has been demonstrated on both high- and low-sulfur coals in developed countries, the applicability to local coals in developing countries, such as China, sometimes requires further adaptation.

Although wet scrubbers can also be utilized in particulate removal, they are most effective when coupled with ESP or filters (EPA, 2002b). Wet scrubbers consist of a spray tower or absorber where flue gas is sprayed with a calcium-based water slurry. The calcium and SO_2 react to form calcium sulfite or sulfate, which can then be thickened, dewatered, and mixed with fly ash for disposal in landfills. Alternatively, the waste can be turned into gypsum for reuse. Limestone with forced oxidation is the process by which the calcium sulfite created in the wet scrubber is oxidized by bubbling compressed air through the slurry to produce wallboard-grade gypsum. In the United States, this technology is useful when connected to local markets for wallboard, cement, and other applications.

Dry scrubbers are an alternative application for SO_2 removal. Instead of saturating the flue gas, dry FGD uses little or no moisture and thus eliminates the need for dewatering. Dry FGD's efficiency is slightly lower than wet FGD (70-90 percent), but capital costs are also lower, and the scrubbers are easier to operate and maintain. Dry scrubbers have been proven with low-sulfur coal in the United States and elsewhere, but their applicability for use with high-sulfur coals has not been widely demonstrated (World Bank, 2006d). Moreover, as wet scrubbers become more competitive in terms of ease of use and cost, dry scrubbers lose their competitive advantage. More than 20 percent of coal-fired utility boiler capacity in the United States uses wet FGD; roughly half that amount uses dry FGD technology (EPA, 2002b). The United States has been a leader in deploying FGD technology, mostly as a result of stringent regulations which required their use (Rubin et al., 2003). Figure 6-1 shows projected U.S. capacity of coal-fired burners with some sort of FGD technology installed. The figure also projects the potential impacts of retrofits based on CAIR, CAMR, and CAVR—legislation which places caps on SO_2, NO_x, and mercury. Though only about one-third of coal-fired capacity currently has FGD installed, this share is projected to increase to over 70 percent by 2020.

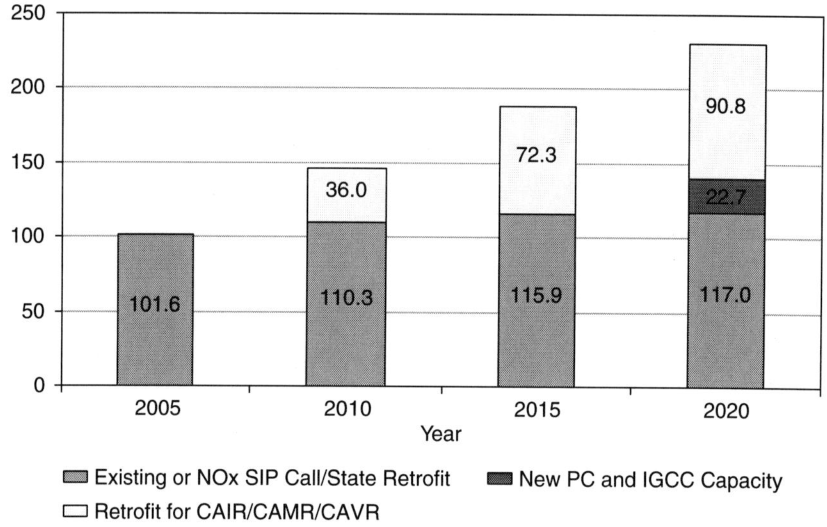

FIGURE 6-1 Projected U.S. capacity of coal-fired burners with FGD technology. SOURCE: EPA, 2005.

China has recently begun adopting and adapting FGD technologies. New power plants are now required to install FGD equipment. There are more than ten kinds of FGD technologies, which have been applied in power stations throughout China. At the end of 2003, the total power capacity equipped with FGD was 45,315 MW, or about 16 percent of the total (including power plants completed, under construction. and invited for public bidding). Limestone-gypsum wet desulfurization was 84 percent of the total FGD capacity. China possesses independent intellectual property rights on the limestone-gypsum desulfurization process, which has cut additional costs to $25/kW, with a desulfurization efficiency above 95 percent—thereby significantly decreasing the proportion of desulfurization technologies in terms of total investment. The electric power industry has made the limestone-gypsum desulfurization technique the main technique for thermal power plant FGD. But a report from the National Development and Reform Commission (NDRC) estimated that no more than 60 percent of scrubbers are operating because, lacking financial incentives and supervision from the State, plant officials turned them off (NDRC, 2006).

Selective Catalytic Reduction

Selective catalytic reduction (SCR) is effective in reducing NO_x emissions, making it an attractive technology for countries with emissions limits, and Germany and Japan are currently utilizing SCR on coal-fired power plants. SCR technology is widely available for low-sulfur coal, but high capital costs and operations and maintenance costs have made it difficult to implement in developing countries (World Bank, 2006b). An even larger impediment to widespread implementation in countries such as China is the lack of restrictions on NO_x emissions. However, in 2005 the Chinese government began recommending that new power plants with a capacity of more than 300 MW apply denitrifying technologies to reduce the exhaustion of NO_x. Following this, some power plants have installed SCR, and more may follow if the government enacts stricter NO_x control regulations.

SCR works by injecting ammonia into the flue gas, converting NO_x to N and H_2O. Selective non-catalytic reduction (SNCR) is a similar technology, the major difference being the presence of a catalyst in SCR, and the temperatures at which the processes take place. SCR requires much lower temperatures (340°-380°C) than SNCR (870°-1,200°C). When utilized in low-sulfur boilers (<1.5 percent sulfur content), SCR has been demonstrated to remove 70-95 percent of NO_x emissions. The United States is currently focusing efforts on demonstrations for medium- and high-sulfur coal as well.

Capital costs depend on required NO_x reductions, unit layout, and type of SCR. SCR systems can either be "hot-side," that is, located between the economizer and the air heater, or "cold-side/post-FGD"—installed downstream of the particulate control. Hot-side systems are severely limited by space constraints and require extensive modifications that result in an average outage of 2-3 months. Cold-side systems are less constrained in terms of space and require shorter outages on average (3-6 weeks), but the flue gas must be reheated and costs are significantly higher than those for hot-side systems.

The United States has lagged behind other countries in SCR deployment and did not install its first units until 1993. However, recent legislation to curb NO_x emissions in the United States has led to an increase in capacity of boilers with SCR technology. Figure 6-2 shows projected U.S. capacity of coal-fired burners with SCR technology, and the potential impacts of retrofits based on CAIR/CAMR/CAVR legislation. Again, SCR-equipped capacity is projected to increase from just under one-third to over 56 percent of total capacity by 2020.

Mercury Control Issues and Technologies

Trace amounts of mercury are present in coal, and when electric utilities burn coal to generate electricity, mercury is released. In 2000, the total mercury content of the coal received at power plants in the United States was approximately 75 tons. Because of fuel processing and other environmental control equipment, total mercury emissions from coal-fired power plants in the United States were approximately 45 tons—representing a 40 percent reduction relative to "as received" coal.[1]

[1]Approximately 30 percent of the mercury in eastern bituminous coal is typically removed by coal washing before shipping to the plant.

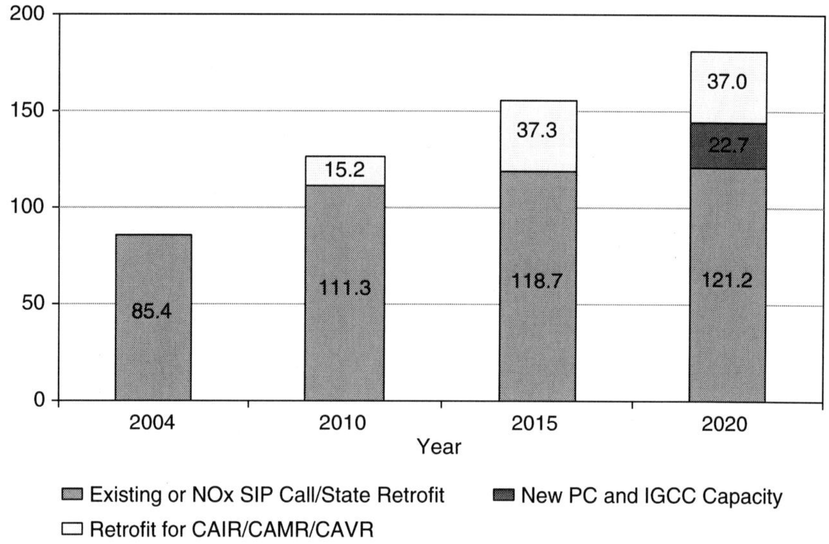

FIGURE 6-2 Projected U.S. capacity of coal-fired burners with SCR technology.
SOURCE: EPA, 2005.

Mercury is known to bioaccumulate in fish as methylmercury—its most toxic form. Human exposure to methylmercury, which occurs primarily from fish consumption, is associated with serious neurological and developmental effects. The U.S. Centers for Disease Control and Prevention estimates that roughly 6 percent of American women of child-bearing age have blood levels of mercury that are above the reference dose set by EPA to represent a safe level.

Since the 1990s, methods for capturing mercury from coal-fired power plant flue gases have been the subject of considerable R&D and demonstration initiatives in the United States. Up until now, control of mercury emissions from coal-fired plants has been achieved primarily as a co-benefit of existing pollution controls. Fabric filters and ESPs, FGD systems, and SCR systems contribute to mercury capture (Srivastava et al., 2005). On average, these pollution controls are estimated to remove 36 percent of the mercury emitted from U.S. coal-fired boilers. Due to mercury speciation, or the partitioning of elemental mercury into various forms (elemental, Hg^0, ionic, Hg^{2+}, and particle-adsorbed mercury, Hg_p), mercury control approaches and success rates vary. In general, bituminous coal produces a majority of Hg^{2+}, while sub-bituminous and lignite coals produce a majority of Hg^0. Figure 6-3 illustrates that units burning sub-bituminous and lignite coals exhibited significantly inferior mercury capture in cold-side electrostatic precipitators (CS-ESP) compared to similarly equipped units burning

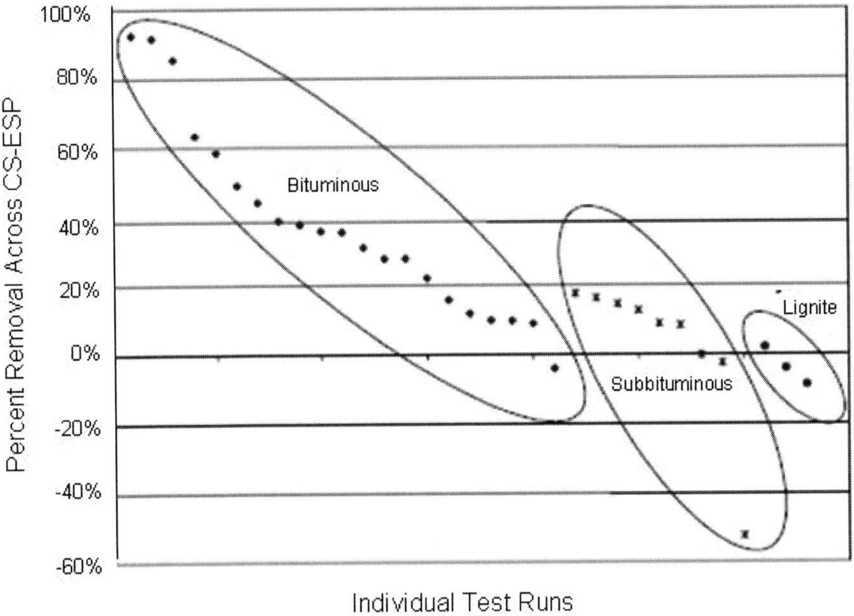

FIGURE 6-3 Mercury capture across cold-side ESP.
SOURCE: DOE, 2003.

bituminous coal (DOE, 2003). In the case of sub-bituminous coals, a maximum mercury capture of 35 percent has been reported for units with ESPs. However, plants that burn sub-bituminous coals and are equipped with fabric filters report relatively high mercury removal levels of 60 to 99 percent (Grover et al., 1999; Butz et al. 2000).

Chlorine in coal can be detrimental to boiler performance because of its corrosive nature. However, chlorine content plays an important role in the control of mercury. The presence of chlorine results in the conversion of mercury to $HgCl_2$, a relatively soluble form that can be effectively captured in a wet FGD system. Elemental mercury is not removed by FGD. SCR systems can also convert elemental mercury to Hg^{2+}, enhancing control by a FGD system. In general, the higher the chlorine content, the higher the degree of mercury removal. Limited tests suggest that the use of wet FGD in combination with SCR may be able to capture 80 to 90 percent of mercury from some high-chlorine bituminous coals, which tend to have a high percentage of ionic mercury. Deactivation of the SCR's mercury oxidation effect with time has been observed in plants burning sub-bituminous PRB coals but remains to be determined for bituminous coals.

In contrast to bituminous coals where a synergism or co-benefit between FGD and mercury control has been found, the opposite may be true for sub-bituminous coals. This was first discovered in an analysis of data from the EPA's 1999 mercury Information Collection Request (ICR). The data showed that mercury removal dropped when a spray dryer absorber was part of the air pollution control train (Kilgroe et al., 2001). The average mercury removal for ICR plants burning sub-bituminous coal with a fabric filter was 72 percent. However, for units burning sub-bituminous coal that had both a spray dryer absorber for SO_2 control and a fabric filter, the average mercury removal dropped to 25 percent.

Injection of powdered activated carbon (PAC) represents the most mature add-on technology for reducing mercury emissions from coal-fired boilers. This sorbent binds with the mercury and is subsequently captured in the ESP or fabric filter, depending on which device is installed. To date, activated carbon injection (ACI) has primarily been used to control mercury emissions from municipal and medical waste incinerators. However, due to state and federal mercury control requirements, commercial contracts have recently been placed for ACI or brominated-ACI equipment for a number of new and existing electrical generating units in the United States (Feeley, 2006). Jones et al. (2007) recently reported results from sorbent injection tests at six full-scale plants, which demonstrated that injection of PAC (for bituminous coal) or brominated-PAC (for lignite or sub-bituminous coal) is capable of reducing mercury emissions by 70 to 90 percent, depending on the coal type, sorbent injection rate, and type of particulate control equipment. Additional short-term field tests and long-term demonstrations of sorbent injection technologies are being conducted at a number of plants representing a range of plant designs, operating characteristics, and fuel types.

Complicating mercury emission controls is the need to dispose of mercury-containing wastes generated from the removal of mercury from flue gases. In addition, injection of activated carbon can impact the ability to sell coal combustion by-products such as fly ash. Although the cost attributed to mercury controls would increase significantly if it rendered fly ash unmarketable (Jones et al., 2007), the majority of existing coal plants in the United States presently landfill their fly ash as waste, and modest quantities of carbon in fly ash may be tolerated (Srivastava et al., 2005). The use and presence of activated carbon may also complicate SCR operation and generate concerns that depleted catalyst be handled as special wastes. It may also complicate FGD system operation, increasing the quantity of FGD by-products requiring disposal in a landfill.

Carbon Capture and Sequestration

Though not widely regulated as a pollutant, governments and energy producers are increasingly seeking methods to capture CO_2 over concerns about global climate change. To achieve further CO_2 reductions (beyond reductions achieved through efficiency improvements), CO_2 must be removed from the gas streams,

TABLE 6-3 Performance and Economic Impacts of Carbon Capture Technologies

	IGCC	Supercritical PC
Net plant output (pre-CO_2 capture), MW	425	462
Plant output derating, percent	14	29
Heat rate increase, percent	17	40
Total capital cost increase, percent	47	73
Cost of electricity increase, percent	38	66
CO_2 capture cost, \$/ton	24	35

NOTE: Based on 90 percent capture.
SOURCE: EPA, 2006.

concentrated, and compressed for transportation to the storage and sequestration location. Commercially available technologies can capture over 90 percent of the CO_2, but are capital intensive, impose an electric power output reduction, and cause energy efficiency penalties. The current costs for carbon capture for all coal technologies are substantial and would significantly increase the cost of electricity. Examples of advanced technologies for post-combustion capture include advanced imines, ammonia scrubbing, and a variety of other promising solvents, and these technologies are in the process of RD&D. Advanced CO_2 capture for IGCC includes improvements to the water-gas shift reaction and hydrogen separation.[2]

Table 6-3 shows the projected performance and economic impacts of applying currently available technologies to IGCC and PC plants. This comparison points to the cost-competitiveness of IGCC, if carbon capture technologies are applied. Near-term goals for U.S. RD&D are: by 2007, developing two capture technologies limiting energy cost increases to 20 percent for pre-combustion, 45 percent for post-combustion; and by 2012, developing two capture technologies limiting cost increases to 10 percent for pre-combustion, 20 percent for post-combustion (NETL, 2006).

Sequestration can take many forms. Biological or terrestrial sequestration relies on long-lived biomass, such as forests, which sequester carbon until reaching equilibrium as trees die and decay, thereby re-releasing carbon back into the atmosphere. Longer-term storage options include geologic and oceanic sequestration. Geologic sequestration is the only method currently available for seques-

[2]Studies by DOE, EPRI, the Canadian Clean Power Coalition, and others indicate that advanced capture processes, applied to SCPC or SCCFB, could result in competitive electricity costs with carbon capture. With the commercialization of these advanced capture processes, the selection of a coal power technology with carbon capture would be based on fuel, operational, and site specifics, thus providing generation companies with a portfolio of proven options for near-zero emission power.

tering CO_2 from large stationary sources (such as power plants) (NETL, 2006). Carbon can be sequestered in a number of geologic formations, including depleted oil and gas reservoirs, unmineable coal seams, and saline formations. The first two options have the attendant benefit of producing additional energy resources, which might otherwise not be recovered. These processes are described in more detail in Appendix B.

REFERENCES

Butz, J. and J. Albiston. 2000. Use of fly ash fractions from western coals for mercury removal from flue gas streams. Proceedings of the Air Quality II: Mercury, Trace Elements, and Particulate Matter Conference. Arlington, VA, September 19-21.

CEC (China Electricity Council). 2004. Electric Power in China.

DOE (U.S. Department of Energy). 2003. A Review of DOE/NETL's Mercury Control Technology R&D Program for Coal-Fired Power Plants. DOE/NETL Hg R&D Program Review.

EIA (Energy Information Administration). 2005. Average Flue Gas Desulfurization Costs. Washington, D.C.: U.S. Department of Energy.

EPA (U.S. Environmental Protection Agency). 2002a. Clean Alternative Fuels: Fischer-Tropsch. Washington, D.C.

EPA. 2002b. Engineering and Economic Factors Affecting the Installation of Control Technologies for Multipollutant Strategies. Washington, D.C.: Office of Research and Development.

EPA. 2005. Multi-Pollutant Regulatory Analysis: The Clean Air Interstate Rule, The Clear Air Mercury Rule, and The Clean Air Visibility Rule, October.

EPA. 2006. Environmental Footprints and Costs of Coal-Based Integrated Gasification Combined Cycle and Pulverized Coal Technologies. Washington, D.C.: Office of Atmospheric Research.

Feeley, T.J. 2006. U.S. DOE's Mercury Control Technology R&D Program Review. 2006 Mercury Control Technology Conference, Pittsburgh, PA, December 11-13, 2006.

Grover, C., J. Butz, S. Haythornthwaite, J. Smith, M. Fox, T. Hunt, R. Chang, T. D. Brown, and E. Prestbo. 1999. Mercury measurements across particulate collectors of PSCO coal-fired utility boilers. Proceedings of the Utility Mega Conference. Atlanta.

IEA (International Energy Agency). 2006a. Clean coal technologies. Retrieved June 9, 2006, from http://www.iea-coal.org.uk/.

IEA. 2006b. Economics of Retrofit Air Pollution Control Technologies. London: IEA Clean Coal Centre.

Jones, A.P., J.W. Hoffmann, D.N. Smith, T.J. Feeley, and J.T. Murphy. 2007. DOE/NETL's Phase II Mercury Control Technology Field Testing Program: Preliminary Economic Analysis of Activated Carbon Injection. Environmental Science and Technology 41(4):1365-1371.

Kilgroe, J.D., C.B. Sedman, R. K. Srivastava, J. V. Ryan, C.W. Lee, and S.A. Thorne. 2002. Control of Mecury Emissions for Coal-fired Electric Utility Boilers: Interim Report with Errata Pages. U.S. Environmental Protection Agency, EPA-600/R-01-109.

Marr, C. 2006. Under federal pressure, Georgia Power and others are spending billions to cut back air pollution. Rome News-Tribune, November 12, 2006, Rome, GA.

NCC (National Coal Council). 2006. Coal: America's Energy Future. Washington, D.C.

NDRC (National Development and Reform Commission). 2005. China Medium and Long Term Energy Conservation Plan.

NDRC. 2006. Perfect Policy, Strengthen Management, Guide and Promote Industrial FGD for Healthy and Rapid Development, available at http://www.ndrc.gov.cn/hjbh/huanjing/t20060731_78633.htm (in Chinese).

NETL (National Energy Technology Laboratory). 2005. Gasification: World Survey Results 2004. Washington, D.C.: Office of Fossil Energy, U.S. Department of Energy.

NETL. 2006. Carbon Sequestration Technology Roadmap and Program Plan—2006. Washington, D.C.: Office of Fossil Energy, U.S. Department of Energy.

NRC (National Research Council). 2003. Review of DOE's Vision 21 R&D Program—Phase I. Washington, D.C.: The National Academies Press.

NRC. 2004. Urbanization, Energy, and Air Pollution in China: The Challenges Ahead. Washington, D.C.: The National Academies Press.

NRC. 2007. Implementing the Stockholm Convention on Persistent Organic Pollutants: Summary of a Workshop in China. From the series: Strenghtening Science-Based Decision Making in Developing Countries. Washington, D.C.: The National Academies Press.

Rubin, E.S., M.R. Taylor, S. Yeh, and D.A. Hounshell. 2003. Experience Curves for Environmental Technology and Their Relationship to Government Actions, Presented at EXTOOL-EXCETP6 Workshop, IEA, Paris, January 22-24.

Srivastava, R.K., J.E. Staudt, and W. Jozewicz. 2005. Preliminary estimates of performance and cost of mercury emission control technology applications on electric utility boilers: An update. Environmental Progress 24:198-213.

World Bank. 2006a. Pressurized Fluidized Bed Combustion. Environmental Management for Power Development Program. Retrieved August 16, 2006, from http://www.worldbank.org/html/fpd/em/power/EA/mitigatn/pfbcsubs.stm.

World Bank. 2006b. Selective Catalytic Reduction (SCR). Environmental Management for Power Development Program. Retrieved August 16, 2006, from http://www.worldbank.org/html/fpd/em/power/EA/mitigatn/aqnoscr.stm.

World Bank. 2006c. Supercritical Coal Fired Power Plants. Environmental Management for Power Development Program. Retrieved August 15, 2006, from http://www.worldbank.org/html/fpd/em/supercritical/supercritical.htm.

World Bank. 2006d. Wet Flue Gas Desulfurization (FGD). Retrieved August 15, 2006, from http://www.worldbank.org/html/fpd/em/power/EA/mitgatn/aqsowet.stm.

World Bank. 2007. Environmental Health and Safety Guidelines. Retrieved June 5, 2007 from http://www.ifc.org/ifcext/policyreview.nsf/Content/EHSGuidelinesUpdate.

7

Renewable Energy Resources

As noted at the outset of Chapter 2, energy usage in the United States and China is presently dominated by hydrocarbon sources and will continue to be for decades. In the case of petroleum, both countries are highly dependent on imported sources. In the case of coal, they have rich domestic sources, but its use for the generation of electricity presents many challenges for minimizing its contribution to air pollution. The desire to reduce both pollution and foreign energy dependence has driven efforts at development of renewable energy sources. Their benefits are well known: in most cases, they produce little or no pollution emissions and because the resources are renewable, they reduce dependencies on foreign supplies and more importantly on finite resources of any kind. This chapter provides a brief discussion of the candidate renewable sources/technologies that could have an important impact in the medium to long term. It also explores the capacities, consumption, and forecasts (where available) for both countries. Finally, it will discuss the challenges of increased use of renewable energy resources.

RENEWABLE ENERGY TRENDS AND CURRENT USE

Renewables currently provide 9 percent of U.S. electricity supply, although this varies greatly by region. In China, renewables accounted for 7 percent of total primary consumption (excluding traditional biomass) in 2004, down from 7.8 percent in 2002, and this too varies greatly by region. China's renewable energy industry is growing rapidly, and in recent years, wind and solar power generation have grown 25 percent (NBS, 2005). In order to better assess the potential of these two resources, China opened a new center, the Center for Wind and Solar Energy

Assessment, which is part of the Chinese Meteorology Administration. Growth in the renewable energy industry is even more difficult to forecast than for more conventional energy sectors, as it is still a relatively small sector in both countries; it is largely dependent upon conventional fuel prices, and is increasingly being driven by government mandates. In China, much of the recent R&D has focused on rural applications for converting its abundant resources, but there are also important urban applications. This section will examine the history of renewable energy consumption in the United States and then discuss the resources available to each country and prospects for future use. A separate discussion of renewable liquid fuels (biofuels) will follow.

Figure 7-1 shows trends in renewable energy consumed by source in the United States, 1949 to 2004, and illustrates that:

- Renewable energy has traditionally been dominated by hydroelectricity and wood, although over the past two decades other renewable sources have begun to appear.
- Through 2030, hydro and wood will continue to dominate, although the other renewable sources will gradually increase in importance.

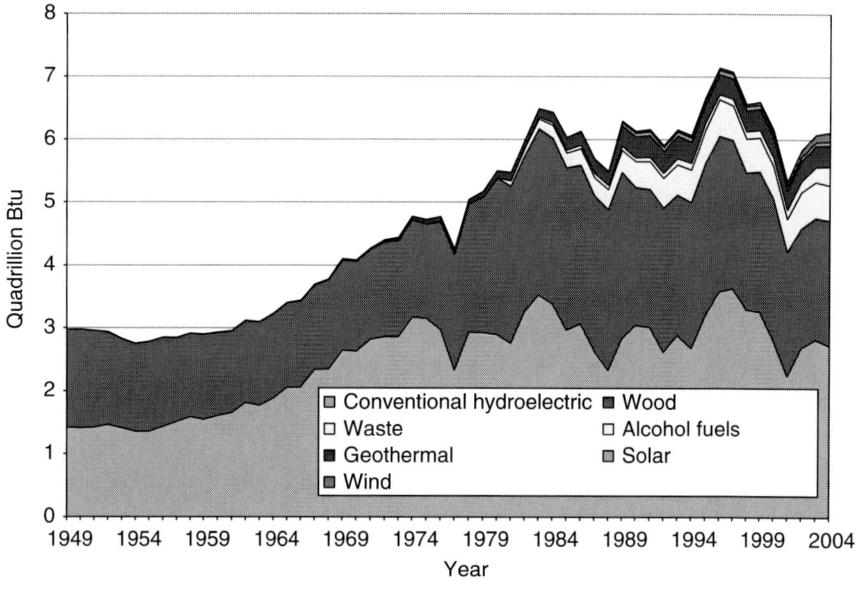

FIGURE 7-1 U.S. renewable energy consumption, by source, 1949-2005.
NOTE: Data for renewables other than hydro and wood N.A. prior to 1980s.
SOURCE: EIA, 2006a.

Figure 7-2 shows another representation of the sources of renewable energy in the United States in 2005, and illustrates that:

- Hydro accounts for 45 percent of renewable energy, and wood accounts for 31 percent.
- Much smaller shares consist of biomass, alcohol fuels, solar, and wind.

Figure 7-3 shows trends in renewable energy consumed in the United States by sector, 1949-2005, and illustrates that:

- Most renewable energy is consumed in the industrial sector, followed by consumption in the residential sector.
- Consumption in the transportation sector (alcohol fuels) was negligible for most of this period, but has increased its share recently.

A closer look at statistics from 2005 indicates that:

- Approximately 60 percent of renewable energy was used in the industrial sector and 21 percent was used in the residential sector.
- Fourteen percent was used in the transportation sector, and 5 percent was used in the commercial sector (EIA, 2006a).

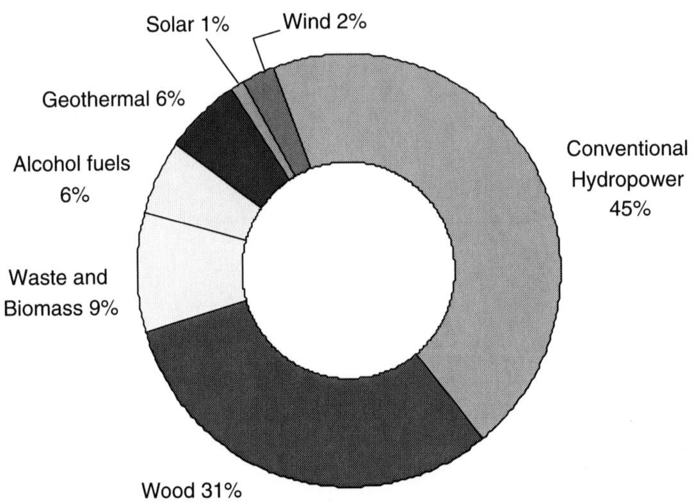

FIGURE 7-2 Relative shares of renewable energy consumption in the United States, 2005.
SOURCE: EIA, 2006a.

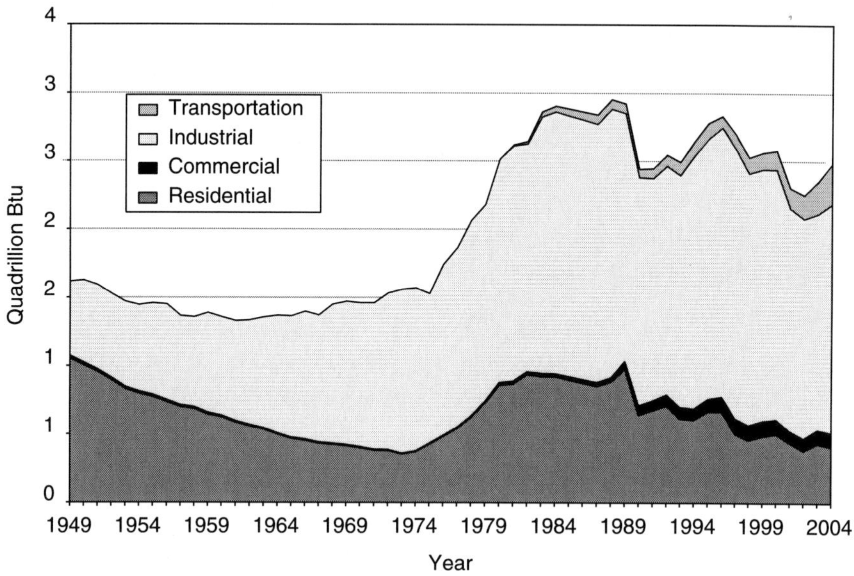

FIGURE 7-3 U.S. renewable energy consumption, by sector, 1949-2005.
NOTE: Data on renewable fuel (ethanol) used in transportation not available prior to 1981.
SOURCE: EIA, 2006a.

Figure 7-4 shows the renewable energy used in the electric power sector in 2005, and illustrates that:

- Hydro dominates, providing over 70 percent of renewable electric power.
- Waste[1] and geothermal are the next largest sources.

RENEWABLE ENERGY FOR ELECTRICITY AND HEATING

Hydropower

Hydroelectric generation is a well-established, base load option with significant air quality advantages. In both countries, conventional hydroelectric power is the dominant renewable resource, although the scale of plants differs in many cases. In the United States, most hydroelectric plants are large-scale plants; in

[1]Municipal solid waste, landfill gas, sludge waste, tires, agricultural by-products, and other biomass.

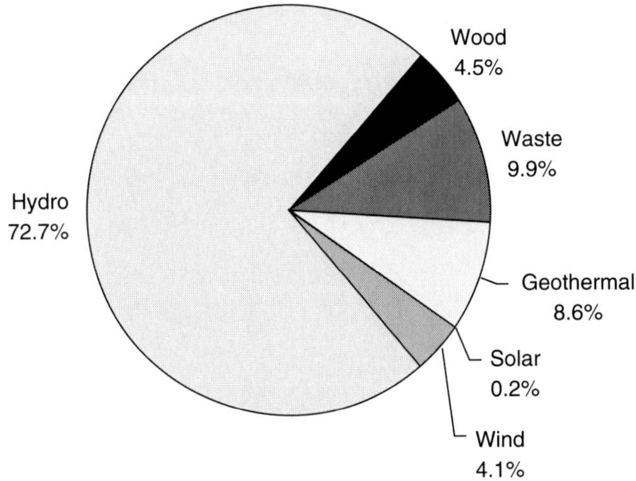

FIGURE 7-4 Renewable energy for electricity, by source, 2005.
SOURCE: EIA, 2006a.

China, although major projects such as the Three Gorges Dam garner the most attention, roughly one-third of China's capacity is classified as small hydro power (<50 MW). In the United States, hydro power accounts for almost 7 percent of total generation, but in the future its growth will be limited. It is projected to decrease to 5 percent due to public concerns over environmental impact, as well as because most sites with large-scale potential to generate power have already been put into use (EIA, 2007).

In China, growth potential for hydropower appears to be more promising. China has abundant water resources for power production, with theoretical reserves of 688 GW and annual energy production of 6,040,000 GWh. The technically feasible reserves are 540 GW, with annual energy production of 2,140,000 GWh; and the economically developable reserve is 400 GW, with annual energy production of 1,280,000 GWh (NDRC, 2005). About 74 percent of water resources are located in southwest China, as shown in Figure 7-5. As of 2004, domestic hydropower generation (including small-scale hydropower) capacity was 108,260 MW, with total energy production of 328,000 GWh (15 percent of total production), and that would reach 401,000 GWh in 2005 (Figure 7-6). In 2004, small hydropower produced 110,000 GWh, providing electric power for one-third of the country and one-fourth of the population. China has recently been making great progress in programming, exploring, designing, constructing, and researching hydropower. Chinese hydropower construction is approaching the world-class

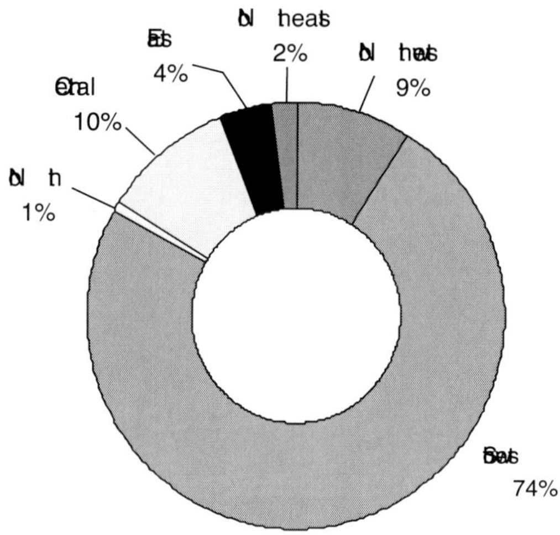

FIGURE 7-5 China's hydropower resources by region.

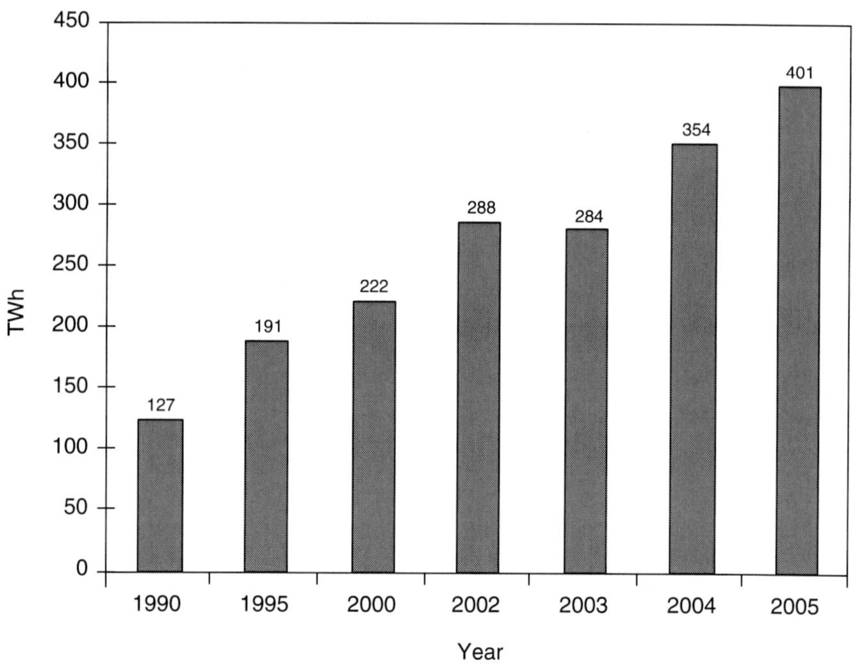

FIGURE 7-6 China's hydropower generation, select years, 1990-2005.
SOURCE: China Electric Power Yearbook, 2005.

level, as evidenced by the Ertan hydropower plant and other large plants, such as Sanxia (Three Gorges), Xiaowan, Longtan, and Shuibuya.

Wind

Wind energy is an electric generation option that has received considerable interest in recent years, and is a well-established proven technology that generates few environmental emissions. Wind is one of the most rapidly growing energy technologies in the United States. Areas with high potential include the Mountain West region, the northeast, and certain offshore areas such as the Great Lakes and portions of the North Atlantic coast.

Wind power is projected to experience the largest growth of all renewables in the United States through 2030, thanks to state mandates, but also as a result of federal tax credits. Wind growth will likewise depend on fossil fuel prices as well as on environmental impacts; apart from public concern over aesthetics, wind turbines have been criticized for their impacts on migratory bird populations. Nevertheless, through the combination of federal tax credits and state programs, wind power is projected to provide at least 93 percent of capacity additions in the renewable sector (see Figure 7-7). Current federal tax credits are set to expire at the end of 2008, thus favoring technologies with short lead times, such as wind, and placing at a disadvantage, biomass, geothermal, and hydropower technologies (EIA, 2007).

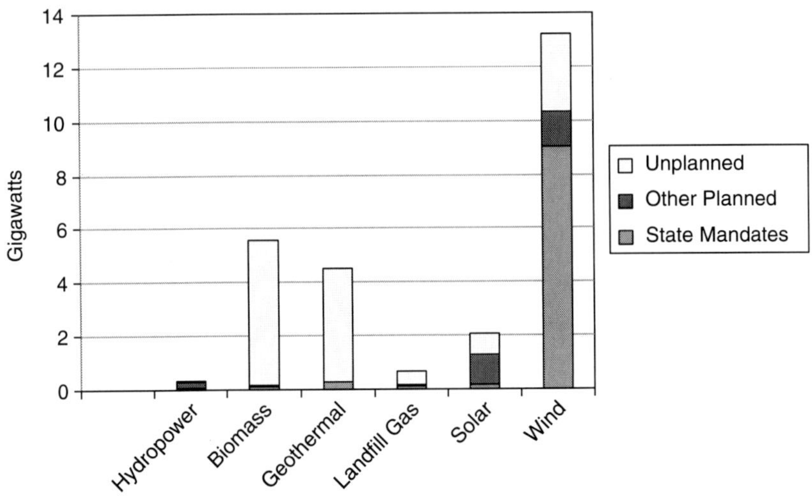

FIGURE 7-7 Projected additions to U.S. renewable generating capacity, 2004-2030.
SOURCE: EIA, 2007.

Wind resources are most abundant throughout northern China, along with some offshore potential west of Hainan Island. According to the calculation of the Chinese Meteorology Academy of Science, the wind reserves at 10 m are 3,226 GW, with a recoverable quantity of 253 GW on land and 750 GW offshore. By 2004, China had built 44 incorporated wind farms totaling 764 MW of capacity, and by the end of 2005, China had increased this capacity to 1.26 GW. There are about 200,000 small wind-driven generators used by herdsmen, totaling 30 MW of capacity. China has been mass-producing 600 kW wind-power generators, and already 96 percent of such units installed in China are domestically produced. China has also begun domestically manufacturing 750 kW wind-power generators. China has set specific targets for wind-powered generation capacity of 5 GW by 2010, and 30 GW by 2020 (NDRC, 2007).

Solar

Solar energy can be converted into heat (thermal) or electricity (photovoltaic or PV). Thermal power has applications in heating buildings and water, as well as in creating steam to generate electricity. The southwestern United States has the highest potential for solar thermal use and solar PV use, although larger sections of the Midwest and southeast have only slightly less potential for PV use. In China, most of the potential for solar energy use is in northern and western China. The Qinghai-Tibet Plain is a high-value center of solar energy, where the total quantity of radiation reaches as high as 6,700 to 8,400 MJ/m^2 per year, and the sun shines 3,200 to 3,300 hours annually.

China is the top producer and consumer of solar-powered water heaters in the world. Solar energy's primary use in China is to heat water for household use, and the price of heating with solar energy is often less than the price of heating with electricity or gas. The total amount of installed capacity was over 80 million m^2 as of 2005, and China's strategic goal for the industry is to cover 400-750 million m^2, or 30 percent of the domestic market for water heaters, mostly in rural areas. China intends to focus first on meeting domestic demand, but the industry's growth could eventually expand to the international market, and beyond conventional use to other applications such as heating, air conditioning, and seawater desalination. Rizhao City (which means City of Sunshine in Chinese) provides one example of solar's potential. As a result of a combination of provincial government support (policy and financial), local industrial capacity, and local political will, the city has increased household solar water heater use to 99 percent in the central urban district, and most traffic signals and outdoor lighting are powered by PV cells. It appears that one key element in Rizhao's reliance on solar power has been the provincial government's support for R&D, which helped improve the technologies and ultimately to lower costs to consumers, while strengthening the local manufacturing capacity (Bai, 2007).

Solar PV technologies receive considerable federal and state government support in the United States and small-scale solar energy applications are growing rapidly there. Solar generators operate during peak loading periods and, without incentives, it is generally too costly to tie them to the grid. Building these technologies from the design stage can reduce costs (relative to retrofits), offset energy costs, and offer savings over the longer term. Rooftop PV modules/tiles and curtain wall or other façade panel installations are examples of building-integrated applications. Second-generation "thin-film" technologies, which utilize a fraction of the active materials and thereby reduce manufacturing costs, may be promising. Some thin films can be incorporated into glass manufacture and provide another opportunity to include solar PV in a building's envelope. Efforts are under way in both countries to tie these building-scale technologies into the grid, but require policy incentives and improvements in the technologies.

China's solar PV installed capacity was 65 MW in 2004, which provided electricity in more than 700 villages and towns totaling 3 million people, mostly in remote areas (NDRC, 2005). The government began providing support for PV development in the 1980s and since then, with assistance from international organizations such as the World Bank, and partnerships with foreign corporations, annual production is approaching 100 MW. China is the largest manufacturer of PV products, and domestic programs such as the National Brightness Program have helped to electrify remote villages off-grid through solar PV. China and the United States have similar manufacturing capacities (just over 8 percent of the world share) but this far outpaces China's current domestic demand, which is still less than 1 percent of worldwide demand (NDRC, 2006). However, this may change as the national government has set ambitious targets for growth in the solar PV power generation sector; by 2010 the installed capacity is to be 700 MW, growing to 1.8 GW by 2020 (NDRC, 2007).

Geothermal

Geothermal power draws its energy from the earth's core; its applications in power generation are limited geographically. In the United States, Nevada, California, and Utah represent the areas where geothermal power is a viable source of renewable energy. However, more limited uses for home heating/cooling and water heating are available throughout the United States via geothermal heat pumps.

In China, geological surveys suggest that geothermal energy resources are abundant and widely distributed. However, of the estimated 5,800 MW of potential capacity, only 30 MW is currently being exploited (CREIA, 2007). Heat pumps, however, are beginning to play a role in China and are being implemented in new building projects (NDRC, 2005).

Biomass

Biomass sources include wood, agricultural waste, industrial organic waste, municipal solid waste, and various other sources and are available in all regions of the United States and China. Its largest use is in co-firing, but it can also fuel power plants and combined heat and power (CHP) plants and can be used as a feedstock in gasification plants. It is difficult to determine the specific cost of biomass feedstocks, as there are significant differences between the various feedstocks, and transportation distance plays a major role in the delivered price of biomass such as forest thinnings and agricultural residues. Some biomass is produced on-site as a by-product of other operations (e.g., mill residues), and the costs of these materials are determined by the presence or absence of competing market outlets. Biomass delivered costs also exhibit considerable variability depending on the resource moisture content and energy content, in addition to collection and transportation costs.

In April 2005, Oak Ridge National Laboratory (ORNL) completed a major assessment of biomass for the U.S. Department of Energy and for the U.S. Department of Agriculture to determine whether the United States is capable of producing a sustainable supply of biomass feedstocks that could be used to displace 30 percent of petroleum consumption by 2030 (ORNL, 2005). Table 7-1 shows the breakdown of potential biomass coming from U.S. forestlands. ORNL makes several important assumptions, including that all forestlands not presently accessible by roads are excluded, all environmentally sensitive areas are excluded, equipment limitations are considered, and recoverable biomass is allocated to both conventional forest products industries and bioenergy and bio-based products. Table 7-2 shows the ORNL breakdown of potential biomass coming from the agricultural sector. The largest contributing source is estimated to be residues from annual crop production (e.g., wheat straw and corn stover), followed by production of perennial energy crops. Thus, based on the ORNL analyses, it appears that the potential exists for sufficient biomass resources to be produced for use as feedstock in an integrated biofuels industry.[2]

Agricultural waste is widespread in China and provides an important energy source; crop residues (primarily stalks and manure) represent roughly 170 Mtce (5.1 EJ). The theoretical quantity of biogas production from industrial organic wastewater is 35 Mtce (1.1 EJ). Forestry residue resources are over 70 Mtce (2.1 EJ). As such, China has been a world leader in developing and implementing biogas technologies, specifically anaerobic digestion (Wu et al., 2007). In 2005, approximately 8 billion m^3 (0.32 EJ) of biogas was utilized, providing energy to 14 million rural residents.

In China, direct burning is the primary use for biomass; much of this takes place in central and western rural China, and its low thermal efficiency (10-

[2]Additional infrastructure will be required to assure adequate consumer access to biomass derived liquid fuels.

TABLE 7-1 Projection of Biomass from U.S. Forestlands for Bioenergy Production

Resource	Representative Moisture Content	BTU as Received (Btu/lb)	BTU (dry basis, Btu/lb)	Quantity (million bdt/yr)[a]
Urban wood wastes including construction and demolition	10 percent-50 percent	4,000-8,000	7,600-9,600	47
Fuelwood harvest from forest lands	40 percent-60 percent	4,000-6,400	7,600-9,600	52
Undergrowth removal for fire protection	40 percent-60 percent			60
Logging and land clearing	40 percent-60 percent	~ 4,500	7,600-9,600	64
Mill residues including pulp and paper	10 percent->50 percent	4,500-8,000	8,000-9,600	145
Total				368

[a]bdt: bone dry tons.
SOURCE: Kitani and Hall, 1989 and ORNL, 2005.

TABLE 7-2 Projection of Biomass from U.S. Agricultural Lands for Bioenergy Production

Resource	Representative Moisture Content	BTU as Received (Btu/lb)	BTU (dry basis, Btu/lb)	Quantity (million bdt/yr)
Grains for biofuels	25 percent-30 percent	4,300-7,300	6,500-9,500	87
Animal manure, process residues, and miscellaneous	85 percent	1,000-4,000	4,000-8,500	106
Perennial energy crops	40 percent-60 percent	4,500-6,500	6,500-9,500	377
Annual crop residues	10 percent-60 percent	4,500-6,500	6,500-9,500	428
Total				998

SOURCE: Kitani and Hall, 1989; ORNL, 2005.

20 percent) and high pollution emissions (significant source of particulate matter) make it a less than desirable fuel source. Anaerobic and gasification technologies are improvements but are not widely used at present. Thus, a major challenge for China will be capitalizing on its abundant biomass resources by upgrading the conversion technologies used, in order to limit the damages caused to human health and the environment. The government has set a target of 3 GW of biomass

power generation by 2020, which will clearly necessitate improved technologies. As of 2006, 39 direct combustion projects totaling 1,284 MW of capacity had been approved, most relying on imported technology (Wu et al., 2007). Research and development of a 1 MW biomass gasification internal combustion engine is also under way; smaller-scale gasification power generators have been in use for decades, mainly utilizing rice husks, but scaling up biomass gasification technologies for commercial use will require large improvements in the overall conversion efficiency (Wu et al., 2002). In Dalian, a pilot project at Sanjiapu is converting agricultural waste into coke, black liquor, and gas which supplies 1,000 local homes and businesses with gas for cooking. A recent technical and economic analysis suggests that decentralized, medium-scale (1-10 MWe) gasification plants may be preferable to larger-scale combustion plants, particularly in rural areas (Wu et al., 2007). These technologies may also benefit from a feed-in tariff resulting from the 2006 Renewable Energy Law, which establishes a subsidy of 0.25 RMB/kWh (~$0.03/kWh) over the price of desulfurized coal.

BIOMASS FOR LIQUID FUEL PRODUCTION

Biomass comprises the largest single source of renewable carbon on the planet, and starch and sugars from biomass currently form the basis for a large and growing renewable liquid fuel industry. While China has begun pilot projects in Jilin and Henan provinces to commercialize grain ethanol, the United States is already a major ethanol producer, recently overtaking Brazil as the world's largest producer (Table 7-3). Ethanol production from grain and biodiesel production from fats and oils are commercial industries and will continue to make significant contributions to liquid fuels production.[3] Fuels produced from cellulosic biomass are not currently (as of 2006) cost-competitive with petroleum fuels, or with conventional biofuels. However, research attention is increasingly focusing on low-cost cellulosic materials as well as on municipal waste materials as next-generation feedstocks (Farrell et al., 2006).

The U.S. corn-to-ethanol industry produced more than 4 billion gallons of alcohol fuel in 2005, and is on track to significantly increase that in 2006. However, use of starch-based biomass fuels has an upper limit, because of the use of food crops as the starting substrate and the inherent competition with the food markets. The agricultural sector estimated that it can produce between 15 and 17 billion gallons of ethanol from crop-based starches before significant impacts to the food market occur. To meet the growing demand for liquid fuels, it is apparent that lignocellulosic forms of biomass will need to supplant the

[3]It should be noted that there continues to be debate over the issue of net energy gain or loss with the production of liquid fuels from biomass. A recent study suggested that, from a life-cycle energy balance perspective, ethanol derived from corn is only slightly positive; ethanol from sugarcane shows a slightly better net positive balance, but wide-scale use of ethanol as an alternative to petroleum-based liquid fuels will almost certainly require cellulosic technology (Farrell et al., 2006).

TABLE 7-3 Top Five Ethanol-Producing Countries, Millions of Gallons, All Grades

Country	2004	2005	2006
United States	3,535	4,264	4,855
Brazil	3,989	4,227	4,491
China	964	1,004	1,017
India	462	449	502
France	219	240	251

SOURCE: RFA, 2007.

current starch substrates. ORNL's recent study suggested that over 95 percent of the U.S. biomass resources available on a sustained basis in 2030 would be cellulosic resources (ORNL, 2005). Lignocellulosics are what comprise woody types of biomass and include the stalks and leafy material of agricultural biomass. Converting these materials is where the real challenge of biomass to liquid fuel production remains.

The Energy Policy Act of 2005 (EPACT) requires a minimum annual renewable fuels consumption in the United States of 6 billion gallons by 2006, and 7.5 billion gallons by 2012. Beyond this, the Biomass R&D Technical Advisory Committee, a panel established by Congress to guide the future of biomass R&D efforts, envisioned that 30 percent of petroleum could be replaced by biofuels by the year 2030—"30 by 30"; however, current production levels are only a small fraction of this target.

Current U.S. ethanol production capacity is 4.4 billion gallons from 97 ethanol refineries, and planned capacity expansions and new capacity under construction total another 2.1 billion gallons (USDA, 2006). Thus, ethanol production will soon exceed the 2006 EPACT target. Ethanol production consumed 1.6 billion bushels of corn in 2005 (about 14 percent of U.S. corn production), and 2.6 billion bushels of corn are expected to be used by 2010 (about 22 percent of an 11.9 billion bushel crop). Despite the rapid increase in production, ethanol consumption has exceeded production for the past few years, which has led to increased imports. Current production costs for the U.S. ethanol industry average about $1.09 per gallon (CARD, 2006). Most U.S. production is based on corn, although other feedstocks include wheat, sorghum, and waste beer.

Though China is the third largest ethanol producer in the world, it is important to note that most of the ethanol (at least 60 percent) is used in the beverage industry. Small-scale operations are well distributed, but are not producing fuel ethanol; China's four state-owned large-scale fuel ethanol manufacturers produced 1.02 million tons (340 million gallons) of ethanol using stale grain as a feedstock (Wu et al., 2007). The National Development and Reform Commission (NDRC) has set targets of 5 million tons of fuel-grade ethanol by 2010, and 10 million tons by 2020. Presently in China, 27 cities utilize ethanol in their public transportation systems, and cities such as Dalian have been blending ethanol into gasoline since 2000. Plans for expanded production and consumption tend to favor non-food feedstocks. Research is under way to expand the use of sweet sorghum stalks in Northeastern China, and cassava in Southern China, as well as other crops suitable for marginal or alkaline land—though storage and pretreatment present obstacles which must be overcome (Wu et al., 2007). Additionally, the Ministry of Science and Technology is funding research on cellulosic conversion, in order to make use of China's abundant cellulosic resources.

BOX 7-1
Reformulated Gasoline (RFG) and Air Quality Goals

Reformulated gasoline (RFG) is increasingly being used to reduce emissions from motor vehicles. What began as a program to prompt refineries supplying nine metropolitan areas of the United States in non-attainment for ozone levels has expanded to a nationwide effort to blend petroleum with additives in order to reduce emissions of volatile organic compounds, NO_x, and other toxics. During its incipient phase in the 1990s, the price differential was noticeable in the metropolitan areas required to sell RFG, but as petroleum prices have risen and more states and metropolitan areas encourage RFG use, price differentials between conventional gas and RFG have become almost negligible.[a] Moreover, since RFG is mandated for certain areas, the element of choice for consumers is eliminated.

RFG consists of petroleum mixed with a cleaner-burning fuel such as ethanol or methyl tertiary-butyl ether (MTBE). Although peak ozone levels in major U.S. metropolitan areas decreased through the late 1990s, a review of the RFG program indicated that little of the reduction seemed to be attributable to use of RFG, and in the case of ethanol, may actually increase O_3 levels (NRC, 1999). Ethanol was first utilized in select markets in the Midwest as the favored additive in RFG over the more effective and generally less costly MTBE. Ethanol use benefited many Midwestern farmers and, because of low transportation costs (if produced and consumed locally), the price differential made it roughly competitive with MTBE for those markets. More recently, MTBE has been determined to pollute local water resources, so many in the industry have been moving away from MTBE due to liability issues and states have taken action to phase out and at least partially ban its use (EIA, 2006a).

---

[a]According to EIA statistics, when introduced in 1995, price differential between RFG and conventional gasoline was greater than 60 cents per gallon. By 1999 the price disparity had settled slightly above 20 cents per gallon, and has decreased since. Recent figures for 2006 show price differentials of less than 10 cents per gallon, representing a 3-5 percent difference from conventional gasoline.

Although ethanol currently accounts for approximately 90 percent of total biofuel production, world biodiesel production has been increasing rapidly. Biodiesel can be produced from waste grease and recycled cooking oil, and thus it provides opportunities to recycle waste materials, though it can also be produced from dedicated crops, such as soybeans. A recent study indicated that a 20 percent biodiesel blend (B20) has no net effect on NO_x emissions; earlier studies had pointed to an increase in NO_x emissions (NREL, 2006).

The National Biodiesel Board (NBB) estimates that U.S. production of biodiesel reached 75 million gallons in 2006, compared to 25 million gallons produced in 2004. The NBB estimated in October 2006 that total production for 2006 would triple yet again, approaching 250 million gallons, aided in large part by state and federal tax credits and grants. U.S. on-highway use consumed 37.1 billion gallons of diesel in 2003 and, at that level of consumption, 2005 biodiesel production represents 0.2 percent of supply. NBB estimates that current U.S. biodiesel manufacturing capacity is 290 million gallons per year—180 million gallons from dedicated biodiesel plants and 110 million gallons within the oleochemical industry (NBB, 2005).

China's biodiesel production represents less than 10 percent of its overall biofuel supply (less than 30 million gallons). Most of this is currently being utilized as industrial fuel oil, particularly in areas lacking access to conventional petroleum-based fuels. Primary feedstocks are rapeseed and soybeans, but both produce edible oils and thus their potential for expanded production is limited. Restaurant waste resources are small and might serve niche markets (e.g., a city's public transportation system), but jatropha (an inedible tree-grown oil seed) has shown promise, particularly in southwest China, where it grows rapidly on marginal lands (Wu et al., 2007).

Taken together, the combined capacities of the U.S ethanol and biodiesel industries are approximately 95 million boe (0.6 EJ) per year. However, in 2004, U.S. petroleum consumption for transportation was approximately 13.86 million barrels per day, or just over 5 billion barrels (31 EJ) per year (EIA, 2006a). Thus, the current biofuels industry provides less than 2 percent of annual U.S. consumption for transportation, or about 1 percent of total petroleum consumption. ORNL projected that by 2030, the United States could sustainably produce over 1.3 billion tons of biomass per year, measured on a bone dry ton (bdt) basis (Figure 7-8), and this would be sufficient feedstock to produce about one-third of U.S transportation fuels. The total resource potential is based on an increase of over seven times current biomass production levels. However, the authors believe that the 1.3 billion tons can be produced with relatively modest changes to land use and agricultural and forestry practices.[4]

[4]The values in the report should not be thought of as upper limits, but just one scenario for a set of assumptions. Over the coming years, significant additional research will be undertaken.

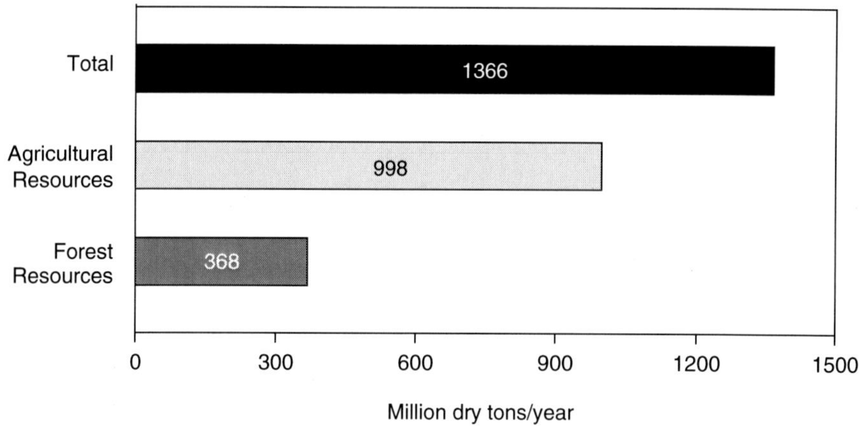

FIGURE 7-8 U.S. annual biomass potential.
SOURCE: ORNL, 2005.

CHALLENGES TO INCREASING RENEWABLE ENERGY USE

Renewable energy generation is expected to grow in China and in the United States in the future. In the United States, total renewable generation, including CHP and end-use generation, is projected to grow by 1.5 percent per year, from 357 billion kWh in 2005 to 519 billion kWh in 2030 (EIA, 2007). China's estimates are less certain, but the government's goals are a generation capacity of 30 percent renewable sources by 2020, supplying 400-500 Mtce (12-15 EJ), accounting for 15 percent of primary energy consumption (NDRC, 2007). In both countries, hydropower will continue to be the dominant renewable source for electricity production, though wind-power generation is expected to enjoy the largest annual increase in terms of percentage. Nevertheless, a realistic technological and economic assessment of the characteristics of these options for generating electricity reveals challenges, as well as the more advertised opportunities.

Most major hydroelectric capacity options in the United States have long since been exploited, and proposals for new or expanded hydro facilities often encounter resistance. Further, hydro facilities can have serious negative consequences for land use and regional ecosystems, particularly fish populations.

Wind energy conversion systems (WECS) present two major problems. First, wind is intermittent and unreliable, and basically operates in a fuel-saver mode. WECS has a capacity factor of about 20 percent, and without adequate storage and/or conventional power generation backup, wind is simply not a long-term, viable option. Second, wind is not entirely without environmental problems. Aside from aesthetic and noise complaints, avian mortality is a challenge, and it is noteworthy that the National Audubon Society has successfully intervened to prevent the development of some wind projects in California due to the danger they pose to endangered condors.

Biomass power has the great advantage of being able to supply reliable, base load power and liquid fuels. However, it has at least three major concerns to address. First, virtually all current biomass plants in the United States use waste from the wood products industries as feedstock. But to become a major source of electricity generation, a large number of dedicated silviculture biomass plantations will have to be developed and maintained to provide fuel for closed-loop biomass systems. Such plantations have yet to be successfully developed, and the commercial and economic feasibility of such plantations has not been established. Second, a prodigious amount of land may be required for biomass energy plantations. This poses challenges in terms of not only land use, but the water use and ecosystem impacts it entails.

Biomass combustion is also an environmentally degrading energy source. Though it is widely used in co-firing operations (along with coal) in order to reduce SO_2 emissions, it still emits various other criteria pollutants, notably $PM_{2.5}$. Thus the vastly increased use of biomass combustion to generate electricity in an increasingly emission-constrained environment must be carefully assessed. Biomass gasification significantly reduces air pollution emissions and provides opportunities to extract high-value chemicals and other by-products, but it is not yet commercial in the United States or in China, and it will face challenges in terms of feedstock uniformity and gas cleanup.

The remaining two renewable energy options for large-scale electricity generation are central receiver solar thermal power and utility-scale PV. Both have serious, unresolved issues:

- Solar PV has not yet been proven technologically viable for substantial, long-term, reliable electricity generation.
- Both are far from being cost competitive.
- Both suffer from intermittency and unreliability problems worse than for even those of WECS (e.g., the wind may blow at night but the sun does not shine at night).
- Due to their high solar radiation requirements, both options are unsuitable for the climatic and weather conditions in much of the United States.
- The land requirements for both options are substantial, and for 1,000 MW could range between 40,000 and 70,000 acres.

In addition, the environmental effects of the large amounts of materials required in solar thermal and PV systems are, at present, incompletely understood. For example, there are substantial environmental, health, and safety hazards associated with the manufacture, use, and disposal of solar PV cells. Some feedstock materials used in PV cells are toxic, carcinogenic, pyrophoric, or flammable, and the actual hazards to health posed by these materials depend on their inherent toxicological properties and the intensity, frequency, and duration of human exposures. Widespread utilization of PV technologies, such as the installation of 130 km^2 of PV cells to approximate a generic 1,000 MW power plant, will require that serious attention be given to these hazards as they relate to the sources, processing, usage, and end-of-product-life disposal.

Overall, costs and distribution of resources are two overriding challenges to wide-scale use of renewable energy resources. With regard to costs, wind power is an exception, as it has become cost-competitive in some markets without the help of subsidies. However, Figure 7-9 projects that levelized costs would stay above avoided costs for other sources of power generation through 2015 (aside from limited areas with geothermal resources), and that solar thermal generation may still be prohibitively costly in 2030 (EIA, 2007). Due to restructuring in the U.S electrical power sector, more technology R&D will likely need to be underwritten by state and federal governments (NRC, 2000).

Programs exist in many U.S. cities which allow individuals and commercial businesses to purchase electricity from renewable resources, albeit at a slightly higher cost, referred to as a green pricing program. A 2000 National Research Council (NRC) report suggested that the U.S. DOE could play a role in encouraging a public demand for "green power" (NRC, 2000). As of 2004, U.S. electrical

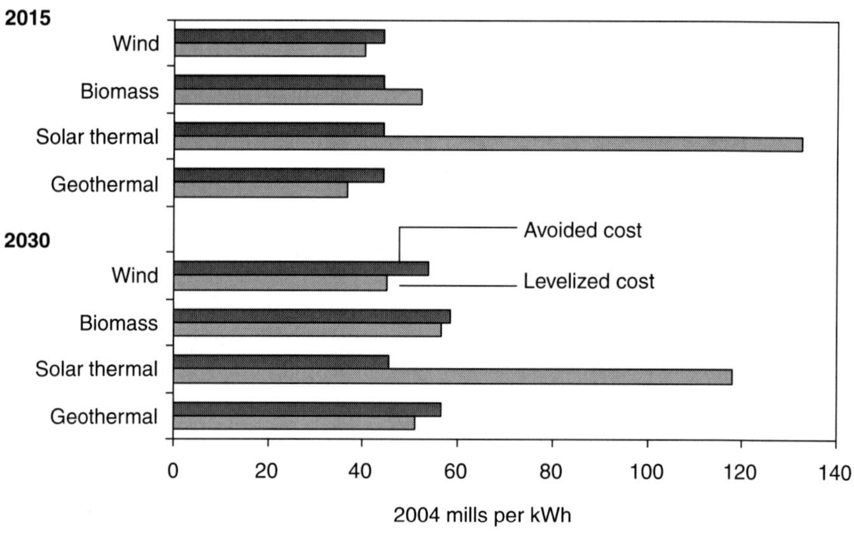

FIGURE 7-9 Levelized and avoided costs for new renewable plants in the northwest, 2015 and 2030.
SOURCE: EIA, 2007.

industries reported 928,333 customers participating in green pricing programs, and 93 percent of these customers were residential customers (EIA, 2006b). Shanghai has experimented with a similar program, the "Shanghai Green Power Program." However, renewable power generation accounted for only 0.02 percent of Shanghai's total generation, and with costs nearly double that of coal-generated electricity, the program has received only limited support (CASS, 2006).

Distribution of resources is another difficulty, particularly in China's case. There is a reverse distribution of the most promising renewable energy resources, relative to population. This is beneficial for remote areas in early stages of urbanization, for they have an opportunity to meet their rapidly increasing energy needs with clean renewable sources. However, renewable resources appear to have far fewer large-scale applications in eastern and coastal China, home to the majority of the population. Nonetheless, smaller-scale applications can supplement and thereby reduce coal consumption in cities, as well as create opportunities for local industries (e.g., manufacturing or biorefineries).

Domestic production capacity for renewable energy technologies is an important consideration. The international market provides increasing opportunities for renewable energy technologies, both as a source of imported technologies, as well as a larger market for domestically produced technologies (NRC, 2000). In some instances, China is already able to satisfy local demand and is poised to enter the international market. However, in other instances, it could benefit from international cooperation (Zeng, 2005). The Clean Development Mechanism (CDM) provides one such opportunity to foster international collaboration on renewable energy technologies. CDM projects, a result of the Kyoto Protocol to reduce greenhouse gas emissions, have been undertaken in China since 2002, and investments totaled nearly US$1 billion through 2005 (CASS, 2006). The CDM is an opportunity for developed countries to work towards meeting their reduction goals by investing, generally at lower cost, in projects to reduce emissions in developing countries, relative to a business-as-usual scenario. While the United States has not ratified the Kyoto Protocol (as of 2007) and is therefore not eligible to partake in CDM projects, it nonetheless has a history of cooperating with China to develop renewable energy resources (Box 7-2).

Hydrogen

As a resource, hydrogen is the third most abundant element on earth, but most of this is contained within H_2O and organic compounds and thus must be extracted for use as a fuel. Most hydrogen is currently produced by applying heat to natural gas to extract the hydrocarbons. While effective for limited applications, given the high price of natural gas, this does not appear to be feasible as a long-term strategy, should hydrogen become a more commonly used fuel source. Electrolysis is another means to obtain hydrogen and, in this process, an electrical current separates water into its elemental components. This process is plagued

BOX 7-2
United States-China Renewable Energy Development and Energy Efficiency Protocol

Renewable energy collaboration between the U.S. Department of Energy (DOE) and China's Ministry of Science and Technology (MOST) began in early 1995. In December of 2006, the protocol was renewed to continue to facilitate programs that create and implement energy efficient and renewable energy technologies using "solar, wind, biomass, geothermal, and hydrogen energy" (DOE, 2006a). Five sections of the protocol relate to renewable energy. First, a plan for rural, village-scale renewable energy projects focuses on providing rural areas throughout China with energy and electricity. The project's preliminary steps have begun in rural villages of China, where these energy needs are being met by companies such as the Asia Pacific Economic Cooperation and the Tibet Solar Electrification Project. Another aspect of the protocol relates to wind energy development that includes both grid-connected and off-grid power. A pilot project that uses a "wind/diesel/battery systems" is currently being tested and has the ability to provide electricity to 120 households on the island of Xiao Qing Dao in the Yellow Sea (DOE, 2006b). The protocol has also encouraged United States-China cooperation on a series of workshops and outreach activities that facilitate U.S. renewable energy companies' interests in doing business in China. Additionally, geothermal production has increased as a result of United States-China collaborative efforts to provide a market for this energy source in China. Geothermal production occurs when high temperatures are used to generate electricity and low to medium temperatures are needed for heating/cooling. The protocol has also increased the number of programs that promote renewable energy policy and planning. Projects, such as the National Township Electrification Program, gather policy makers and technical experts from around the world to discuss how to provide electricity to townships and villages in China that are currently lacking.

by inefficiencies, particularly if the initial electricity is produced by fossil fuel combustion, which is often the case. More efforts are needed to extract hydrogen using renewable energy (e.g., wind), thus realizing the full benefits of hydrogen as a "clean" fuel. Hydrogen from coal is yet another means to utilize an abundant domestic resource while significantly reducing emissions. Much of the research taking place is a result of public-private RD&D in cooperation with the U.S. DOE's National Energy and Technology Laboratory.

Hydrogen as a fuel source is appealing because it holds the potential to replace gasoline in the transportation sector, exclusively using domestic resources, thereby significantly reducing dependence on foreign oil. In addition, it can potentially eliminate almost all criteria pollutants and greenhouse gases from vehicular emissions. These factors led the United States to announce in 2003 its intention to transition to a hydrogen economy. In some instances, notably in California, states have pushed ahead with this initiative. California is currently developing the Hydrogen Highway Network, using partnerships between government and the private sector. This Hydrogen Highway would be comprised of hydrogen fueling stations located approximately every 20 miles along state and federal highways, effectively giving all Californians access to hydrogen for fuel-cell vehicles by 2010. This is an ambitious target and well ahead of national forecasts of a transition taking place by 2050.

In order for hydrogen to become a viable alternative fuel source and to replace fossil fuels, it must overcome a number of challenges. First, it must prove to be cost-effective. Hydrogen is currently produced at reasonable cost for industrial purposes, but large-scale use would require production using renewable energy, which is still often prohibitively costly. Transmission and storage are key cost issues as well; costs increase drastically when the H_2 must be distributed over dispersed locations (NRC, 2004). Therefore, it is difficult to imagine a centralized network for production and distribution, particularly in the short term.

REFERENCES

Bai, X. 2007. Rizhao: Solar-Powered City. State of the World 2007: Our Urban Future. Washington, D.C.: World Watch Institute.

CARD (Center for Agricultural and Rural Development). 2006. Policy and Competitiveness of U.S. and Brazilian Ethanol. Iowa Ag Review Online 12(2), retrieved from http://www.card.iastate.edu/iowa_ag_ review/spring_06/article3.aspx.

CASS (Chinese Academy of Social Sciences). 2006. Understanding China's Energy Policy: Economic Growth and Energy Use, Fuel Diversity, Energy/Carbon Intensity, and International Cooperation. Background Paper prepared for Stern Review on the Economics of Climate Change.

CREIA (Chinese Renewable Energy Industries Association). 2007. http://www.creia.net/ (accessed January 30, 2007).

DOE (U.S. Department of Energy). 2006a. U.S. and China Announce Cooperation on FutureGen and Sign Energy Efficiency Protocol at U.S.-China Strategic Economic Dialogue.

DOE. 2006b. Fact Sheet: Cooperation with the People's Republic of China.

Editorial Board of the China Electric Power Yearbook. 2005. China Electric Power Yearbook. Beijing: China Electric Power Press.

EIA (U.S. Energy Information Administration). 2006a. Annual Energy Review 2006. Washington, D.C.: U.S. Department of Energy.

EIA. 2006b. Renewable Energy Annual 2004. Washington, D.C.: U.S. Department of Energy.

EIA. 2007. Annual Energy Outlook 2007. Washington, D.C.: U.S. Department of Energy.

Farrell, A.E., R.J. Plevin, B.T. Turner, A.D. Jones, M. O'Hare, and D.M. Kammen. 2006. Ethanol can contribute to energy and environmental goals. Science 311(5760):506-508.

Kitani, O. and C.W. Hall, eds. 1989. Biomass Handbook. New York: Gordon and Breach Science Publishers.

NBB (National Biodiesel Board). 2005. U.S Biodiesel Production Capacity.

NBS (National Bureau of Statistics—China). 2005. China Statistical Yearbook. Beijing: China Statistics Press.

NDRC (National Development and Reform Commission). 2005. Overview of Renewable Energy Development in China: Recent Progress and Future Prospects.

NDRC. 2006. Report on the Development of the Photovoltaic Industry in China. China Renewable Energy Development Project Office.

NDRC. 2007. Medium- and Long-Term Renewable Energy Development Plan.

NREL (National Renewable Energy Laboratory). 2006. Effects of Biodiesel Blends on Vehicle Emissions, Fiscal Year 2006 Annual Operating Plan: Milestone 10.4 Report, October. Washington, D.C.: Office of Energy Effficiency and Renewable Energy.

NRC (National Research Council). 1999. Ozone-Forming Potential of Reformulated Gasoline. Washington, D.C.: National Academy Press.

NRC. 2000. Renewable Power Pathways: A Review of the U.S. Department of Energy's Renewable Energy Programs. Washington, D.C.: National Academy Press.

NRC. 2004. The Hydrogen Economy: Opportunities, Costs, Barriers, and R&D Needs. Washington, D.C.: The National Academies Press.

ORNL (Oak Ridge National Laboratory). 2005. Biomass as Feedstock for a Bioenergy and Bioproducts Industry: The Technical Feasibility of a Billion Ton Supply, prepared under contract to the U.S. Department of Energy and the U.S Department of Agriculture.

RFA (Renewable Fuels Association). 2007. http://www.rfa.org.

USDA (U.S. Department of Agriculture). 2006. Ethanol Reshapes the Corn Market. Amber Waves. http://www.ers.usda.gov/AmberWaves/April06/Features/Ethanol.htm.

Wu, C.Z., H. Huang, S.P. Zheng, and X.L. Yin. 2002. An economic analysis of biomass gasification and power generation in China. Bioresource Technology 83:65-70.

Wu, C.Z., X.L. Yin, Z.H. Yuan, Z.Q. Zhou, and X.S. Zhuang. 2007. The development of bioenergy technology in China. Int. J. Energy, in press.

Zeng, P.Y. 2005. Keynote Remarks on Beijing International Renewable Energy Conference, November.

8

The Pittsburgh Experience

HISTORY

Pittsburgh is located in a region rich in coal and is also near where oil was first discovered in the United States. Pittsburgh has perhaps the longest documented history of air pollution and of efforts to ameliorate air pollution of any city in the United States. With an abundant supply of coal, used for fuel since the mid-18th century, Pittsburgh established itself as the "smokiest" city in the country, although this was not always considered unpleasant. One difficulty in addressing air pollution was that it was actually thought to be a sign of bountiful resources and industriousness. Presence of these natural energy supplies defined Pittsburgh's evolution.

By the early 19th century heavy industry picked up rapidly as Pittsburgh suffered from competition in other sectors such as trade. In particular, the abundance of coal gave rise to the industry soon to be synonymous with Pittsburgh: steel production. Producing steel requires a great deal of energy, which at the time could readily be provided by coal. Further, coke, used in the smelting process and therefore an important resource for metal industries, is made from coal. Steel production in Pittsburgh started in the 1700s, and by 1820 had become the major industry in the city. The resultant air pollution was viewed as requisite to the city's economic survival. Soon, not only did the industry dominate the city, but also the city dominated the industry, taking advantage of the region's coal and also the ability to transport steel via its rivers and railroads. Myriad steel mills and coke ovens would soon line Pittsburgh's three rivers, particularly the Monongahela, stretching from below Pittsburgh, upstream past the smaller mill towns that developed near the plants and up the smaller valleys leading in to the larger river valley. By the turn of the century, some of the wealthiest individuals in the world lived in Pittsburgh, including steel barons such as Andrew Carnegie and bankers such as Andrew Mellon, a major investor in the growth industries of coke, coal, and iron.

Air pollution was a particular problem associated with steel production. As detailed below, smelting and refining iron using coal can and did produce significant amounts of soot. Sulfur in the coal and iron and zinc ores produce sulfur dioxide, and combustion can lead to NO_x formation. Coking leads to large emissions of carbon monoxide and air toxics such as benzene. As steel production increased, so did air pollution. While the rivers were vital for transporting coal, the river valleys had a negative impact on air quality, and during periods of atmospheric stagnation, the factories' emissions were trapped. Office workers would take two shirts to work as one would get discolored during the day. Because of the dark polluted air, Pittsburgh famously used its streetlights from morning right on through the night; photographers capturing downtown in the early 20th century noted the time of day to illustrate this point, as Figure 8-1 indicates.

FIGURE 8-1 Photograph taken at 9:20 am in downtown Pittsburgh, 1946.
SOURCE: Carnegie Library of Pittsburgh.

Ordinances pertaining to chimney height date back to Pittsburgh's founding in 1816. Although they are linked to abating the unpleasantness of air pollution, it was not until the 1860s and 1870s that air pollution became a worrisome issue (Mershon and Tarr, 2003). While lawsuits were filed seeking redress for damage caused by air pollution, the Pennsylvania Supreme Court continually ruled in favor of industry. It was not until the 1880s that air quality temporarily improved. From 1884 to 1892, the city enjoyed a brief respite from the smoke, while natural gas replaced coal as the city's major fuel. By the end of the 19th century, however, the gas reserves were depleted. It was during this period that the Ladies Health Protective Association was formed to combat smoke, among other health-related environmental issues; and newspapers began targeting air pollution as not just an unpleasant circumstance but as an all-out health risk. In 1911, the Mellon family founded an institute of industrial research at the University of Pittsburgh, which carried out a number of smoke studies.

In October of 1948, a particularly severe stagnation episode occurred, lasting 4 days (Lipfert, 1994a). While the impact was more widespread, it had a particular effect on Donora, a mill town on the Monongahela River, 15 miles from Pittsburgh. Zinc is an ingredient in some steels; Donora was home to about 14,000 inhabitants and the Donora Zinc Works. As the stagnation episode continued, pollution built up, and the impact on health became noticeable. People started to experience severe respiratory problems with some deaths. Doctors recommended evacuation of those with respiratory problems, though this was hampered by traffic congestion and the severe smog. By the end of the episode, about 20 deaths occurred and about 7,000 additional individuals were ill. This episode became a catalyst for action.

As a result of actions taken in the years prior to the incident, Pittsburgh was not so severely affected as Donora. Pittsburgh had gotten a jump on the problem. In 1945, the mayor of Pittsburgh, along with the city elite, had begun to identify actions to improve air quality. Actions included reducing the use of bituminous coal as part of a 1941 smoke ordinance (Tarr, 1981). Natural gas was piped into homes for heating. Diesel engines began replacing coal-fired engines in locomotives and riverboats.

Socially and economically, Pittsburgh grew up as a steel town, and in the 1800s attracted European immigrants to work in the mines and mills. Initially, such jobs did not pay well, leading to attempts at unionization and resulting in strikes, and deadly conflict[1] (Tarr, 2002). As unionization took hold early in the 20th century, pay improved, and a rising middle class emerged. The region's economy suffered badly during the 1970s as steel prices plummeted. While some have suggested that the shuttering of the mills was related to the environmental regulations of the 1970s, in fact the mills used older technologies, and newer

[1]The most noteable was the Homestead Strike of 1892, one of the most serious disputes in U.S. labor history. Fighting broke out between union steelworkers and private security guards, resulting in deaths on both sides, severely tarnishing millowner Andrew Carnegie's legacy.

mills in Japan, and elsewhere, used less labor and had a less expensive workforce. (Mills in Japan suffered a similar fate a couple of decades later, and were displaced by steel manufacturing in other countries with a less expensive work force and newer mills.) Unemployment skyrocketed and the region was forced to find alternatives.

Today, while steel is still a component of the region's economy, it is no longer so dominant. Coal continues to play a major role in the region as the primary fuel for producing electricity, and both the coal and electricity are used locally and exported. The rivers are still used for transport, but are not as central in that role. Oil production is virtually gone. Pittsburgh has developed a more diverse economy, in particular by making a push in high-technology areas such as robotics.

Air pollution in Pittsburgh is not nearly as severe as it once was, and is more regional in scale (Bergin et al., 2005; Farrell and Keating, 2002; Millet et al., 2005; Polidori et al., 2006). Ozone and acid deposition are major concerns, along with particulate matter (including soot), though the composition and levels have changed.

PHYSICAL, ECONOMIC, AND SOCIETAL SETTING

Pittsburgh lies in southwestern Pennsylvania, where the Monongahela and Allegheny Rivers combine to form the Ohio River. Like many historically industrial cities in the United States, Pittsburgh developed in a river valley, surrounded by forests. The Allegheny Mountains lay to the east, and continue southward, with the Great Lakes to the north and west. Topography has had an effect on Pittsburgh's air quality: since the city sits at the confluence of the three rivers, the majority of the population resides in the valley or at the base of the slopes of the outlying hills and mountains. Weather patterns sometimes cause polluted air from Pittsburgh and from power plants in the Midwest or industrial centers on the East Coast to hover over the region. While coal is no longer mined within the city, mining operations are still quite active in the region, particularly in West Virginia to the south.

Pittsburgh still suffers from severe atmospheric inversions, though emissions are not so high that these episodes raise major alarms (this is not to say that there are no health impacts), nor are street lights needed during the day. The fact that the region lies in an air pollution transport corridor causes greater problems today. Winds typically blow from west to east (approximately), leading Pittsburgh to be impacted by emissions from states such as Ohio, Illinois, Indiana, and beyond, and emissions are transported from the region to the heavily populated Atlantic coast. In particular, a relatively frequent occurrence during the summer is a "Bermuda high" leading to warm temperatures in the north-central to north-eastern states, with air masses slowly moving over the region towards the coast. High temperatures can increase emissions of volatile organic compounds (VOCs), particularly from trees. Along with NO_x and SO_x emissions from automobiles and power plants, the resulting air mass is not only warm, but laden with ozone and sulfate. As it moves slowly eastward, these pollutants can build up. Summer smog has replaced winter stagnation episodes as the major concern.

In comparison with its historical focus on the steel industry, Pittsburgh now has a more diverse economy. Steel is not gone, as a few plants have modernized and now produce specialty steels. Approximately 10 percent of the region's workforce is still employed in manufacturing. But health services and education now employ about 20 percent of the workforce. High-tech industries such as robotics and informatics have developed from research activities at the local universities. The *Wall Street Journal* nicknamed the Pittsburgh region "Robo-burgh" in 1999, and in recent years the area has embraced that distinction, with more than 100 firms specializing in robotics (Kara, 2006). Pittsburgh's strong industrial and engineering heritage helped it become a leader in the field, with the Robotics Institute at Carnegie Mellon University providing much of the talent. Nevertheless, household income remains somewhat depressed; Pittsburgh's median annual household income was $31,910 in 2005, compared to a national median of $44,400.[2]

While much of the early growth in Pittsburgh was fueled by immigrants during the 1800s, the region added to the immigrant-based population by an influx of residents after the Civil War and World War II. The population now remains rather stable with relatively little emigration. Today, Pittsburgh has a population of nearly 340,000, down from a peak of more than 675,000 in 1950. Allegheny County has a population of 1.24 million, down somewhat from its peak of 1.45 million in 1980. There has been a shift in population from the urban core to the suburbs; in response the area recently started a limited subway service to complement the existing bus service.

While historically a large fraction of the population was employed as labor in the manufacturing sector, there was also a significant amount of local wealth. The families of the company owners and financiers lived in the area, most notably Andrew Carnegie of US Steel and Andrew Mellon of the Mellon Bank. Their wealth and influence has had a major, continuously evolving impact on the city. They started universities (Carnegie Institute of Technology and the Mellon Institute), headquartered their companies in the city (Pittsburgh, until the 1980s, was the headquarters of a large number of Fortune 500 companies), and they supported efforts to improve air quality. This, as mentioned earlier, may have helped save Pittsburgh from the same catastrophe that hit Donora.

SOURCES AND LEVELS OF AIR POLLUTION

Air pollution in the Pittsburgh area has changed in character over the last century, evolving from the localized smog plaguing coal-dependent industrial

[2]According to the U.S. Census Bureau, and American Community Survey, 2005.

cities to a more regional problem. America's post-war prosperity led to a tremendous increase in demand for electricity, with coal as a major fuel. Power plants sprang up in the region, again benefiting from the local coal reserves. While pulverized-coal power plants tend to emit relatively smaller amounts of soot (per kilogram of fuel burned) than other types of combustion, the high temperatures lead to NO_x formation. The sulfur in the coal also comes out of the stack as SO_2. Emissions of sulfur and nitrogen oxides from power plants grew throughout the latter part of the 20th century, though opacity regulations limited the amount of soot (EPA, 1991).

Coal, particularly soft coal, when burned inefficiently (e.g., for heating), led to significant emissions of soot, blackening the atmosphere. This was accompanied by less visible emissions of SO_2 and toxics. Coke production produced air toxics, in particular benzene. Smelting led to metal emissions. As the region cleaned up the industries, and when foreign competition shuttered many as well, such emissions dropped (Figure 8-2).

While factory emissions may have been decreasing, automobile use was increasing. Although generally not as visible, automotive emissions led to an

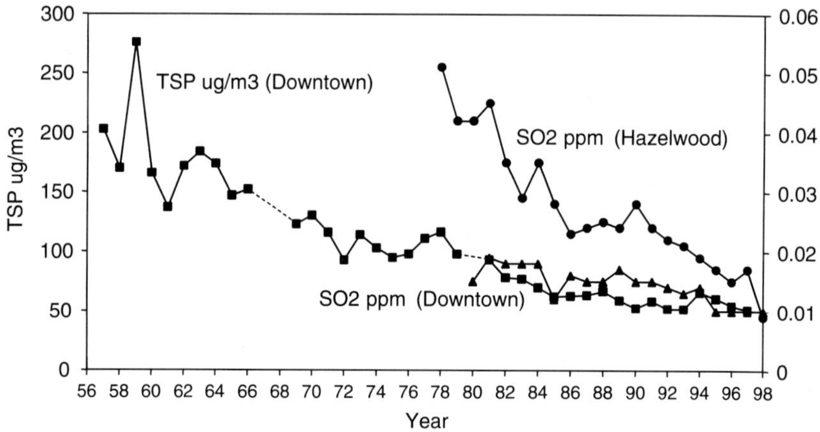

FIGURE 8-2 Annual arithmetic average concentrations of total suspended particles (TSP) and SO_2 in or near downtown Pittsburgh. The TSP measurements were made with high-volume samplers at two downtown locations: the County Office Building (1957-1982) and Flag Plaza (1983-1997), as part of the National Air Sampling Network and the Air Quality Program of the Allegheny County Health Department. Reliable data are not available for 1967, 1968, and 1980. The SO_2 measurements were made with continuous monitors at Flag Plaza downtown (1980-1998) and in the Hazelwood section of the city (1978-1998).
SOURCE: Reprinted with permission from Davidson et al., 2000.

increase in nitrogen oxides, carbon monoxide, and VOCs. Automobile use in the United States grew faster than the population, as did automotive emissions (Tables 8-1 and 8-2), leading to a new concern for air pollution control in the Pittsburgh region and in many other metropolitan areas.

Controls on automobiles and power plants have had a dramatic effect on emissions, and further reductions are planned. The regulations resulting from the Clean Air Act Amendments (CAAA) of 1990 were especially important for the Pittsburgh region. Title IV targeted acid deposition, and required significant reductions in sulfur dioxide. Title II required stricter automotive controls and Title I focused on ozone-forming emissions, although NO_x emissions were underemphasized as research was just coming out to suggest the importance of NO_x to regional ozone formation (Chameides et al., 1992; Milford et al., 1989, 1994; NRC, 1991; Sillman and Logan, 1987).

It was not only Pittsburgh's local air quality problem (though it was in nonattainment of the ozone standard), but also problems in the large cities along the eastern seaboard that led to greater consideration of the region's emissions. In the

TABLE 8-1 U.S. Population and Vehicle Use Trends, 1960-1980

	Year			Percent Increase	
	1960	1970	1980	1960-1970	1970-1980
Population[a] (millions)	178.6	203.3	226.5	13.8	11.4
Passenger cars[b] (thousands)	61,671	89,244	121,601	44.7	36.3
Vehicle miles traveled[b] (millions)	587,000	917,000	1,112,000	56.2	21.3
Fuel consumed[b] (million gallons)	41,171	67,819	69,982	64.7	3.2

[a]U.S. Census Bureau, based on resident population.
[b]U.S. Department of Transportation.

TABLE 8-2 Key Vehicular Emissions in the United States, 1970

	Highway Vehicles (1970)	United States Total (1970)	Percent of Total Emissions
CO emissions (thousand short tons)	163,231	204,043	80.0
NO_x emissions (thousand short tons)	12,624	26,883	47.0
VOC emissions (thousand short tons)	16,910	34,659	48.8

SOURCE: U.S. Environmental Protection Agency National Emissions Inventory 1970-2002.

1980s, virtually all of the major coastal cities from Washington, D.C. to Boston had ozone levels above the National Ambient Air Quality Standards (NAAQS), and they pointed the finger upwind, not only at the other coastal cities, but inland to the Midwest. Given the emerging research that showed that biogenic VOC emissions led to relatively rapid oxidation of NO_x, and that the upwind cities also had ozone problems, the claim was made that much of the ozone was being transported into their cities—therefore upwind sources needed to be controlled, whether the upwind areas were in attainment or not. Controversy around this issue led to the creation of a large ozone analysis program: the Ozone Transport Assessment Group (OTAG), with representatives from the affected states, industries, research organizations, and the U.S. and Canadian governments.

OTAG (Bergin et al., 2005; Farrell and Keating, 2002) and groups they funded analyzed available air quality data, conducting a major air quality assessment effort to understand ozone formation and transport in the eastern United States. They also assessed available emissions control options for the region. This effort also motivated other stakeholders to perform their own analyses. Via graphical and modeling results (Figure 8-3), OTAG convinced policy makers of the role of regional NO_x emissions, leading to what is commonly called the "NO_x SIP Call" in which EPA mandated state-by-state reductions in NO_x (Farrell and Keating, 2002). Like the acid rain provisions of the 1990 CAAA, the NO_x SIP Call used emissions trading to promote more efficient control choices than would be used under a traditional command-and-control policy. While aspects of how OTAG conducted parts of their studies were controversial, the program has led to emissions reductions throughout the eastern United States (Table 8-3), including the power plants around Pittsburgh, which installed low-NO_x burners and selective catalytic reduction (SCR) control devices, or switched fuels (e.g., to natural gas and cleaner coal). Both the NO_x SIP Call and the acid rain controls led to reductions not only locally, but upwind as well—thus the region's air quality has benefited, with reduced ozone, particulate matter, and acid deposition (Figure 8-4).

While concentrations have been coming down, Pittsburgh, like most of the region, still violates the new, stricter ozone standard (Figure 8-5), and the annual $PM_{2.5}$ standard. Recent data suggest that the area would not meet the new 24-hour standard as well (EPA, 2005, 2006). Recognizing that further regional controls are needed, the U.S. EPA proposed the Clean Air Interstate Rule (CAIR), which calls for further reductions in SO_2 and NO_x emissions from power plants. These controls are expected to bring the area into attainment with the ozone NAAQS, but not with the annual $PM_{2.5}$ standard (EPA, 2005). In January 2006, the Allegheny Board of Health approved an expenditure of up to \$840,000 to develop a SIP for $PM_{2.5}$. The Allegheny County Health Department (ACHD) is responsible for developing the SIP, which is due in April 2008 (ACHD, 2006b).

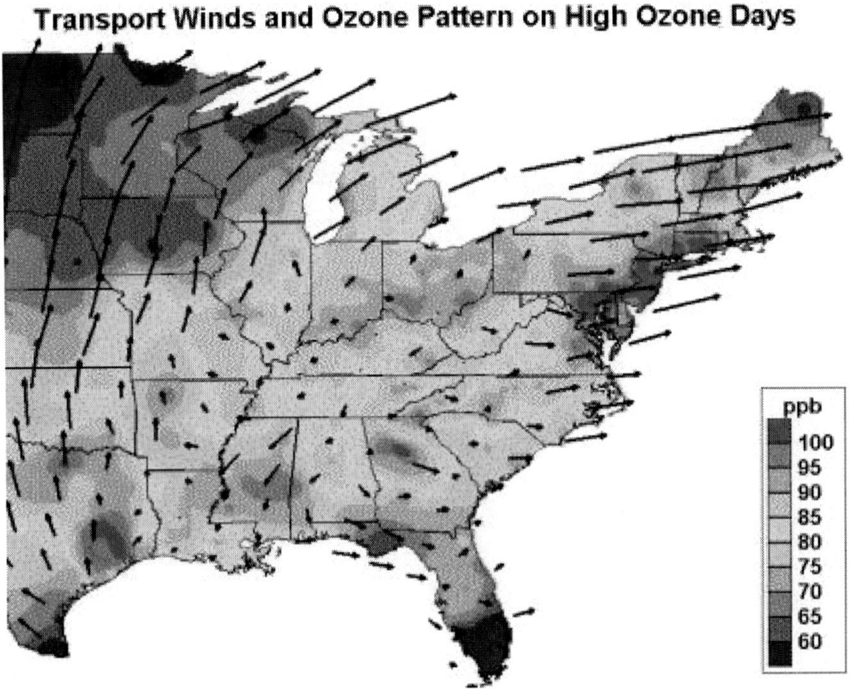

FIGURE 8-3 Ozone concentrations and wind vectors showing regionally high ozone levels across the Midwest and Northeast, and transport from Texas northward and the Ohio River Valley to the east. This graphical representation suggests the build-up of ozone as the air masses moved from the Midwest to the "Amtrak" corridor between Washington, D.C., and Boston.

TABLE 8-3 Estimated Point-Source Criteria Air Emission Change in Allegheny County (tons per year emitted or percent of 1996/1999 baseline year)

	Pollutant					
1996 Base Year	CO	NO_x	$PM_{2.5}$	PM_{10}	SO_2	VOC
(short tons/yr)	10,259	24,141	1,768	4,205	46,789	4,762
1997	–2.1%	2.3%		–0.7%	7.6%	2.4%
1998	–9.2%	–18.0%		–16.2%	–18.1%	–20.2%
1999	–12.7%	–20.2%		–24.1%	–8.0%	–21.3%
2000	–9.6%	–21.7%	–12.5%	–32.3%	7.3%	–30.9%
2001	–15.2%	–27.0%	–31.1%	–40.3%	16.0%	–40.2%
2002	–16.7%	–32.8%	–21.2%	–40.1%	0.9%	–44.6%
2003	–14.3%	–40.1%	–12.2%	–36.6%	8.7%	–48.1%

SOURCE: ACHD, 2004.

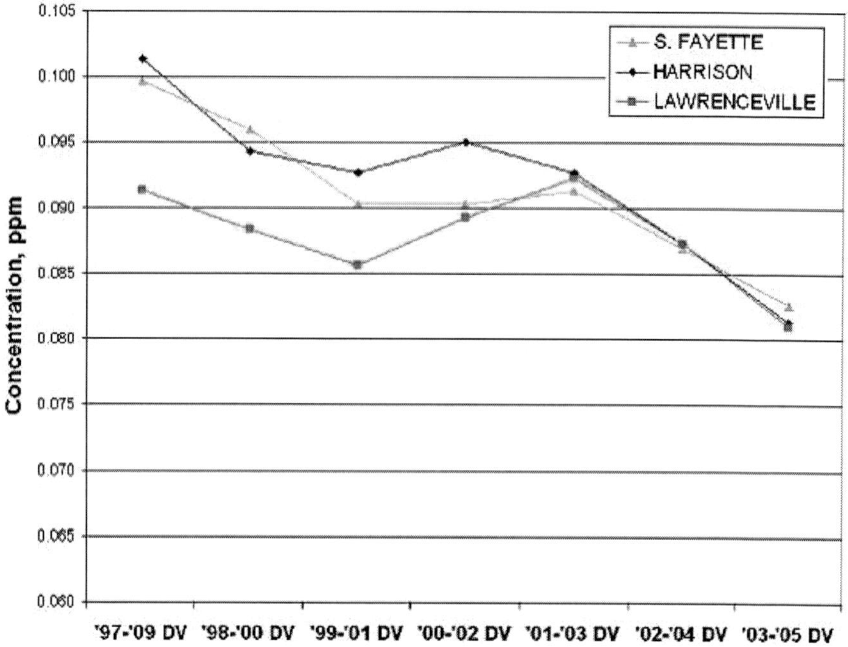

FIGURE 8-4 Ozone 8-hour design values, ACHD sites, 1997-2005.
SOURCE: ACHD, 2006a.

ENERGY RESOURCES AND USE

As noted earlier, a major source of energy in the Pittsburgh region continues to be coal, which is used primarily for electricity production (Figure 8-6). Coal accounts for about two-thirds of the total electricity produced, but is also used for heating (e.g., combined heat and power plants) and is still essential in coking for steel production.

Table 8-4 provides more detail on coal production and consumption in Pennsylvania. In spite of the abundance of coal, electricity is also produced by nuclear power at the Beaver Valley I and II units (about one-fourth of the total production). Indeed, Pittsburgh was the first city to benefit from nuclear power for electricity production. Oil is no longer produced in the area. Hydropower and other renewable forms of electricity production have been limited. Major uses of energy in Pennsylvania include transportation, domestic and commercial lighting, heating and cooling, and industrial production. Transportation relies on petroleum-based fuels, with limited use of alternatives (though gasoline does include some ethanol). Heating relies largely on natural gas, though fuel oil, electricity, and district heating are used as well (see Box 8-1). Thus, except for coal, much of the energy

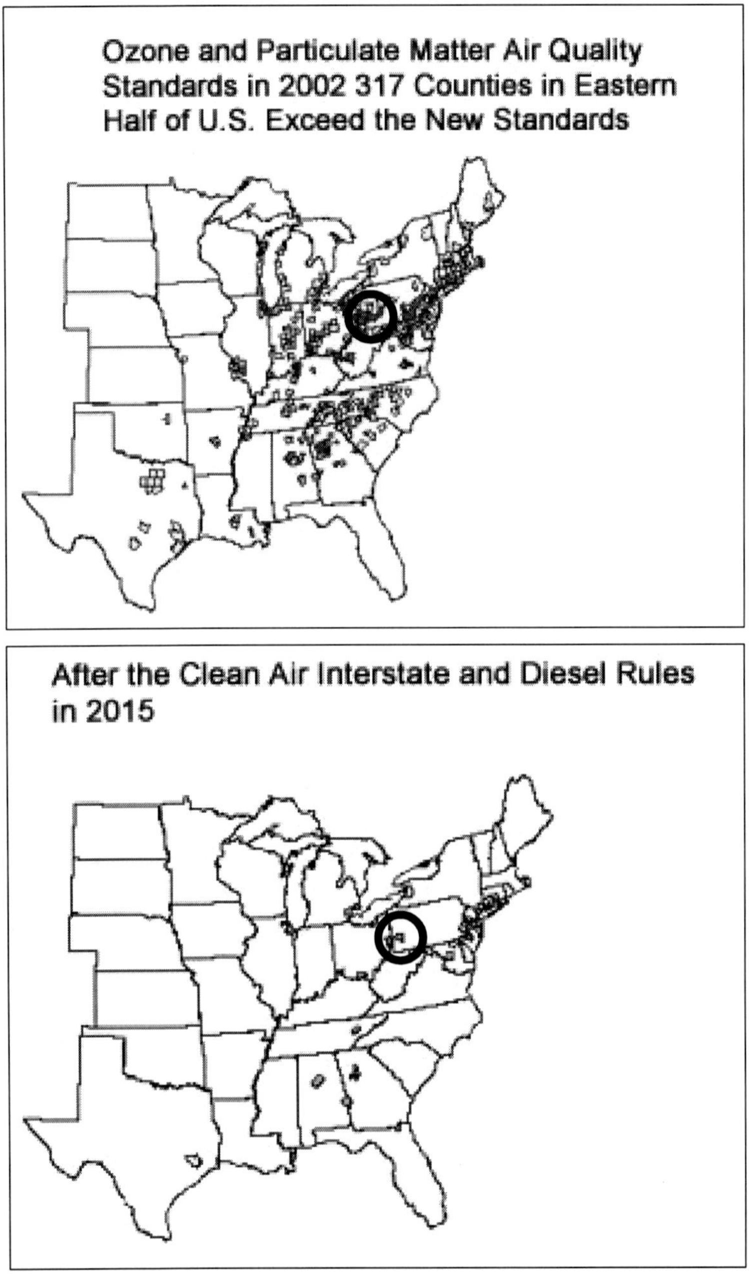

FIGURE 8-5 Areas in non-attainment in 2002, and projected non-attainment following implementation of CAIR.
SOURCE: EPA, 2005.

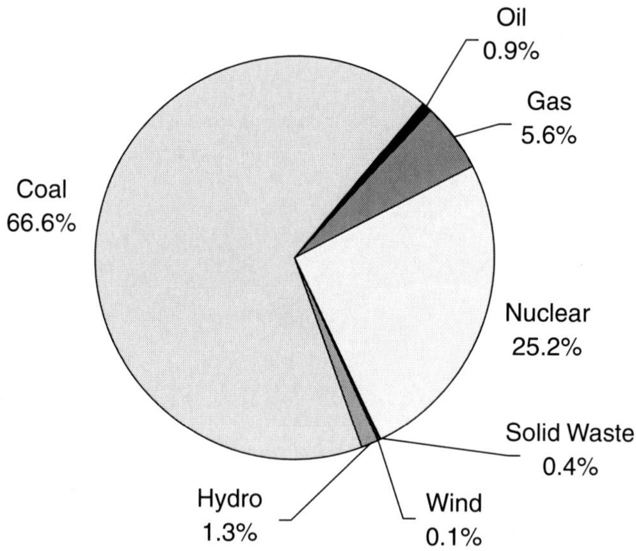

FIGURE 8-6 Regional electricity generation by energy source, 2005.
SOURCE: PPUC, 2006.

TABLE 8-4 Pennsylvania Coal Statistics

		Electrical Generation	Industrial Plants	Residential/ Commercial
Production[a]	**State Total, 2004**	47,728	11,425	796
	(thousand short tons)			
	By rail	23,761	718	53
	By water	14,968	940	29
	By tram, conveyor, or slurry pipeline	1,742	4,584	—
	By truck	7,257	5,183	715
	Used within PA	24,796	9,906	612
Consumption[b]	State Total, 2005	54,464	2,937	623
	State Total, 2004	51,698	3,337	680

[a]Distribution of U.S. coal by origin state (EIA, 2006).
[b]U.S. coal consumption by end use sector, by census division and state (EIA, 2006).

BOX 8-1
District Heating and the Bellefield Boiler Plant

The Bellefield Boiler Plant was built by Andrew Carnegie in 1907. Located in the Oakland section of Pittsburgh, Pennsylvania, it supplies steam for heating and refrigeration to the University of Pittsburgh, the Pittsburgh Medical Center, Carnegie Mellon University, The Carnegie Library, the Pittsburgh Board of Education, and other institutional sites (PDEP, 2006; CLP, 2003). The plant has six boilers that release exhaust to one of two stacks, which heat fuel oil, coal, natural gas, or a combination of coal/natural gas. The Bellefield Boiler Plant meets air pollution standards by using a mixture of 80 percent coal and 20 percent natural gas (CLP, 2003). Some of the boilers do not have emissions' controls in place, yet others have burners that are designed to lower the nitrogen oxide in natural gas production. The facility is a major source of air pollutants such as nitrogen oxides and carbon monoxide emissions, as well as a minor source of particulate matter, sulfur dioxide, and volatile organic compounds (ACHD 2004). Annually, over 60,000 tons of low-sulfur Kentucky coal is brought in daily during the winter months (CLP, 2003).

is imported to the region, and electricity is exported. Although renewables do not make up a large contribution of the Pittsburgh area's energy portfolio, the bulk of new power plants under consideration in Pennsylvania are wind-power plants (43 percent of total additional MW), while new coal plants would account for approximately 26 percent of additional MW (PPUC, 2006).

There currently is very limited city-wide data on energy consumption. As will be described shortly, a task force in 2006 was formed to compile much of this data, in order to help the city develop recommendations on reducing energy use and subsequent greenhouse gas emissions. However, the region is served almost exclusively by one energy company, Duquesne Light Company (Duquesne), and therefore Duquesne's data may serve as a proxy for residential, commercial, and industrial energy consumption (the notable omission is transportation). Total demand increased by 1.2 percent annually from 1990 to 2005, with projected increases through 2010 (Figure 8-7). Although no longer a power station operator, Duquesne was instrumental in the 1970s in implementing some of the first full-scale, plant-wide scrubber systems, as well as in putting fly ash to use as fill for highway embankments in and around Pittsburgh (Duquesne, 2006).

FirstEnergy is a major electric utility providing power in the region. The Bruce Mansfield Plant, located in Shippingport, Pennsylvania, is approximately 25 miles northwest of Pittsburgh. Although it does not supply much electricity to the Pittsburgh metropolitan area, it is nonetheless an important source of regional emissions. It is the largest coal-fired power plant in North America, with three

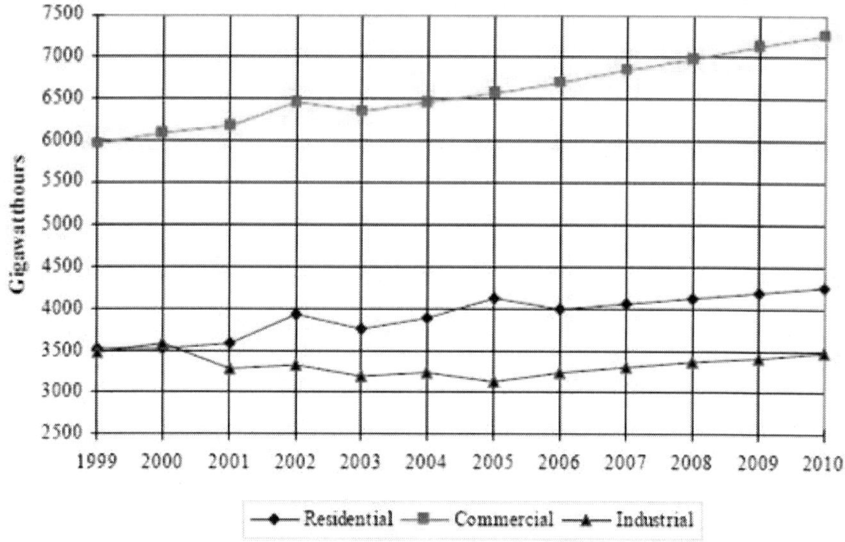

FIGURE 8-7 Duquesne Light Company historic and forecast energy demand.
SOURCE: PPUC, 2006.

boilers producing 2,360 MW of electricity by burning more than 7 million tons of coal annually (FirstEnergy, 2004). The plant is equipped with a suite of environmental control technologies; all units are equipped with SCR to remove NO_x, along with scrubbers and electrostatic precipitators (ESPs) to remove virtually all particulate matter and 92 percent of SO_2 (more than 400,000 short tons annually), with planned upgrades to increase this removal to 95 percent (FirstEnergy, 2005). This emphasis on environmental control technologies has not come without a cost, though. More than one out of every three dollars spent to build the $1.4 billion facility was spent on environmental protection, and similarly, one out of every three employees operates pollution-control equipment (FirstEnergy, 2004). However, the plant has devised a unique way to recover some of those costs, while continuing its efforts in environmental protection.

FirstEnergy developed a process to convert its scrubber by-product, calcium sulfite, into gypsum, a common ingredient in wallboard. Dubbed "North America's largest recycling project," the Forced Oxidation Gypsum plant was launched in 1999 and is able to provide nearly a half a million tons of gypsum each year, which is converted into enough drywall for 70,000 new homes (FirstEnergy, 2005). National Gypsum, the other half of this partnership, built a facility adjacent to the power plant, in order to take advantage of the steady supply of a low-cost raw material to manufacture its drywall. FirstEnergy benefits from the additional

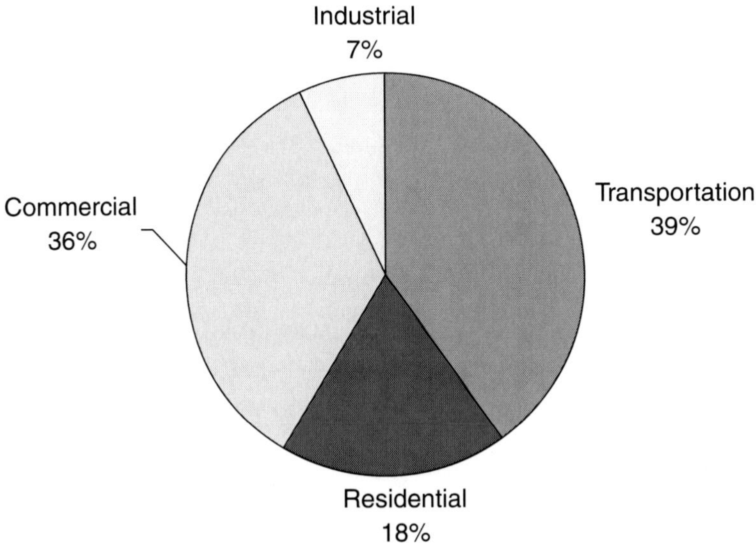

FIGURE 8-8 Pittsburgh's energy consumption by sector, 2003.
SOURCE: Heinz School Research Team, 2006.

revenue that its former "waste product" generates, as well as from the decreased need for waste disposal.

As mentioned earlier, Pittsburgh decided to establish a task force in order to better address issues such as energy efficiency and greenhouse gas emissions. In October 2006, Mayor Luke Ravenstahl initiated the Green Government Task Force, a 30-member panel charged with studying the city's energy use and developing recommendations for a local action plan (Heinz School Research Team, 2006). In order to identify appropriate policies for reducing consumption and emissions, the city must first be able to recognize where there are opportunities to increase efficiency and/or reduce consumption. Thus, led by a group of students at Carnegie Mellon University, the Task Force's initial activity was to assemble an inventory of the city's energy consumption (Figure 8-8). This represents the first comprehensive look at energy consumption for the region and will serve as a useful tool in guiding future decisions.

POLLUTION AND ENERGY POLICIES AND THE APPROACH TO AIR QUALITY MANAGEMENT

Early in the 20th century, efforts to reduce smoke in Pittsburgh met with only marginal success. This changed in the 1940s due to local political pressure on governmental agencies coupled with technological advances in combustion and changing fuels. In particular, "smokeless" fuels or combustion approaches were being developed and shown to be practical (Mershon and Tarr, 2003; Tarr et al., 1980; Tarr and Lamperes, 1981). In 1941, the mayor, who was up for reelection, appointed a Commission for the Elimination of Smoke, and noted that Pittsburgh must, in the interest of its economy, its reputation, and the health of its citizens, curb the smoke and smog. The Commission included industrial and civic leaders, the editor of a local newspaper, a doctor who headed the Pittsburgh Department of Public Health, members of civic clubs, and a member of the Board of Education, among others. In addition to the direct workings of the Commission, the newspapers provided support for anti-smoke regulations, and the Civic Club and League of Women Voters educated the community on the costs of smog, not only on health, but on welfare costs as well (e.g., increased soiling of clothes). Not surprisingly, the Commission's report calling for reduced use of bituminous coal was not fully supported by the local Coal Operators Association, which argued that it would cost jobs in the region. The Commission countered that it would enhance employment by increasing regional demand for the local coal. Furthermore, cleaner air would "bring about a new era of growth, prosperity and well being." The city council adopted a proposed ordinance in July of 1941, with implementation to start in October. As part of the ordinance, the Bureau of Smoke Prevention was formed inside the Department of Health.

Non-governmental organizations (NGOs) have played a major role in promoting air quality improvement in Pittsburgh. The Civic Club, which was founded in 1895 and which had representation on the original Commission, created the United Smoke Council (USC) in 1945. At the county level, the Allegheny Conference on Community Development (ACCD), established in 1943, had the mission of "overall community improvement," and took on smog as a concern. One sponsor of the organization was Richard K. Mellon, nephew of Andrew Mellon. USC merged into the ACCD in 1945; the combined group maintained pressure to continue reductions in smoke emissions.

Recognizing that smog did not obey city boundaries, and that many of the larger mills were located outside the city, the USC, ACCD, and others pushed for a county-wide smoke ordinance. This raised more opposition than regulations limited to just the city, from the mills and also the railroads. It required additional regulatory action, and in 1947, the state of Pennsylvania enacted a bill to give Allegheny County legal authority to regulate all sources of smoke in the county, including those passing through (i.e., the railroads). (While the transfer of this authority from the state to the county may seem minor, it prevails today, and it is unique in the United States.) This led to the formation of the Allegheny County Smoke Abatement Advisory Committee, which was headed by the executive director of the Mellon Institute with representation from various stakeholders, particularly the affected industries. The Committee was charged with developing a plan to reduce smoke in the region. Given the strong representation of the industrial stakeholders, the resulting plan was not as strict as the city's. Nonetheless, it did provide pressure to further improve air quality.

Today, the ACHD continues to lead the local monitoring and regulation of air pollutants, while the Pennsylvania Department of Environmental Protection's Bureau of Air Quality regulates at the state level and maintains a regional office in Pittsburgh. The Health Department has more than 40 staff working on air quality issues. Fifty percent of the department's funding comes from federal funds, 40 percent from emissions fees and about 10 percent from permit fees.

The Group Against Smog and Pollution (GASP), founded in 1969, is a leading NGO that now promotes cleaner air. GASP has had a number of influential members, and also draws expertise from the local universities. Sustainable Pittsburgh is a newer area NGO that focuses on transportation and land-use planning. Sustainable Pittsburgh has sponsored an annual smart growth conference since 2000, and has actively promoted the development of an integrated and advanced public transportation system for the metropolitan area.

As discussed in Chapter 4, in the United States, the federal government sets NAAQS that state and local governments are charged with implementing. The federal government also sets emissions standards for stationary sources and for new motor vehicles, oversees state and local air pollution control activities, and supports them through grant programs. State and local control agencies like the ACHD take the lead in planning for the attainment and for ongoing maintenance of air quality standards, conduct air quality monitoring, track emissions, and develop and enforce the construction and operating permits for sources. The ACHD operates 24 monitoring sites across the Pittsburgh metropolitan area. State and local agencies also undertake their own initiatives, including special monitoring, research, and control initiatives. For example, the ACHD is supporting school bus retrofits to curtail diesel emissions and a trade-out program for wood stoves, and is enforcing a local ordinance that limits idling of diesel equipment.

Source inspection and enforcement are critically important roles performed by state and local agencies. The ACHD has six inspectors who visit every major regional source of air pollution at least once per year. The U.S. Steel's Clairton Coke Plant, a 2.5-mile-long facility on the Monongahela River that is a major source of toxic air pollution and fine particulate matter impacting the nearby Liberty neighborhood is inspected daily. The ACHD has police powers, meaning it can levy fines against non-complying polluters, and in cases of egregious violations, it can impose criminal sanctions.

The ACHD works with the Pennsylvania Department of Natural Resources (PaDNR) on regional issues, and to develop State Implementation Plans (SIPs). The Clean Air Act mandates that states with areas found to be in violation of the NAAQS submit SIPs specifying how they will reach attainment, usually via local and regional control measures. Currently, the Pittsburgh area does not meet the NAAQS for ozone and $PM_{2.5}$.

As part of the effort to address regional haze that impairs visual air quality across the United States, five regional planning organizations (RPOs) have been established by states, tribes, and federal agencies. These groups are staffed largely by members of the state and tribal air quality management agencies in each of the regions. Pennsylvania belongs to the Mid Atlantic, Northeast Visibility Union (MANE-VU) covering Maryland northward and Pennsylvania eastward. MANE-VU was formed to coordinate regional haze-planning activities, with the additional goal of reducing visibility impairment in national parks and wilderness areas (MANE-VU, 2006). MANE-VU provides technical assistance and a forum for discussion, and encourages coordination with other regions as well. Although the RPO encourages a coordinated approach, the individual states retain the authority to set regulations.

Because of its geographic location, Pittsburgh must look beyond the PaDNR and MANE-VU to deal with ozone and PM non-attainment. Pittsburgh is just downwind of Ohio, which belongs to the Midwest Regional Planning Organization, and Ohio and other states further upwind have substantial emissions of nitrogen oxides and sulfur oxides that produce ozone and particulate matter. Hence the area also relies on more widely applicable regulations enacted by the EPA (such as CAIR [EPA, 2005]) for providing the needed emissions reductions.

Local universities have contributed significantly to air quality management in Pittsburgh by advancing scientific understanding of these problems. Pittsburgh has several colleges and universities, the largest of which are the University of Pittsburgh, a state-related university with 34,000 students, and Carnegie Mellon University, a private university with 10,000 students. Researchers at both universities have made significant contributions to understanding local, regional, and global air pollution issues. In particular, researchers at the University of Pittsburgh's School of Public Health and School of Medicine, including Herbert Needleman, Julian Andelman, and Bernard Goldberg, have made seminal contributions to understanding the health effects of air pollution. Professor Lester Lave at Carnegie Mellon University conducted a pioneering study that demonstrated the association between premature mortality and particulate air pollution (Lave and Seskin, 1973). From 2000 through 2004, Carnegie Mellon University hosted the "Pittsburgh Supersite," an intensive air pollution field study conducted with funding from the EPA and the U.S. Department of Energy (DOE). The objectives of the study were to better characterize the particulate air pollution in the Pittsburgh region, develop and evaluate new measurement methods, assess source contributions to pollution concentrations, and investigate relationships between pollutant levels and health impacts.

As mentioned earlier, GASP is a leading environmental advocacy organization active on air quality issues in southwestern Pennsylvania. GASP educates the public on air pollution issues and clean transportation and energy alternatives. The organization is currently working to help area school districts secure funding to retrofit school buses to reduce diesel engine emissions. GASP also has a successful track record of litigation, including working to enforce pollution-control requirements at the Clairton Coke Works, LTV Corporation's Hazelwood Works, and the Shenango Neville Island Coke Plant. Sustainable Pittsburgh is a more recently formed public policy advocacy organization working to advance relevant causes such as energy efficiency, support for public transporation, and smart growth. In addition to hosting convocations, Sustainable Pittsburgh compiles a regional biennial indicators report which incorporates indicators on air quality, energy consumption, and toxic emissions. Its 2004 report raised the issue of the need for comprehensive energy consumption data, and in general it highlights research needs and gaps (Sustainable Pittsburgh, 2004).

Pittsburgh's Green Building Alliance is another organization promoting environmentally responsible practices, and thus having an impact on energy consumption and air quality. The Green Building Alliance is involved, along with the regional NGO Clean Air – Cool Planet, and the International Council for Local Environmental Initiatives, in carrying out Pittsburgh's Climate Protection Initiative. The Green Government Task Force mentioned earlier will also be instrumental in the early stages of this initiative. Together, these organizations will assist local leaders in shaping policy, obtaining available funds (statewide and federal) for energy efficiency projects, and taking advantage of state and federal tax incentives for doing so.

LESSONS LEARNED

Pittsburgh's evolution provides several lessons about energy use and air pollution—some positive, some negative. As a region, it was able to capitalize on its abundant resources, though the abundance and attractiveness of those resources have varied over the years. Local oil is no longer a major contributor to the regional economy. Coal remains abundant, but its desirability as a fuel has decreased, due to the recognition of its impact on air quality and climate change. Damage to the environment from coal mining, as well as the health and safety of mine workers, also continue to be issues of concern (NRC, 2005). The cleaning up mines can be prohibitively expensive, leading to an environmental legacy lasting longer than the mine's period of operation. Such impacts and costs need to be considered.

One of Pittsburgh's initial successes in air quality management is the adoption of pollution-control measures in the 1940s, which may have helped the city avoid severe and acute episodes of pollution such as the episode that caused thousands of illnesses and numerous deaths in Donora, just a few kilometers away. As noted by the *Monessen Daily Independent*, "the Zinc Works may have cost the valley more jobs than it ever supplied, and the cost to the Donora-Webster area in terms of general community welfare is probably incalculable." The failure to control emissions from the Zinc Works may have delayed some costs, but a tremendous price was paid and the controls eventually were required anyway.

Individuals, often associated with NGOs, were critical in getting area governments to enact air quality regulations. In Pittsburgh, organizations such as the Civic Club and the League of Women Voters pushed for reducing smoke, bringing various stakeholders to join in the planning. It was through the efforts of such organizations that the city produced a report, which the Mine Workers and other coal industry and labor representatives signed, that said smoke elimination would "bring about a new era of growth, prosperity and well being," with "little or no additional burden on low income groups" (Tarr, 2002), thus facilitating progress in improving air quality. Now, GASP is a driving force. Such NGOs, at the local and national levels, continue to pressure for decreased emissions of air pollutants, including carbon dioxide.

Andrew Carnegie and Andrew Mellon started two companies that had a tremendous impact on the region. As well, one of their more lasting contributions appears to be the two educational institutions that they founded—the Carnegie Institute of Technology and the Mellon Institute. While the steel industry contributed to very prosperous periods, there were periods of significant hardship when that industry had downturns in the 1930s and 1980s. The most recent downturn in the steel industry appears final, as most of the mills have been razed and replaced by a diverse set of uses, including "high technology" companies. The region's economic transition has been tied to the research conducted at the local universities, leading to the diversification of the economy, with many of the new companies being relatively non-polluting. The region was very dependent upon steel through much of the 20th century. Consequently, the economic impacts were severe when Pittsburgh's steel mills were closed due to overseas competition. In part, Pittsburgh's mills were unable to compete due to the continued use of older technologies. It was not environmental controls that made the industry less competitive. The fate of Pittsburgh's steel industry suggests the need to continually update the technologies being used in manufacturing. Indeed, more modern facilities can be both more cost competitive and less polluting.

The region's coal-fired power plants are now required to retrofit their facilities to reduce emissions of sulfur and nitrogen oxides. These requirements are due in part to the regional nature of ozone and particulate matter pollution, as downwind areas have pushed for these reductions. As explained in Chapter 6, retrofitting is an expensive process and is generally brought about as a result of tight regulations and strict enforcement. In some cases, the need to retrofit leads to plant closures. The costs of retrofit controls argue strongly for taking more aggressive action to lower emissions as facilities are designed. Looking to the future, high levels of SO_x and NO_x reductions and measures to control mercury and CO_2 should be targeted by new facilities.

Pittsburgh has a unique regulatory structure for addressing air pollution problems, due in large part to the history of air pollution issues. Locally, the Allegheny Pollution Control Division takes charge. Having a local agency lead pollution-control efforts worked well for addressing pollution problems with a limited spatial scope. Local agencies were more intimately familiar with the types of sources, and were more responsive to local pressures. However, as the more local "smoke" problems were reduced, the need to address regional air pollution grew. Ozone, secondary fine particulate matter, and acid deposition are not as effectively addressed by local agencies working in isolation. Pennsylvania's Department of Natural Resources takes the lead in developing control strategies for ozone and particulate matter, and must develop plans and regulations to meet the NAAQS and regulations to deal with regional haze. However, current air pollution problems extend well beyond state boundaries, as Pennsylvania both receives pollution from other eastern states and contributes to their pollution burdens. Recognition of the regional character of current air pollution problems led to regional initiatives (e.g., OTAG), the establishment of RPOs, and federal requirements for region-wide controls. Regional programs are successfully reducing ozone and fine particulate matter in the eastern United States. Effective programs across states appear to benefit from organizations operating at similar scales, and which are given the appropriate authority.

Most recent regulations in Pittsburgh and elsewhere have been the result of federal intervention, rather than resulting from cooperative actions at the local level (Davidson and Davis, 2005). Thus, multistakeholder groups such as OTAG are instrumental in addressing complex, multifaceted issues. Moreover, the regional nature of these challenges requires geographically broad standards (typically federally established), but also opens up the opportunity for regional emissions trading, which in many cases appears to be the most effective way to address specific issues (e.g., acid deposition). Even as local air quality improves, downwind areas may continue to demand reductions, again necessitating regional and perhaps federal standards and intervention.

FUTURE DIRECTIONS

In the future, it is reasonable to expect further reductions in emissions. In the 1990 Clean Air Act (CAA), coke production was given added time to reduce emissions, and by 2015, controls on the coke batteries are to be in place to reduce air toxics such as benzene. While current trends suggest that coal will continue to be the most common fuel used for electricity generation, there will be continuing pressure to further reduce emissions. First, the CAA and other recent regulations require continuing reductions in sulfur and nitrogen oxides emissions (EPA, 1998, 2005). Further, downwind areas are not expected to come into attainment with the NAAQS, and regulatory mechanisms and political pressures allow them to look upwind (EPA, 2005).

In November 2004, Pennsylvania Governor Edward Rendell signed Act 213 into law, requiring electric distribution companies and suppliers to include a specific percentage of electricity from alternative resources in the generation that they sell to Pennsylvania customers. This Act, the Alternative Energy Portfolio

Standards Act, without mandating specific resources or quantities, established a 15-year schedule of gradual increases, with minimum thresholds for Tier I, Tier II, and solar PV.[3] By 2021, companies will be required to provide 8 percent of electricity from Tier I, 10 percent from Tier II, and a full 0.5 percent from solar PV (PPUC, 2006). The Pennsylvania Public Utilities Commission is tasked with implementing the act, and will work in conjunction with the Pennsylvania Department of Environmental Protection, which is primarily responsible for ensuring compliance. Although recent plans for future power plant construction indicate a shift toward increased wind power (likely to ensure compliance with the Act), a number of alternative sources (coal mine methane, waste coal, and IGCC) may also be employed, allowing area energy providers to meet the rising standards, while still utilizing coal.

Pittsburgh is still undergoing a renaissance of sorts, as it positions itself to be a leader in green technologies. In keeping with the determination to shed itself of the image of a smoky steel city, Pittsburgh's city leaders see this as an opportunity to be an innovator, rather than merely following what other cities have done. Pittsburgh is among the top five cities nationwide in the number of certified green buildings, including the first certified green convention center in the world, as well as the largest mixed-use green building in the United States (built by PNC Bank), slated for completion in 2008 (Heinz School Research Team, 2006). Additional actions include purchasing hybrid vehicles for the city's fleet, using biofuels in the more than 1,000 city-owned vehicles (Mayor's office press release, 2006), and training maintenance workers throughout the city to comply with the DOE's EnergyStar program, which requires buildings to reduce power consumption (Boren, 2006).

REFERENCES

ACHD (Allegheny County Health Department). 2004. Air Quality Program. Review of Application: Title V Operating Permit. June 17, 2004.

ACHD. 2006a. 2005 Air Quality Annual Report. Pittsburgh, PA.

ACHD. 2006b. Eco-Currents 6(1). Pittsburgh, PA.

Bergin, M.S., J.J. West, T.J. Keating, and A.G. Russell, 2005. Regional atmospheric pollution and transboundary air quality management. Annual Review of Environment and Resources 30:1-37.

Boren, J. 2006. City conservation panel eyes savings. Pittsburgh Tribune-Review. October 20.

Chameides, W.L., F. Fehsenfeld, M.O. Rodgers, C. Cardelino, J. Martinez, D. Parrish, W. Lonneman, D.R. Lawson, R.A. Rasmussen, P. Zimmerman, J. Greenberg, P. Middleton, and T. Wang. 1992. Ozone precursor relationships in the ambient atmosphere. Journal of Geophysical Research-Atmospheres 97(D5):6037-6055.

CLP (Carnegie Library of Pittsburgh). 2003. Oakland: Bellefield Boiler Plant. Pittsburgh, PA.

[3]Tier I includes solar thermal energy, wind, low-impact hydro, geothermal, biomass, biologically derived methane gas, coal mine methane, and fuel cells. Tier II includes waste coal, distributed generation systems, demand-side management, large-scale hydro, municipal solid waste, pulping process and wood manufacturing by-products, and coal integrated gasification combined cycle (IGCC) technology.

Davidson, C.I. and D.L. Davis. 2005. A Chronology of Airborne Particulate Matter in Pittsburgh. History and Reviews of Aerosol Science.

Davidson, C.I., W. Tang, S. Finger, V. Etyemezian, M.F. Striegel, and S.I. Sherwood. 2000. Soiling patterns on a tall limestone building: Changes over 60 years. Environmental Science & Technology. American Chemical Society 34(4).

Dequesne Light Company. 2006. http://www.duquesnelight.com/OurCommunity/Environment.

EIA (Energy Information Administration). 2006. State Energy Profiles. Pennsylvania.

EPA (U.S. Environmental Protection Agency). 1998. Finding of Significant Contribution and Rulemaking for Certain States in the Ozone Transport Assessment Group Region for Purposes of Reducing Regional Transport of Ozone. Code of Federal Regulations, U.S. Environmental Protection Agency.

EPA. 2005. Clean Air Interstate Rule: 40 CFR Parts 51, 72 et al. Federal Register, U.S. Environmental Protection Agency.

EPA. 2006. National Ambient Air Quality Standards for Particulate Matter; Final Rule. 40 CFR Part 50, U.S. Environmental Protection Agency.

Farrell, A.E. and T.J. Keating. 2002. Transboundary environmental assessment: Lessons from OTAG. Environmental Science & Technology 36(12):2537-2544.

FirstEnergy Corporation. 2004. Bruce Mansfield Plant: Facts at a Glance.

FirstEnergy Corporation. 2005. Air Issues Report, December 1.

Heinz School Research Team. 2006. Pittsburgh Climate Protection Initiative: Greenhouse Gas Emissions Inventory. Carnegie Mellon University, December.

Kara, D. 2006. The private sector: All signs point to Roboburgh. The Pittsburgh Post-Gazette, June 20.

Lave, L.B. and E.P. Seskin. 1973. Analysis of association between United-States mortality and air-pollution. Journal of the American Statistical Association 68(342):284-290.

Lipfert, F. 1994a. Air Pollution and Community Health: A Critical Review and Data Sourcebook. New York: International Thomson Publishing.

Lipfert, F.W. 1994b. Filter artifacts associated with particulate measurements: Recent evidence and effects on statistical relationships. Atmospheric Environment 28(20):3233-3249.

MANE-VU. 2006. http://www.manevu.org.

Mershon, S.R. and J.A. Tarr. 2003. Strategies for clean air: The Pittsburgh and Allegheny County smoke control movements, 1940-1960. In Devastation and Renewal: An Environmental History of Pittsburgh and its Region, J.A. Tarr, ed., University of Pittsburgh Press.

Milford, J.B., A.G. Russell, and G.J. McRae. 1989. A new approach to photochemical pollution-control—implications of spatial patterns in pollutant responses to reductions in nitrogen-oxides and reactive organic gas emissions. Environmental Science & Technology 23(10):1290-1301.

Milford, J.B., D.F. Gao, S. Sillman, P. Blossey, and A.G. Russell. 1994. Total reactive nitrogen (No(Y)) as an indicator of the sensitivity of ozone to reductions in hydrocarbon and NO_x emissions. Journal of Geophysical Research-Atmospheres 99(D2):3533-3542.

Millet, D.B., N.M. Donahue, S.N. Pandis, A. Polidori, C.O. Stanier, B.J. Turpin, and A.H. Goldstein. 2005. Atmospheric volatile organic compound measurements during the Pittsburgh Air Quality Study: Results, interpretation, and quantification of primary and secondary contributions. Journal of Geophysical Research-Atmospheres 110(D7).

NRC (National Research Council). 1991. Rethinking the Ozone Problem in Urban and Regional Air Pollution. Washington, D.C.: National Academy Press.

NRC. 2005. Superfund and Mining Megasites: Lessons from the Coeur d'Alene River Basin. Washington, D.C.: National Academies Press.

PDEP (Pennsylvania Department of Environmental Protection). 2006. Allegheny Health Department Announces Agreement to Curb Air Pollution. Daily Update, May 4.

Polidori, A., B.J. Turpin, H.J. Lim, J.C. Cabada, R. Subramanian, S.N. Pandis, and A.L. Robinson. 2006. Local and regional secondary organic aerosol: Insights from a year of semi-continuous carbon measurements at Pittsburgh. Aerosol Science and Technology 40(10):861-872.

PPUC (Pennsylvania Public Utility Commission). 2006. Electric Power Outlook for Pennsylvania 2005-2010.

Sillman, S. and J.A. Logan. 1987. Ozone Production in Rural Areas—a Photochemical Model Study. Abstracts of Papers of the American Chemical Society 193, 196-ENVR.

Sustainable Pittsburgh. 2004. Southwestern Pennsylvania Regional Sustainability Indicators Report 2004.

Tarr, J.A. 1981. Changing fuel use behavior—the Pittsburgh Smoke Control Movement, 1940-1950. Technological Forecasting and Social Change 20(4):331-346.

Tarr, J.A. 2002. The metabolism of the industrial city—the case of Pittsburgh. Journal of Urban History 28(5):511-545.

Tarr, J.A. and B.C. Lamperes. 1981. Changing fuel use behavior and energy transitions—the Pittsburgh Smoke Control Movement, 1940-1950—a case-study in historical analogy. Journal of Social History 14(4):561-588.

Tarr, J.A., G.D. Goodman, and K. Koons. 1980. Coal and natural-gas—fuel and environmental-policy in Pittsburgh and Allegheny County, Pennsylvania, 1940-1960. Science Technology & Human Values (32):19-21.

9

The Huainan Experience

PHYSICAL, ECONOMIC, AND SOCIETAL SETTING

Huainan is an industrial city and an important Chinese energy base relying on the coal, electric power generation, and chemical industries. It is located in north-central Anhui Province, roughly 100 km from the provincial capital of Hefei, and 250 km from the city of Nanjing in neighboring Jiangsu Province. It has a mild climate and the topography is mostly plains, located around the central part of the Huai River, which traverses the city from west to east. The northern shore of the Huai River is the Huaibei plain; the southern shore is a hilly area. In 2004, the total area of the city was 2,585 km^2, of which the urban area was 1,489 km^2, and the population was 2,335,800, of which 1,625,100 were urban.

In 1984, together with Chongqing, Dalian, Qingdao, and other cities, the State Council approved Huainan as a county-level city, granting it local legislative power. In 1985, the State Council approved the city as an open city (i.e., open to foreign investment). As a county-level city, Huainan administers five districts (Figure 9-2), a national-level experimental zone, and a provincial-level economic development zone. Huainan's ecology has also earned it the distinction of being a provincial Garden City, which is a measure of environmental improvement, and is awarded by the Ministry of Construction. Huainan has also endeavored to become a National Model City for Environmental Protection (see Chapter 4).

Huainan is in a warm temperature zone with a monsoon climate and with four distinctive seasons. The annual average temperature is 15°C, and the annual average rainfall is 970 mm. Huainan's four distinct seasons are also characterized by different pollution characteristics. In the fall and winter, the air is dry and cold and the pollution level is worse than that in the spring or summer due, in large

FIGURE 9-1 Huainan in China.

part, to a higher rate of coal-burning for commercial and residential heating. Rain and wind during the monsoon season help to reduce local air pollution as well.

Huainan's leaders have been espousing the concept of the "Three Bases," a goal that by 2010 Huainan will (1) produce 100 million tons (3 EJ) of coal per year; (2) be an important supplier of electrical power in Eastern China; and (3) establish itself as the regional base for the chemical industry. The concept of the Three Bases was first put forward by former President Jiang Zemin, and has since become the guiding strategy in the city's development.

Industry dominates Huainan's economy, particularly primary industries like energy raw materials (coal) and agriculture. The chemical industry is also a prominent part of the local economy, producing large amounts of ammonia, fertilizer (401,000 tons and 222,000 tons in 2004 respectively), and other chemicals. The city's GDP reached 21.5 billion RMB in 2004, an increase of 16.2 percent over the previous year, reaching the 10th Five-Year Plan (FYP) target a full year in advance. GDP in 2005 was estimated at 26.0 billion RMB, another increase of over 16 percent. Urban residents had a per capita income of 8,530 RMB (US$1060) which represented a similar increase of 15 percent from the year prior,

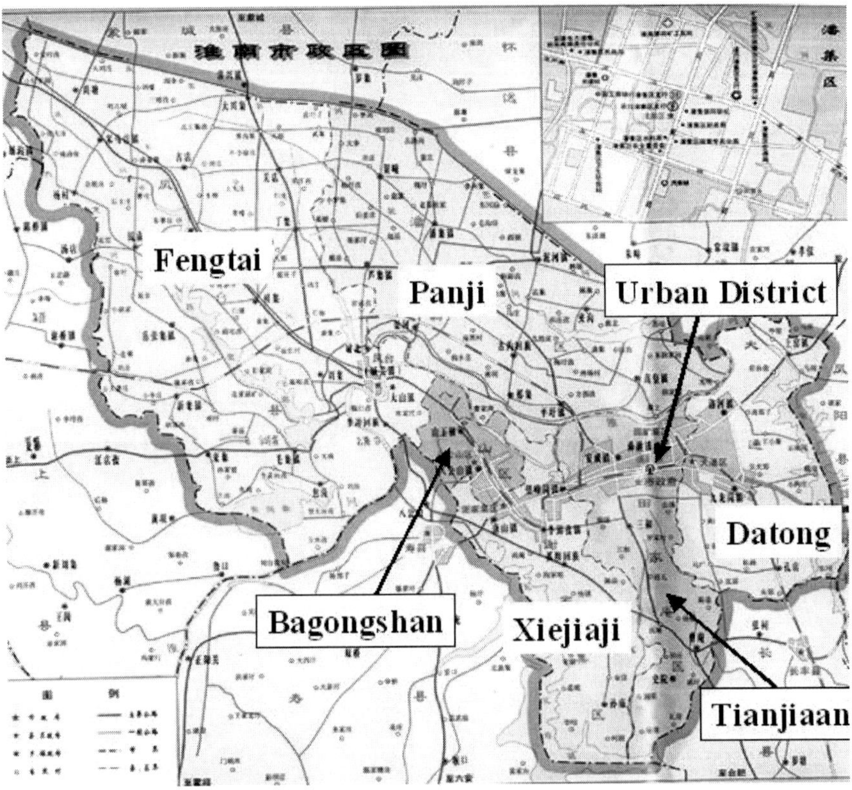

FIGURE 9-2 Huainan municipality administrative map.

while rural farmers had a lower income of 2,700 RMB, increasing only 5 percent from the previous year (Huainan Municipal Government, 2005).

Anhui University of Science and Technology is the major regional university, playing a key role in mining research, with a strong emphasis on environmental engineering. Industry also plays an important role in Huainan's research capacity. The Huainan Chemical Group maintains the Research Institute of Chemical Designing with a research staff of 140, who focus much of their work on coal technologies, in particular, coal gasification technologies utilizing locally mined coal. Within the Huainan Mining Group is the Development Center of Science and Technology, containing 6 laboratories and 98 research personnel. The Development Center has been instrumental in researching, developing, and disseminating technologies for capturing and utilizing coalbed methane (CBM) (HBST, 2005). In 2005, China's National Development and Reform Commission approved a

National Engineering Center for Coal Gas Control, to be partially based in Huainan. This center will focus on coalmine disaster control and prevention, geology, and safety technology (Xinhua Net, 2005).

SOURCES AND LEVELS OF AIR POLLUTION

Air pollution, as well as contamination of other media in Huainan, is believed to be dominated by industrial activity, most notably by the energy raw materials sector. In 2005, the total amount of waste water, gas, and solids emitted by energy materials production was estimated to be 81-95 percent of all industrial emissions. Energy consumption per unit GDP was 49 percent higher than for other industrial sectors. The emission of waste water and gaseous pollutants were 27-49 percent higher, and the air emissions from electrical power generation and the coal chemical industry were 50 percent higher than other industries. SO_2, smoke dust, coal waste, and fly ash from these industries amounted to 99.9 percent, 99.47 percent, 60.5 percent, and 34.5 percent, respectively, of Huainan's total air emissions (HEPB, 2005). Industrial waste gas emissions outpaced residential-sector emissions by a ratio of approximately 40 to 1, and of those industrial emissions, roughly 92 percent were from fuel combustion, as opposed to other manufacturing processes.

In 2005, total SO_2 and dust emissions were 119,000 and 33,000 tons, respectively, 28 percent and 12 percent higher than those of 2000 (HEPB, 2005). The resulting ambient annual average concentrations for SO_2 and NO_2 in the urban area in 2005, based on four monitoring sites, were each 0.030 mg/m^3, satisfying Class II standards for those pollutants according to China's Ambient Air Quality Standards (see Chapter 4). The PM_{10} average value was 0.104 mg/m^3, which satisfied Class III standards. The monthly average dust deposition through 2005 was 7.0 ton/km^2, which did not exceed the standard. The annual average pH value for precipitation was 6.98 and there was no evidence of acid rain detected in samples. Detailed data on specific pollutants follow.

Criteria Pollutants

Sulfur Dioxide (SO_2)

From 2001 to 2005, the annual average SO_2 concentrations in the urban district steadily increased. Figures 9-3 and 9-4 show annual average SO_2 concentrations from four monitoring sites in the urban district and in three administrative districts, respectively.

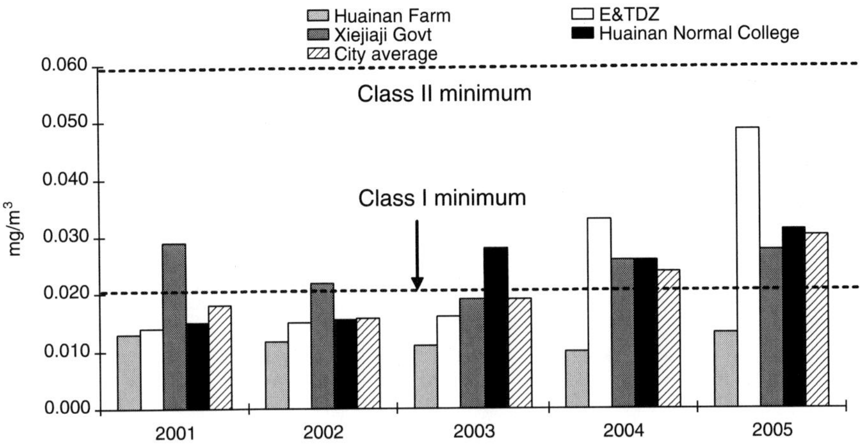

FIGURE 9-3 Annual average SO_2 concentrations at monitoring sites in the urban district of Huainan, 2001-2005.

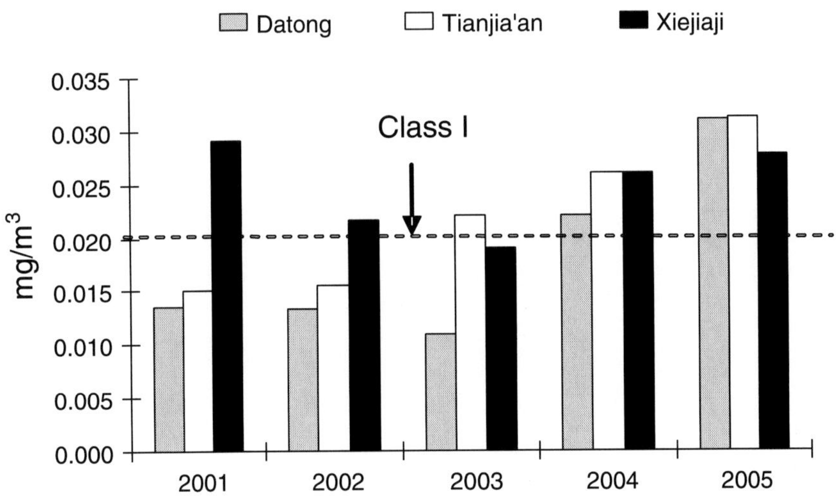

FIGURE 9-4 Annual average SO_2 concentrations at monitors in three administrative districts of Huainan, 2001-2005.

Nitrogen Dioxide (NO_2)

From 2001 to 2005, the annual average NO_2 concentrations increased in the urban district, notably in the E&TDZ and Xiejiaji areas, though the overall average for 2005 still satisfied Class I standards (Figure 9-5).

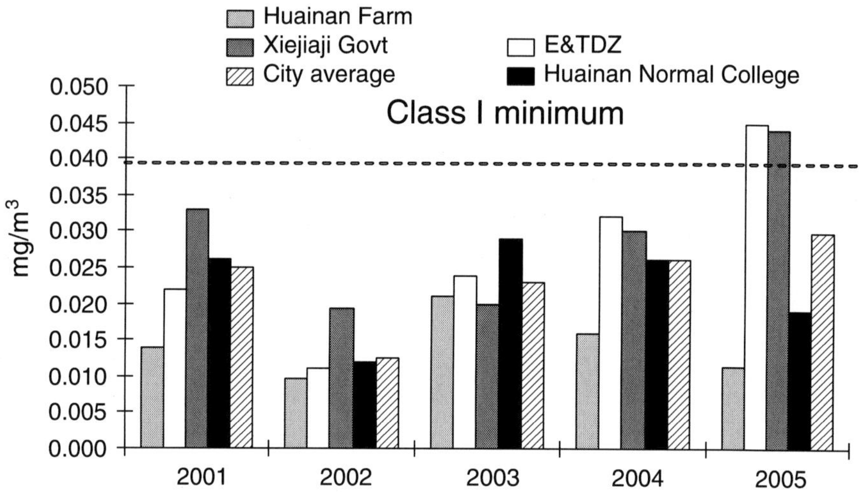

FIGURE 9-5 Annual average NO_2 concentrations at monitoring sites in the urban district of Huainan, 2001-2005.

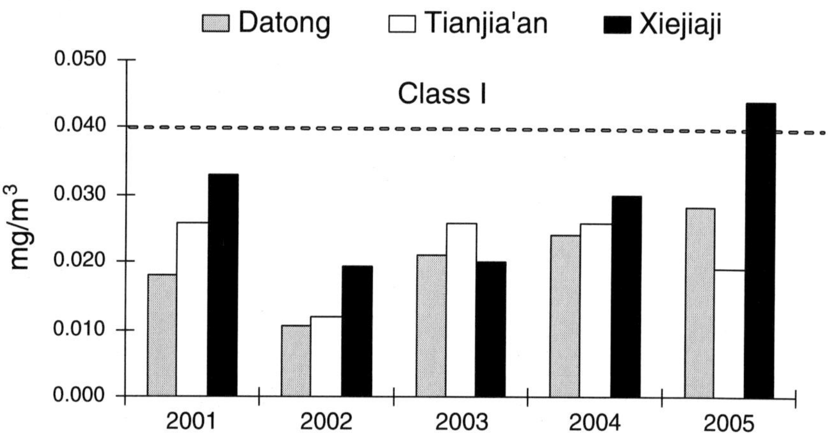

FIGURE 9-6 Annual average NO_2 concentrations in three administrative districts of Huainan, 2001-2005.

Particulate Matter (PM_{10})

Between 2002 and 2005, PM_{10} concentrations in the urban district declined slightly, though they still exceed the Class II standard. The PM_{10} annual averages for 2002-2005 are shown in Figure 9-7, as measured in the urban district. Figure 9-8 shows PM_{10} annual averages from three administrative districts and

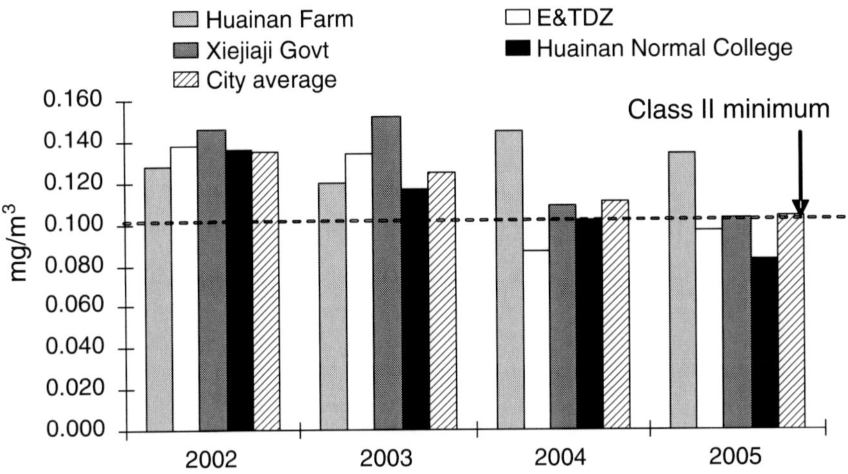

FIGURE 9-7 Annual average PM_{10} concentrations at monitoring sites in the urban district of Huainan, 2002-2005.
NOTE: Prior to 2002, TSP was measured rather than PM_{10}.

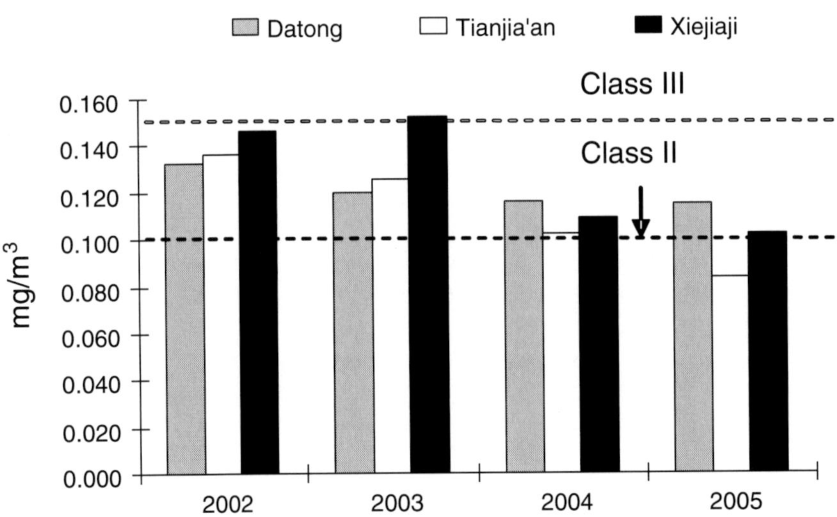

FIGURE 9-8 Annual Average PM_{10} concentrations in three administrative districts of Huainan, 2002-2005.

indicates that the Xiejiaji area in particular experienced the most dramatic decline in PM_{10} concentrations over 4 years—though it is still not quite in attainment of the Class II standard.

Dust Fall

From 2001 to 2005, the average dust fall in the urban district ranged from 6.2 to 7.6 ton/km^2 per month, averaging 7.0 ton/km^2 per month over those 5 years. The maximum value of 7.6 ton/km^2 per month occurred in 2001 and the minimum value of 6.2 ton/km^2 per month occurred in 2003. A closer look at dust fall rates at various locations reveals that the range was much greater within Huainan. The Huainan Steel Factory and Xinzhuangzi Mine in particular experienced high rates (in excess of 11.0 ton/km^2 per month) during this time period (Figure 9-9). Figure 9-10 shows that the Bagong Hills area experienced the highest rates of dust fall while Xiejiaji experienced a significant decline between 2002 and 2003.

Air Quality Trends

From 1996 to 2000, the comprehensive index in Huainan gradually decreased, signaling that air quality was improving. The comprehensive index for 1996 was above five (poorest air quality, according to China's Air Pollution Index), while the 5 index steadily decreased from 1997 to 2000. The average comprehensive index of the 10th FYP (2001-2005) was 2.4, 38 percent lower than that of the 9th FYP (1996-2000). The quantitative amount of this trend may be impacted by the change in analysis parameters, such as the transformation from TSP to PM_{10}, and from NO_x to NO_2, and the change in monitoring method from manual monitoring to automatic monitoring, which improved the quality of the data. Qualitatively, it can be noted that the comprehensive index of the air pollution was kept at a relatively low level in the 10th FYP, even as the city's GDP increased annually; but it also should be noted that certain pollutants, namely SO_2, gradually increased after a period of decline from 1996 to 2000.

From 2001 to 2005, SO_2 and NO_2 levels increased, while PM_{10} and dust fall decreased. Table 9-1 summarizes these trends. Specifically, the daily average values of SO_2 and NO_2 in 2005 increased by 66.7 percent and 20.0 percent, respectively. The average values of PM_{10} and dust fall decreased by 23.0 percent and 17.1 percent over the same period. The pH value in the precipitating rain in 2005 decreased 0.8 percent, though it was still not considered acidic. Indeed, the pH tends to suggest an abundance of alkaline material.

Comprehensive Index Trends, 1996-2005

Figures 9-11, 9-12, and 9-13 show the monitored results of SO_x, NO_x, and PM in Huainan from 1992 to 2006. In most cases, pollution levels have decreased

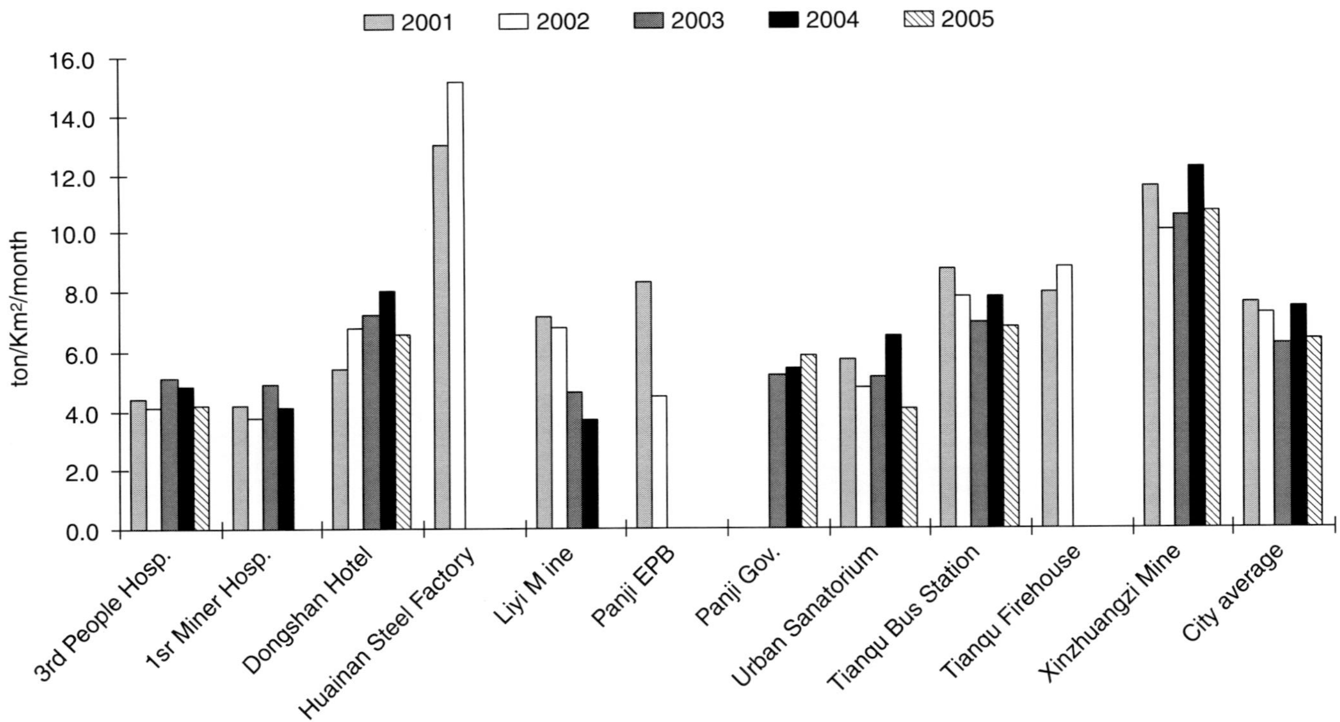

FIGURE 9-9 Dust fall distribution with time variation at various monitoring places.

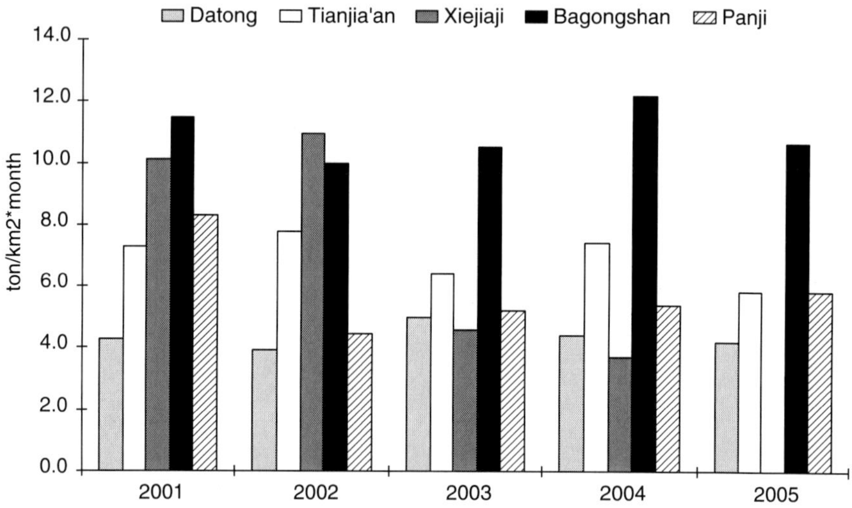

FIGURE 9-10 Dust fall distribution with time variation at various administrative areas, 2001-2005.

TABLE 9-1 Average Annual Values for Various Air Pollutants in 2001-2005

Year	SO_2 (mg/m^3)	NO_2 (mg/m^3)	PM_{10} (mg/m^3)	Dust (t/km^2·mo.)	Rain pH value
2001	0.018	0.025	—	7.6	7.38
2002	0.016	0.013	0.135	7.2	6.65
2003	0.019	0.023	0.125	6.2	7.08
2004	0.024	0.026	0.111	7.4	6.47
2005	0.030	0.030	0.104	6.3	7.32
2005/2001± percent	66.7	20.0	–23.0	–17.1	–0.8

steadily since 1992, in part due to government regulation and closing of highly polluting industries. For example, the local government closed 8 boilers (totaling 111 MW of capacity), rebuilt a 300 MW power unit during the reconstruction of the Tianjia'an power plant, and closed some cement factories with annual outputs under 1.5 million tons. All of these measures played an important role in local air quality improvement. Prior to 1992, there were no boilers with desulfurization equipment, but facilities have begun installing scrubbers, particularly for power plant boilers.

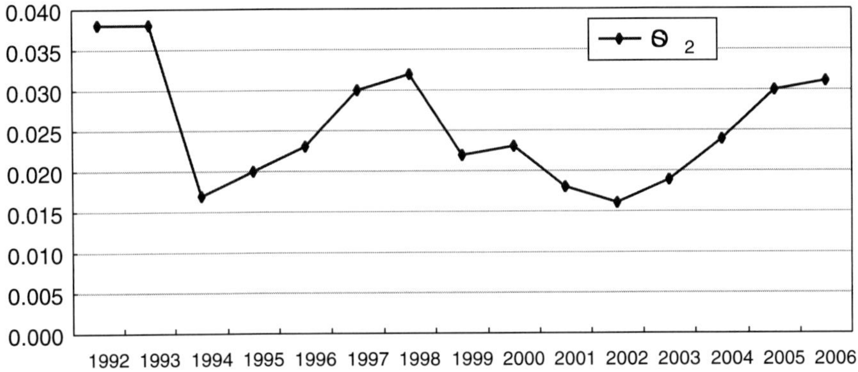

FIGURE 9-11 Monitored SO_x results (mg/m^3), 1992-2006.

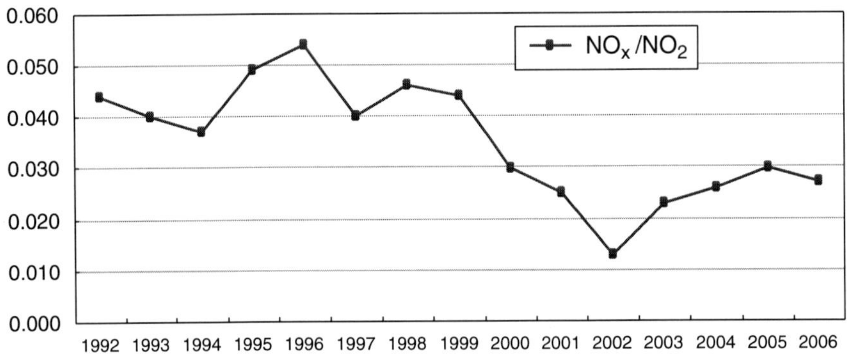

FIGURE 9-12 Monitored NO_x/NO_2 results (mg/m^3), 1992-2006.

Table 9-2 shows the 2000-2005 air emissions trends in more detail and Table 9-3 shows the environmental bearing capacity of Huainan city. Cities calculate their environmental bearing capacity, alternatively referred to as carrying capacity, in order to determine the amount of certain emissions the local environment might bear without further compromising environmental quality (according to the State Environmental Protection Agency's classification system). The bearing capacity is used as a benchmark by which cities might achieve their daily air quality targets 90 percent of the year. For atmospheric bearing capacity,

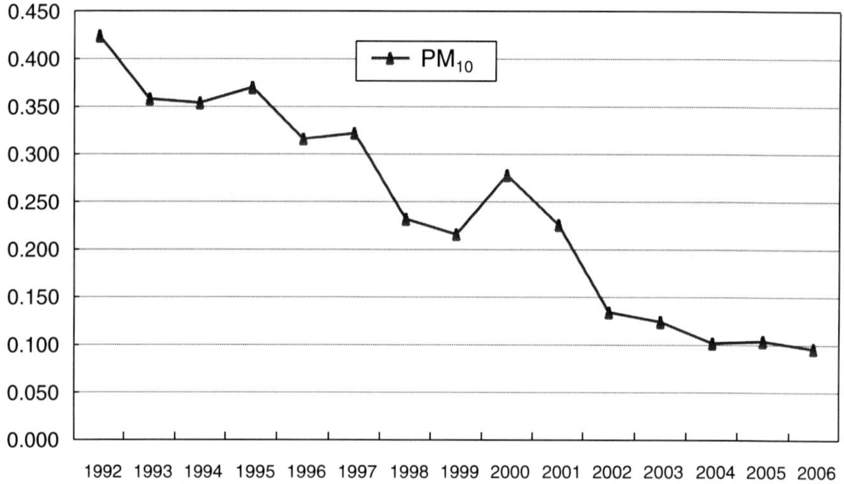

FIGURE 9-13 Monitored PM results (mg/m^3), 1992-2006.

TABLE 9-2 SO_2 and Dust Emissions (in tons), 2000-2005

	2000	2001	2002	2003	2004	2005	Percent Increase
SO_2 emissions	92598	94603	94791	91326	114910	118730	28.2
From industry	91473	91698	90734	90466	112857	117930	28.9
Dust emissions	29339	31353	31412	27264	35669	32818	11.9
From industry	27249	27823	26482	25674	27970	31618	16.0

TABLE 9-3 Emissions and Theoretical Environmental Bearing Capacity for Huainan in 2003 (unit: ton)

	SO_2	TSP
Estimated emissions	91326	27264
Bearing capacity	117900	172000

cities evaluate stationary and mobile sources of emissions (SO_2, PM_{10}, and NO_x), atmospheric conditions, prevailing air quality, and recent trends in both energy structure/consumption and air quality.

ENERGY RESOURCES AND USE

Huainan is a city of rich coal resources. It was recognized in 1949 as the Huainan Coalmine Special District, and was listed in the 1950s as one of China's five famous coalfields. The estimated total coal reserves are estimated at 44.4 billion tons (1345 EJ), and the identified coal reserve is 15.3 billion tons (464 EJ), amounting to 32 percent of the reserves in eastern China and 19 percent of the country's total coal reserves (Huainan Municipal Government, 2005). In addition, CBM reserves are estimated at 593 billion m^3 (23.6 EJ). Production of CBM in 2005 amounted to 5 million m^3. The coalfield is the largest (7,250 km^2) and most recently developed coalfield among those south of the Yellow River and in the southeast coastal area; additionally, the local coal is generally low in sulfur (<0.5 percent) and low in phosphorous and its geological conditions are very favorable to mining (Huainan Mining Group, 2005). At present, there are 12 pairs of coal production tunnels in the city with an annual production capacity of 37.6 million tons (1.14 EJ) as of 2005 (Table 9-4); and future production in the new mine (Panxie) area is expected to increase dramatically.

In 2005, coal consumption totaled 11 million tons, providing about 80 percent of the total energy consumption of the whole city, though this number was down from nearly 90 percent in 2004 (HEPB, 2005). As coal production has increased, so too has electrical power production. Beginning with the Tianjia'an Power Plant in the mid-1950s and its four 6 MW generators (an historic development for China at the time), Huainan's capacity has increased to 3,400 MW, including a 600 MW generator installed at the Pingwei Power Plant in 1984, which was at that time the largest in China (Huainan Municipal Government, 2005). Electricity generation reached 23.2 million kWh in 2004, almost exclusively derived from coal. The residential gasification rate reached 58 percent in 2004, meaning that fewer residents are depending upon indoor coal combustion for cooking and for other residential uses (HEPB, 2005). The increasing coal and electrical capacity have also underpinned

TABLE 9-4 Coal Output in Huainan (actual and projected)

	Output (million ton)		
Year	Total	From Old Mine Area	From New Mine Area
2003	35.83	10	25.83
2007	70	12	58
2010	100	7	93
2020	150	5	145

the development of other regional industries, including chemicals, pharmaceuticals, building materials, textiles, machinery, electronics, and high-tech industries.

Huainan is working to diversify its economy, while capitalizing on its most abundant resource by moving from being a coal supplier to an electricity producer/supplier. The local government hopes to capitalize on the value of energy production, which is significantly more profitable than the production of coal itself. Both local and regional groups have been investing on a large scale in "coal by wire" projects which involve on-site electricity generation and transmission from the coal mines. This is one potential way to eliminate conflict between coal suppliers and power companies, guarantee sufficient supply, and improve efficiency. Huainan and similar coal-supplying cities have been involved in contentious price battles with power plants, leading to power shortages throughout the country (CERW, 2003). With regional shortfalls of electricity generation predicted for the near future, investors are eager to build on the energy producing potential of Huainan (CCII, 2003).

Huainan has also begun to realize the economic, safety, and ecological benefits of capturing and utilizing methane in the coal mines. Many coal mine accidents in China are gas-related, therefore most mines pump methane out and discharge it. But the Pansan Coal Mine in Huainan has adopted technology which allows the mine to fuel 20,000 homes in the area and to generate local electricity (Xinhua Economic News Service, 2005). Although a small plant (2,400 kW), it is novel in its reliance on methane derived from the local mines. The cleaner burning methane had previously been released into the air, further polluting the region. In 2003, 130 million m^3 of methane were extracted from Huainan's coal mines, of which approximately 10 percent was utilized by local residents, in addition to some use for industrial boilers and power generation. As Huainan and other coal mining areas adopt technologies to harness methane from coal mines, China has mandated that industries exploit and utilize this vast reserve, which by some estimations rivals natural gas reserves.

In addition to air pollution, increasing coal production and consumption has brought about two further challenges, specifically subsidence and waste storage. The total area affected by subsidence related to excavation was 113 km^2 in 2005. Of this total area, over half (58 km^2) was in the old mine area, with most of the remaining (54.3 km^2) occurring in the new the Panxie mine area. The old mine area is south of the Huai River, the east area extends from the Jiulong mountain to the Datongjuren village, and the west area extends from Lier mine to west Kongji mine, forming two large coal excavation sink areas with a total length of 25 km and a total area of 6,342 hectares. Economic losses due to subsidence were estimated at 333 million RMB, including a crop economic loss of 191 million RMB (HEPB, 2005). Compared with data for 2000, the total subsidence area in 2005 was 46 percent larger, increasing at a rate of 7.9 percent per year. Projections for 2010 estimate a subsidence area of 165 km^2 and economic losses of 497 million RMB, including a crop economic loss of 294 million RMB.

Huainan generated 7 million tons of coal solid waste and coal fly ash in 2005. The city was able to recycle 6 million tons of this for use in road construction and cement, for a utilization rate of almost 85 percent and an improvement of 1.5 percent compared to 2000 (HEPB, 2005). Still, the city's total volume of solid waste and fly ash was nearly 49 percent higher than in 2000. The total volume of solid waste from coal has reached 27 million tons and occupies an area of 620 hectares.

Huainan, like many other cities in China, has favored supercritical technologies for future coal-fired power production. However, the city has also been exploring integrated gasification combined cycle technology since the early 1990s. Though still considered too expensive to be used solely for electrical power production, Huainan has acquired General Electric's gasification technology (originally developed by Texaco) and intends to use it to produce methane as well as power, likely in a 50/50 split. This also provides an opportunity for future CO_2 emission controls (see Chapter 6).

Huainan has been involved in a provincial experiment to utilize alternative vehicle fuels, dimethyl ether (DME), and methanol. Anhui, Jilin, and Henan provinces were the first provinces in China to use DME/methanol, which can be produced through coal gasification and used as cleaner-burning vehicle fuels (relative to conventional gasoline or diesel) (HEPB, 2005). However, there is still concern over pollution and toxicity, particularly in the case of methanol.

POLLUTION AND ENERGY POLICIES AND THE APPROACH TO AIR QUALITY MANAGEMENT

In the 10th FYP, from 2001 to 2005, air quality monitoring in the Huainan urban district changed in four aspects. First, automatic monitoring supplanted manual monitoring. Second, more monitoring sites were established. Third, monitoring NO_x was changed to monitoring NO_2, and TSP monitoring was changed to monitoring PM_{10}. Finally, the monitoring frequency and duration were changed. Automatic data monitoring was widely adopted after 2003. The city now operates five monitoring stations, as well as flue gas monitors on the stacks at three local power plants.

In order to better inform policy, Huainan has made a concerted effort to upgrade its monitoring capacity. Environmental monitoring is the responsibility of the Local Environmental Monitoring Center, which is part of Huainan's Environmental Protection Bureau. The provincial and municipal governments have invested 9.6 million RMB to monitor air quality, construct an automatic water quality monitoring system for the Huai River, and create an online monitoring system for major pollution sources. The primary and secondary pollution sources of enterprises and institutions are being monitored. Pollution data is released by the Local Environmental Monitoring Center and can be found in media publications and on the Internet (http://www.hnhb.gov.cn). Daily, monthly, and yearly reports are provided. Table 9-5 provides an example of the daily report; Table 9-6 shows a portion of the annual report from 2002.

TABLE 9-5 Example for the Daily Report of Ambient Air Quality in China

[Air quality] API (air pollution index): 59 Air quality grade: II Dominating pollutant: PM	[Air quality] API (air pollution index): 81 Air quality grade: II Dominating pollutant: PM
Huainan Environmental Monitoring Center 22, Oct., 2005	Huainan Environmental Monitoring Center 23, Oct., 2005

TABLE 9-6 Part of Huainan Annual Environmental Report, 2002

Air Pollution:

The air quality in the urban district was Grade II. The detailed pollutants were:

PM_{10}: 0.227 mg/m^3 (changed less than 1 percent)
SO_2: 0.016 mg/m^3 (compared with the last year, decreased 36 percent)
NO_2: 0.013 mg/m^3 (compared with the last year, decreased 28 percent)
Dust: 7.24 ton/km^2.month (compared with the last year, decreased 5 percent)
No acid rain

Emission of industrial waste

SO_2: 90700 ton (compared with the last year, decreased 1 percent)
Smoke dust: 26500 ton (compared with the last year, decreased 5 percent)
Other dust: 6300 ton (compared with the last year, decreased 6 percent)

Measures to protect the environment

- Two projects to control the waste gas have been finished
- Rebuilt the fly ash removing system in the thermal power plant of Anhui Huainan Chemical Group
- Rebuilt the dust separation system in the cement production process, Anhui Huainan Mine Group
- Implemented "Management regulation of the environmental pollution for small scale boiler/stove in Huainan"
- Comprehensive reuse of solid waste reached 5,782,100 tons, an increase of 7.7 percent compared with previous year, including 2,029,500 tons of coal fly ash and 3,305,000 tons of other coal waste

A system of environmental quality responsibility has been put into practice. Required performance criteria for the construction of an ecological city[1] and the establishment of a National Model City for Environmental Protection have been established for all levels of government and corporations. In order to carry out the nation's environmental regulations and to adhere to local environmental legisla-

[1]Huainan was selected as the site for two circular economy demonstrations focusing on coal mine ecology (Huainan Mining Group, 2005).

tion, the local government is strictly implementing industrial techno-economic policy, including environmental impact assessments. The city government has set up some environmental protection laws and regulations, such as the "Protection Regulation for Huai River Area by Environmental Protection Office" (1992), and the "Environmental Protection Regulation for Developmental Projects in Huainan" (1999).

However, construction of basic environmental equipment, such as the pipe networks for urban natural gas and centralized heat supply, has been slow. In addition, there is no basic environmental equipment construction in the connecting regions between urban and rural areas. These conditions do not currently meet the standards of a National Model City for Environmental Protection, which Huainan is pursuing. With coal washing, coal mixing, briquette, and coal water slurry technologies, raw coal quality has been improved, and coal variety has been increased. Thus, the total pollution of coal production is estimated to have been reduced by 20 percent. Clean production had been brought into effect in the Fengtai Jiuhe fertilizer plant and in the Jiangsu Debang Chemical Ltd., where the gross pollution from chemical plants has been cut down by 50 percent. In recent years, as a result of strengthened regulatory control of particulate emissions, most of Huainan's power plants have installed electrostatic precipitators (ESP). This has led to a marked decrease in total dust emissions, although they are still the major pollutant in urban areas. SO_2 and NO_x are currently considered to be less problematic than particulate matter (PM) control. Boilers above 600 MW capacity must adopt desulfurization equipment with removal efficiency exceeding 90 percent (Huainan Mining Group, 2005). As capacity increases, desulfurization is thought to be efficient enough to remain in attainment of Class II standards. There is no control strategy for NO_x in place, although many facilities are implementing low-NO_x burners. NO_x emission concentrations are not supposed to exceed 500 mg/m^3. For PM, scrubbers are 95 percent efficient at removal, but ESP at new plants is 99.7 percent efficient, though this is still likely not enough to address the rapid increase in coal consumption.

FUTURE DIRECTIONS

In the period of the 11th FYP (2006-2010), the industrial structure will maintain its current status, which will be the main challenge to environmental improvement in Huainan. Huainan's specific goals as part of the Three Bases strategy are producing 100 million tons (3.0 EJ) of coal annually by 2007, and increasing this to 120 million tons (3.6 EJ) by 2010. Additionally, Huainan plans to add 10,000 MW electrical generation installed capacity, much of this added capacity being devoted to regional electricity exports which could total 6,000 MW (Huainan Mining Group).

Table 9-7 shows the electric power production situation and energy consumption into the future.

TABLE 9-7 Electric Power Output and Energy Consumption

Year	Unit	2003	2007	2010	2020
Electric power output	Billion kWh	21	61	82	130
Coal consumption	Million ton	6.8	20	27	45

Shanghai could be a major beneficiary of this increased electrical output—Shanghai Electric and Huainan Mining Corporation have agreed to jointly construct new power plants, and the State Grid Corporation of China is planning to construct a high-voltage grid from Huainan to Shanghai. One proposed power station planned by Huaneng Power International, China Power Investment Group, and Huainan Mining Group is to have a generating capacity of 10,000 MW by 2010, and to produce 30 million tons of coal (0.9 EJ) per year. Nicknamed the "Thermal Power Three Gorges," this project could rival the original Three Gorges project along the Yangtze River. By 2020 developers hope that this coal and power base will have a total generating capacity of 20,000 MW and be producing 60 million tons (1.8 EJ) of coal annually. The Huainan project is intended to ease regional power shortages, stabilize coal and energy prices, decrease pollution associated with long-distance coal transport, as well as bring jobs and income to the province. Estimates of investment for the project are around 100 billion yuan (CCII, 2003).

In the short term, Huainan is planning to focus its investments on supercritical power generation. In 2020, it is estimated that 20,000 MW of supercritical thermal power units will annually save 5.2 million tons of coal (0.16 EJ), reduce SO_2 by 8,410 tons, dust by 5,260 tons, and decrease PM_{10} concentrations, as compared with a conventional subcritical unit (Huainan EPB). Water is an increasingly scarce resource in Huainan, and increasing coal-fired power plant activity will put an even larger strain on this resource. Coal-fired power plants are large consumers of water, for both rinsing and cooling. Therefore, the water used in fly ash rinsing in coal-fired power plants should be fully recovered, and the government has set a target of ≥ 97 percent industrial wastewater reuse in these plants (Huainan Mining Group, 2005). An analysis of another coal-rich city, Zaozhuang, suggests that adopting low-emission coal gasification technologies in only 24 percent of its market by 2020, would yield considerably better emissions reductions (15-60 percent depending on pollutant) than adopting the best end-of-pipe control technologies (Wang et al., 2005). This is significant in that Zaozhuang's health damages from air pollution exposure were calculated to be 10 percent of local GDP in 2000, and would rise to 16 percent by 2020 (Wang and Mauzerall, 2006).

Huainan is a prime candidate for further CBM projects. Increased coal extraction also increases the opportunities for commercially developing CBM. By 2007, the city plans to be extracting 220 million m^3 and locally utilizing 45 percent.

Huainan has plans to construct holding tanks connected to piped networks in order to improve utilization. By 2020, it is projected that Huainan will be extracting 600 million m^3 of CBM, with a utilization rate above 90 percent (HBST, 2005). It has also begun the process of fuel switching for some industrial boilers, with plans to reform 18 boilers to operate on gas by 2010 (Huainan Mining Group). Additionally, it is constructing combined cooling, heat, and power plants which will total 100 MW of capacity by 2010, increasing to 150 MW by 2020, and operating on coal gas.

According to the construction goal of the Three Bases, annual coal consumption could reach 29 million tons (0.88 EJ) by 2010, or 1.7 times higher than that of 2000, increasing another 4.1 times in 2020. Air pollution caused by coal flue gas will still be dominant, which will result in air pollution concentrations increasing in the Huainan urban area, Fengtai County, and the Jianzhi town area. Future energy development and construction will focus on the Panxie mine areas. The continuing pileup of coal waste and coal fly ash will occupy a large amount of farmland, exacerbating air, water, and soil contamination. The increased generation amount of coal waste and coal fly ash is estimated to be 11.6 and 22.7 million tons in 2010 and 2020, respectively—64.5 percent and 2.2 times higher than that of 2005. The increased generation amount of the desulfurization by-product will be 0.4 and 0.96 million tons in 2010 and 2020, respectively. With the development of the new superscale mines in the Panxie mine area and the increasing of the exploitation intensity in the old mine area, the total coal subsidence area will increase 47 percent by 2010. Ecological recovery in the coal excavation sites will be a long-term process, as well as a huge systems engineering problem, because the coal excavation sites are large and increasing rapidly.

Analysis

The story of Huainan is the story of a city in transition. It is a story of dynamic tension between efforts to adopt more stringent air pollution standards, to install basic pollution control equipment, to install more modern combustion technology, and simultaneously to rapidly expand production of coal and coal fired generation of electricity. This chapter cites averages over the 2001-2005 period which for the most part meet the Class II standard, but the trends over that period show continuing increases for some important pollutants, especially SO_2. These trends suggest that the tension between remediation and consumption is unbalanced in favor of consumption, and air pollution will worsen in the short to medium term.

Huainan is a sort of demonstration city, as it attempts to move from a coal production base to an energy production base, realizing the economic benefits of the value-added process. At the same time, it will face the consequences of increased emissions as a result of the steep increase in coal combustion. Huainan has made great strides in closing down inefficient boilers, consolidating mining productions to improve efficiency, reducing pollution, increasing gasification rates in the urban areas, and utilizing alternative resources such as CBM. In that respect, it can be viewed as a model city to a number of industrial cities in China. Although PM_{10} and dust are currently the primary pollutants, increased use of automobiles, along with the increase in coal combustion, will certainly lead to increases in SO_2 and NO_x. Scrubbers on the new power plants will aid in mitigating additional SO_2 emissions, but there is not currently any strategy to address increasing NO_x emissions. Current plans recommend installing low-NO_x burners but are concerned with concentrations and not total emissions. This may present challenges for future NO_x control, especially considering the long life expectancies of coal-fired power plants and the certain increase in automobile use.

Coal provides an abundant local resource, but in addition to its impacts on air quality, the mining and extraction processes can also be environmentally degrading. Coal washing and sieving should be incorporated into the large coal-by-wire projects to reduce SO_2 and Hg, prior to combustion. Future power generation stations, which will almost certainly be coal-based, ought to be sited to minimize human exposure, and consideration should be given to alternatives to supercritical generation. Source inspection and enforcement will also be useful, in order to ensure that industries are in compliance with emissions requirements.

Establishing the environmental monitoring center was a step in the right direction for Huainan's environmental management. The next challenge will be to expand the number of monitoring sites and to make these data accessible to researchers. The regional universities and research centers are an important asset; as in Pittsburgh's experience, these groups ought to be part of the research side of the local air quality management regime. In particular, Huainan will benefit from studies of $PM_{2.5}$, given its disproportionate impact on health. Agricultural burning in the rural districts still appears to impact urban air quality and thus deserves more attention. Improved regional cooperation will also be necessary—beginning with research to understand the regional sources and contributions to local air pollution. This will aid each city in the region, including Huainan, in developing appropriate responses, and will create a framework to address emerging regional air quality challenges.

REFERENCES

CCII (China Coal Information Institute). 2003. Huainan: 'Thermal Power Three-Gorge Project' brings chances to CBM. China Coal Information Institute Newsletter, July 20.

CERW (China Energy Report Weekly). 2003. Survey: Coordinating the Coal and Power Industries. November 14.

HBST (Huainan Bureau of Science and Technology). 2005. Scientific Research on Energy Resources and Introduction of Scientific Input in Environmental Protection in Huainan City. Presentation to committee, October 2005.

HBUP (Huainan Bureau of Urban Planning). 2005. Report of Bureau of Urban Planning of Huainan Municipality. Presentation to committee, October 2005.

HEPB (Huainan Environmental Protection Bureau). 2005. Brief Report of Environmental Protection. Presentation to committee, October 2005.

HEPB Report of air quality in Huainan 2001-2005.

HEPB Report of air quality in Huainan 1996-2000.

HEPB Report of air quality in Huainan 1991-1995.

Huainan Mining Group. 2005. Brief Introduction to Huainan Coal Mining Group. Presentation to committee, October 2005.

Huainan Municipal Government. 2005. On the Present Situation and Future Plans of the Social Economic Development of Huainan Municipality. Presentation to committee, October, 2005.

Huainan Municipal Government. 2006. Official Government website. http://www.huainan.gov.cn.

Wang, X. and D.L. Mauzerall. 2006. Evaluating impacts of air pollution in China on public health: Implications for future air pollution and energy policies. Atmospheric Environment 40:1706-1721.

Wang, X., D.L. Mauzerall, Y.T. Hu, A.G. Russell, E.D. Larson, J.H. Woo, D.G. Streets, and A. Guenter. 2005. A high-resolution emission inventory for eastern China in 2000 and three scenarios for 2020. Atmospheric Environment 39:5917-5933.

Xinhua Economic News Service. 2005. China endeavors to turn coal mine gas into clean energy source. August 1.

Xinhua Net. 2005. Country to establish National Engineering Center for Coal Gas Control. November 14, available at http://www.sina.com.cn (in Chinese).

10

The Los Angeles Experience

HISTORY

Los Angeles was a small quiet town for much of the 19th century. The completion of the first railroad into Los Angeles, in 1876, helped to change all of that by connecting it to San Francisco, and in 1885 another line connected Los Angeles to the eastern United States. Oil was discovered in 1892, which further spurred the population boom and also made Los Angeles one of the world's most important suppliers of petroleum in the early 20th century, by some accounts providing one-quarter of the world's supply. The population surged from just over 5,000 in 1870 to more than 100,000 in 1900, and by 1930, the population was 1.2 million people. Two key interventions allowed the population to swell the way it did. First, an aqueduct was developed (at the time the world's longest) and was completed in 1913, supplying ample water to the region. The Los Angeles Water Department, which was responsible for constructing the aqueduct, went on to become the Los Angeles Department of Water and Power (LADWP), now the city's publicly owned power company. In addition to sufficient water resources, planners sought land, and with a reliable source of water in tow, Los Angeles in 1915 began annexing small towns without their own water resources. Los Angeles had also annexed a small strip of land along the San Pedro Bay, which had been selected by a federal panel to be the site for development of a major port, which in 1907 was officially founded as the Port of Los Angeles (POLA, 2007).

This pattern of development was also influenced by the electric rail car system of Pacific Electric, and by 1932, Los Angeles had grown to a city covering 1,165 km^2, which was incidentally the year it first hosted the Olympic Games. World War II brought about an economic boom, as Los Angeles developed into a center for production of aircraft, war supplies, and munitions. Shipbuilding became the Port of Los Angeles' primary economic activity. But with increased production and activity came an increase in air pollution. The first recognized episodes of smog in Los Angeles occurred in the summer of 1943. Visibility was limited to only three blocks and residents suffered from smarting eyes, respiratory discomfort, nausea, and vomiting. The phenomenon was termed a "gas attack," and was blamed on a nearby butadiene plant, but the situation did not improve when the plant was shut down. Smog events continued to plague Los Angeles throughout the 1940s (Figure 10-1).

The post-World War II economic boom was characterized by lateral development into the San Fernando Valley, enabled by the creation of a freeway system which would grow to become one of the world's largest. Right around this time, many urban regions in North America began phasing out electric streetcars in favor of buses, and as Los Angeles' streetcars went out of business, personal automobiles filled the void. In addition to being a major consumer of automobiles, Los Angeles was also the United States' second largest manufacturer behind Detroit, and also a major manufacturer of tires.

FIGURE 10-1 View of part of the Los Angeles Civic Center masked by smog in 1948.
SOURCE: Los Angeles Times photographic archive, UCLA Library.

During this economic boom, Los Angeles also developed severe air quality problems. The City of Los Angeles began its air pollution control program in 1945, establishing the Bureau of Smoke Control in its health department. On June 10, 1947, California Governor Earl Warren signed into law the Air Pollution Control Act, authorizing the creation of an Air Pollution Control District (APCD) in every county of the state. The Los Angeles County APCD was then established—the first of its kind in the United States. During the 1940s and 1950s, air pollution control focused on obvious sources, such as backyard burning and incinerators, open burning at garbage dumps, and smoke emissions from factories (SCAQMD, 1997). In 1953, the Los Angeles County APCD started requiring controls to reduce hydrocarbon emissions from industrial gasoline storage tanks, gasoline tank trucks, and underground storage tanks at service stations. California officially adopted the Ringelmann System, which measures the opacity of smoke arising from stacks and other sources.

During the 1950s and 1960s, local air quality officials implemented the use of vapor recovery equipment for the bulk transfer of gasoline, regulated petroleum-based solvents, and required permits for rendering plants that processed animal waste. But air quality continued to worsen due to the rapid growth in the number of automobiles and the increased miles traveled as a result of increased urbanization and the layout of the Los Angeles urban area (Figure 10-2). By the mid-1960s, total oxidant (ozone plus NO_2) levels approached 800 ppb and 24-hour-average PM_{10} concentrations exceeded 600 $\mu g/m^3$.

In 1959, the California Legislature established the California Motor Vehicle Pollution Control Board to test emissions and to certify emission control devices. In 1961, the first automotive emissions control technology in the United States, Positive Crankcase Ventilation, was mandated by California to control hydrocarbon crankcase emissions. In 1966, California imposed initial regulations for automobile tailpipe emissions for hydrocarbons and CO, the first of their kind in the United States. The California Highway Patrol began random roadside inspections of vehicle smog control devices. In the 1960s, regulators took the first step to clean up motor vehicle fuels by reducing the amount of highly photochemically reactive olefins in gasoline.

In 1967, California's Legislature passed the Mulford-Carrell Act, which combined two Department of Health bureaus—the Bureau of Air Sanitation and the Motor Vehicle Pollution Control Board—to establish the California Air Resources Board (CARB). Since its first meeting on February 8, 1968, in Sacramento, CARB has worked with the public, the business sector, and local governments to find solutions to California's air pollution problem. The resulting state emission standards set by CARB continue to outpace the rest of the nation, and have prompted the development of new control technologies for industrial facilities and motor vehicles.

Starting in 1970, the federal government phased out lead in gasoline. In 1975, the first oxidizing catalytic converters to reduce CO and hydrocarbon tailpipe

FIGURE 10-2 Photograph of the Pasadena Freeway, 1950s.
SOURCE: EPA.

emissions came into use as part of CARB's Motor Vehicle Emission Control Program. This is the state's first example of "technology forcing" regulations, compelling industry to develop a new pollution-control capability by a set deadline. In 1977, the first three-way catalytic converter to control hydrocarbons, nitrogen oxides, and CO was introduced. During the late 1970s, Los Angeles and later the entire state required vehicle inspections for measuring emissions and inspecting emission control equipment, which in 1984 evolved into the California Smog Check Program administered by the state Bureau of Automotive Repair. CARB has pioneered a motor fuels specification enforcement program since 1977, in response to the adoption of a state Reid Vapor Pressure (RVP) standard. Other regulations were adopted to further control the chemical properties of gasoline by limiting lead, sulfur, phosphorus, and manganese, and to control the sulfur content of vehicular diesel fuel in Los Angeles.

In 1988, CARB adopted regulations effective on 1994 model cars requiring that they be equipped with "on-board diagnostic" (OBD) computer systems to monitor emission performance and emission control equipment. Owners are alerted when there is a problem. All 1996 and newer vehicles less than 14,000 lbs. (e.g., passenger cars, pickup trucks, sport utility vehicles) throughout the United States are equipped with OBD II systems, the second generation of OBD requirements. In 1990, CARB adopted standards for cleaner burning fuels (also called Phase I Reformulated Gasoline) resulting in gasoline composition changes that reduced vehicle emissions and enabled advances in catalytic converter technology. This consisted of lowering previously regulated components (RVP and sulfur); requiring the use of oxygenates year round; and regulating additional components (benzene, total aromatics, and olefins). This was followed by the introduction of Phase II Reformulated Gasoline (also known as cleaner burning gasoline) in 1996. In 1999, the Board amended and adopted Low-Emission Vehicle regulations, known as LEVII, which set stringent emission standards for most minivans, pickup trucks, and sport utility vehicles, to reduce emissions of these vehicles to the level of emissions from passenger cars by 2007.

PHYSICAL, ECONOMIC, AND SOCIETAL SETTING

Today the Los Angeles metropolitan area is the second-most populated urban area in the United States, after the New York Metropolitan Area. California's population more than quadrupled following World War II, growing from under 7 million in 1940 to over 30 million in 1990 (McConnell, 1992). About 16 million people—7 percent of the entire U.S. population, and more than half the population of California—now live in the South Coast Air Basin (SoCAB) (Croes and Fujita, 2003). Los Angeles has a diverse economy of many heavy and light industries, including two major ports, oil production and refining, steel production, aerospace manufacturing, and coal-fired power plants. The combined ports of Los Angeles and Long Beach make up the largest container complex in the

United States. Due in part to its "invention" of the suburb and freeway, it now has the greatest number of cars, trucks, and miles of roadways of any city in the world, altogether 27 intertwining freeways which handle roughly 100 vehicle-miles traveled (160 km) daily.

These mobile and stationary sources of air pollution, emitted into an air basin surrounded by 3,000-meter mountains with prevailing oceanic winds, and capped by a persistent thermal inversion under sunny skies, created the highest levels of photochemical air pollution ever recorded. But, no city has ever made greater progress toward air quality goals, with air pollution levels now less than one-fourth of those in the past, during a period when the population doubled and vehicle miles traveled quadrupled. The economy has grown an average of nearly 5.2 percent annually over the past 10 years, surpassing the U.S. economy's average growth rate in recent years (LAEDC, 2006).

The Los Angeles metropolitan area includes the City of Los Angeles and consists of five counties: Los Angeles, San Bernardino, Riverside, Ventura, and Orange. It is an area of complex terrain bounded by the Pacific Ocean to the west; to the north by narrow coastal mountains and valleys, the San Joaquin Valley, and the Sierra Nevada Mountains; and to the south and east by deserts (see Figure 10-3). The SoCAB consists of most of this area and is the air quality jurisdiction for the South Coast Air Quality Management District. The area of the basin is about 17,500 km^2 of which the city of Los Angeles occupies 1,291 km^2. Although the air basin boundaries were established with topographical features in mind, winds often transport pollutants from one basin to another.

Southern California is in the semi-permanent high-pressure zone of the eastern Pacific. During summer, average temperatures are 25°C, with maximum daily readings often exceeding 35°C. Precipitation events are rare. Frequent and persistent temperature inversions are caused by subsidence of descending air that warms when it is compressed over cool, moist marine air. These inversions often occur during periods of maximum solar radiation which create daytime mixed layers of ~1,000 m thickness, though the top of this layer can be lower during extreme ozone episodes (Blumenthal et al., 1978). During summer, the sea-land breeze is strong during the day with a weak land-sea breeze at night. Owing to the high summer temperatures and extensive urbanization in the SoCAB, the land surface temperature does not usually fall below the water temperature at night, and nocturnal and morning winds are less vigorous than daytime winds. The land surface cools sufficiently to create surface inversions with depths as shallow as ~50 m. Surface heating usually erodes the surface and marine layers within a few hours after sunrise each day. Summertime flow patterns are from the west and south during the morning, switching to predominantly westerly winds by the afternoon. The land/sea breeze circulation moves air back and forth between the SoCAB and the Pacific Ocean, as well as along the coast to other air basins. These wind flow patterns are described by Smith et al. (1972), Keith and Selik (1977), and Hayes et al. (1984).

FIGURE 10-3 Map of the Los Angeles Basin.
SOURCE: California Air Resources Board.

In addition to these general features, many smaller features affect the movement of pollutants within the SoCAB. Heating of the San Gabriel and San Bernardino Mountains during the daytime engenders upslope flows that can transport pollutants from the surface into the upper parts of, and sometimes above, the mixed layer. When the slopes cool after sunset, the denser air flows back into the SoCAB with pollutants entrained in it. These elevated layers can also transport horizontally towards the coast as part of the sea breeze return flow, and can be brought back to the surface in the morning through fumigation and subsidence processes (Lu and Turco, 1994). Convergence zones occur where terrain and pressure gradients direct wind flow in opposite directions, resulting in an upwelling of air. Smith et al. (1984) have identified convergence zones at Elsinore (McElroy et al., 1982; Smith and Edinger, 1984), the San Fernando Valley (Edinger and Helvey, 1961), El Mirage, the Coachella Valley, and Ventura County. Rosenthal (1972) and Mass and Albright (1989) identified the Catalina Eddy, a counterclockwise mesoscale circulation within the Southern California Bight, as a mechanism for transporting air pollution. This eddy circulation transports pollutants from the SoCAB to Ventura County, especially after the SoCAB ozone levels drop, due to wind ventilation caused by an approaching low-pressure trough from the northwest. Transport also occurs from the SoCAB to the Mojave Desert. Winds are directly related to the pressure field, which, in summer, results from a consistent mesoscale component added to a varying synoptic-scale component (Green et al., 1992a, 1992b).

SOURCES AND LEVELS OF AIR POLLUTION

Mobile sources such as gasoline-fueled vehicles (10 million cars and light trucks for 16 million people) and diesel-powered vehicles (250,000 trucks and buses) play a major role in Los Angeles' air quality problems. Because of proximity to the Pacific Ocean and geography, the meteorology is particularly conducive to generating poor air quality. Los Angeles' pollutant formation potential is the worst in the United States, due to its unique combination of recirculation patterns, stagnation, inversions, and topography. The Los Angeles Air Basin's carrying capacity (an estimate of the maximum atmospheric burden a region can have and still attain air quality standards) per capita is five times less than Houston's (36 versus 181 lbs VOC and NO_x/person/year), which has similar ozone peaks.

As a result of the state's poor air quality and large population, California residents receive more than 40 percent of the nation's population-weighted exposure to ozone values above the national 8-hour standard of 0.08 ppm, and more than 60

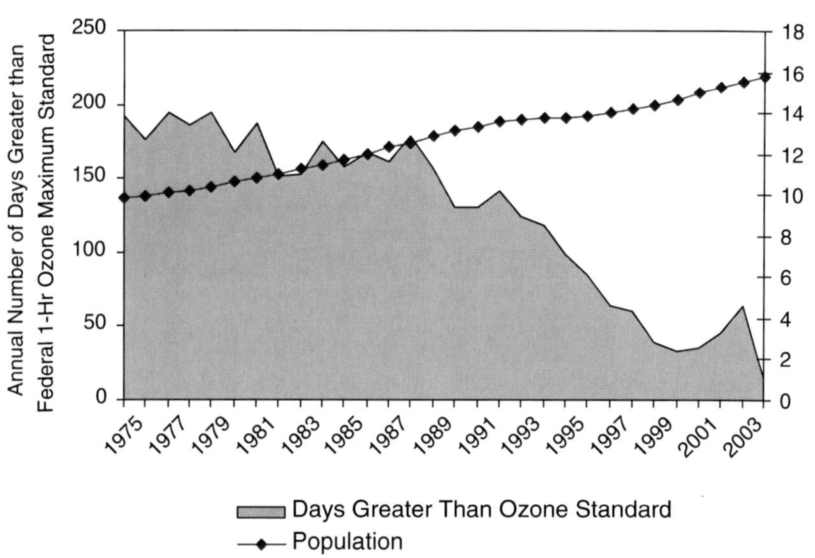

FIGURE 10-4 The South Coast air quality improves despite increased population growth from 1975 to 2004.

percent of the population-weighted exposure to $PM_{2.5}$ values above the previous annual standard of 15 μg/m^3. Southern California residents account for over 70 percent of the total U.S. exposure to exceedances of the 1-hour-average national ozone standard, and particle concentrations in SoCAB exceed California's new annual-average ambient air quality standards for $PM_{2.5}$ (12 μg/m^3) and PM_{10} (20 μg/m^3) by at least a factor of 2. Although air quality in the SoCAB and other air basins in southern California has improved significantly, attainment of U.S. and California ambient ozone standards remains a long-term goal.

In the SoCAB, which has the most extensive history of particle monitoring in California, monthly $PM_{2.5}$ and PM_{10} concentrations vary by less than a factor of 2 throughout the year, and peak during October and November (Dolislager and Motallebi, 1999). $PM_{2.5}$ is dominated by carbon (elemental and organic) and ammonium nitrate, especially in the eastern part of the basin where NO_x emissions are photochemically processed to nitric acid and pass over intensive dairy operations with high ammonia emissions. Sulfate concentrations are low (less than 5 μg/m^3) due to the lack of coal combustion and the use of low-sulfur fuels in California. PM_{10} is composed of roughly equal fractions of $PM_{2.5}$ and coarse material dominated by dust, but also including nitrate particles formed from the reaction of nitric acid with sea salt particles. PM is California's greatest challenge and is responsible for over 6,500 premature deaths per year (about 10 times greater than for ozone and 20 times greater than for cancer cases from known toxic air contaminants). Air pollution is estimated to cost Californians \$51 billion per year—\$4 billion per year in direct medical costs, with the remainder of the value assigned to premature death. CARB calculates that California gains \$3 in health benefits for every \$1 it currently invests in air pollution control.

On-road mobile sources are the single largest source category for ozone precursor pollutants, accounting for about 64, 45, and 69 percent of average daily NO_X, VOC, and CO, respectively. Most of the on-road emissions are due to gasoline vehicles, but diesel vehicles contribute substantially to NO_X emissions. Second to on-road mobile sources, stationary and area-wide sources are significant sources of VOC, while other mobile sources are currently a less important source of VOC. In contrast, other mobile sources generate relatively large emissions of NO_X, while stationary and area-wide sources are less important NO_X contributors. The vast majority of CO emissions are associated with on-road and other mobile sources.

Los Angeles historically experienced the most severe smog in the United States, but air pollution levels have improved dramatically. The health-based standards for lead, CO, NO_2, SO_2, and sulfates have all been attained, and peak ozone levels have dropped 75 percent relative to levels in the mid-1960s. California has also had success with PM_{10} and air toxics. Prior to the implementation of emission reduction measures beginning in the early 1960s, hourly averaged ozone mixing ratios approaching 0.70 ppm were reported in the SoCAB, and Stage III episodes (ozone exceeding 0.50 ppm) were relatively frequent events.

Four decades of progressively more stringent controls on emissions of NO_X and VOC, have significantly reduced the frequency and intensity of excessive ozone levels in the SoCAB. The Basin recorded 167 days exceeding the National Ambient Air Quality Standard of 0.12 ppm maximum hourly average in 1980, 158 days in 1985, 131 days in 1990, 98 days in 1995, and 33 days in 2000 (CARB, 2002). The number of days exceeding the more stringent California Ambient Air Quality Standard of 0.09 ppm for a maximum hourly average declined from 210 in 1980 to 115 in 2000. The maximum hourly average mixing ratios of ozone in the SoCAB declined during this 20-year period from 0.49 ppm to 0.18 ppm. This progress has been achieved despite almost a 50 percent increase in population and a doubling of vehicle miles traveled, and the growth of California into the fifth largest economy in the world. Since 1989, when a permanent particulate monitoring network was deployed, PM_{10} concentrations have declined about 5 percent per year in the SoCAB (Dolislager and Motallebi, 1999).

The air quality improvement was achieved through significant emission reductions (CARB, 2002). Emissions of CO and VOC (and to a lesser extent NO_X) from new passenger vehicles are reduced by a factor of a hundred in comparison to pre-control vehicles in 1963, and the standards are now applicable for 100,000 miles. Stationary source NO_X emissions have been reduced by a factor of 10 since 1980 using low-NO_X burners, selective catalytic reduction, cleaner fuels (i.e., natural gas), vapor recovery, and low-VOC coatings and solvents. From 1980 to 2000, statewide emissions from passenger vehicles (NO_X + VOC) decreased from 5,500 to 2,400 tons/day and CO from 31,000 to 12,000 tons/day, and stationary sources (NO_X + VOC) from 2,800 to 1,200 tons/day (Figures 10-5 and 10-6). California's economy grew by 75 percent despite the $10 billion cost per year for air pollution measures adopted since 1990.

Greenhouse Gases

Over the past century California has seen changes in climate-related conditions such as average temperature (up seven-tenths of a degree Fahrenheit), sea level (up 3 to 8 inches), spring run-off (decreased by 12 percent), and the timing of snowmelt and spring bloom (advanced by 1 to 3 weeks) (Cal/EPA, 2002). Knowles and Cayan (2004) project that the Sierra snowpack that functions as the state's largest reservoir could shrink by a third by 2060, and to half its historic size by 2090. California's famous vineyards, the milk industry, and ski resorts would then no longer be viable by 2100.

Environmental actions are strongly supported by the public. The July 2004 Special Survey on Californians and the Environment, conducted by the Public Policy Institute of California, found that 8 in 10 Californians support the state law that requires automakers to further reduce the emission of greenhouse gases from new cars in California by 2009.

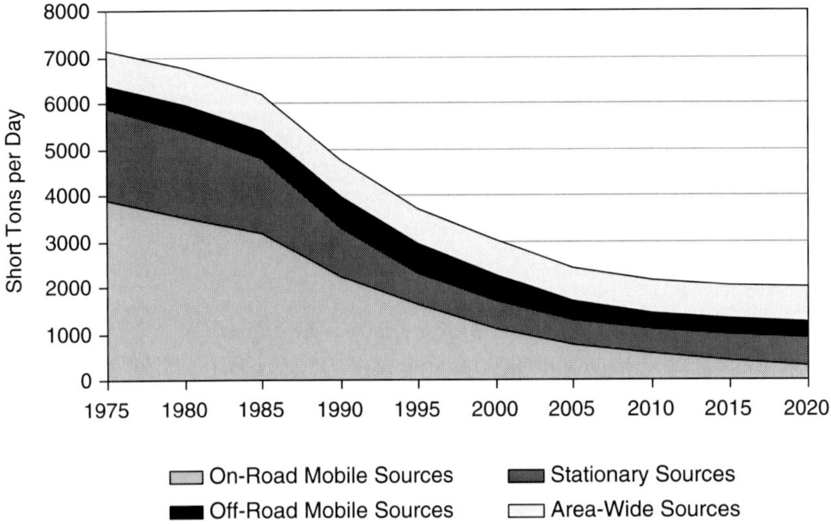

FIGURE 10-5 Statewide ROG emission trend.

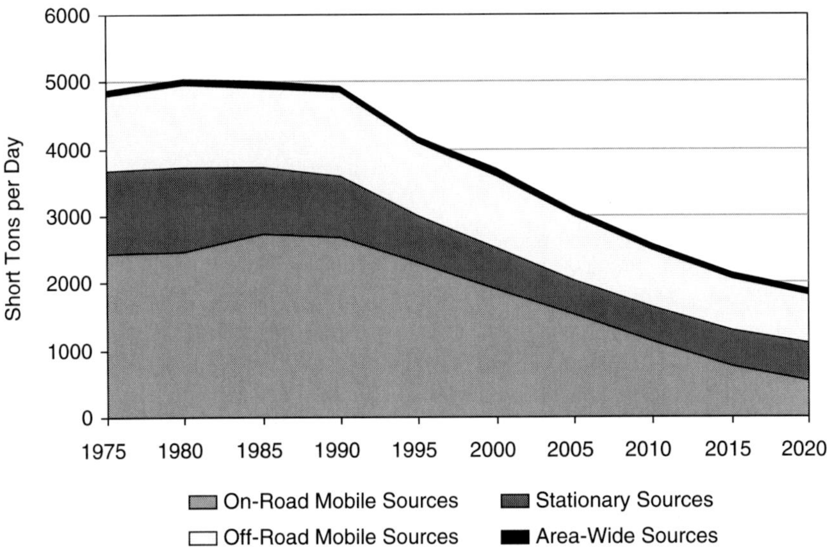

FIGURE 10-6 Statewide NO_x emission trend.

CARB considers greenhouse gases to be ozone and particle precursors as climate change can affect urban air pollution. Transportation is California's largest source of carbon dioxide, with passenger vehicles and light-duty trucks creating more than 30 percent of total climate change emissions. Due to advances in vehicle technology, it is now possible to reduce such emissions without sacrificing performance or other desirable vehicle attributes. In 2004, CARB adopted regulations that reduce greenhouse gases emitted by passenger vehicles and light trucks, although this measure is being litigated by the automotive industry. Reductions in greenhouse gases on the order of 30 percent can be achieved for all vehicle types using technologies already deployed in production vehicles. The costs to consumers are on the order of a few hundred to a thousand dollars per vehicle, and are more than offset by reduced operating costs of up to $5,000. Gas-electric hybrid vehicles and other technologies can achieve greater reductions.

The standards adopted by the Board phase in during the 2009 through 2016 model years. When fully phased in, the near-term (2009-2012) standards will result in about a 22 percent reduction as compared to the 2002 fleet, and the mid-term (2013-2016) standards will result in about a 30 percent reduction. Several technologies stand out as providing significant reductions in emissions at favorable costs. These include discrete variable valve lift or camless valve actuation to optimize valve operation, rather than relying on fixed valve timing and lift as has historically been done; turbocharging to boost power and allow for engine downsizing; improved multispeed transmissions; and improved air conditioning systems that operate optimally, leak less, and/or use an alternative refrigerant.

CARB estimates that the proposed regulation will reduce climate change emissions from the light-duty passenger vehicle fleet by an estimated 87,700 CO_2-equivalent tons per day statewide in 2020, and by 155,200 CO_2-equivalent tons per day in 2030. This equates to an 18 percent reduction in climate changes emissions from the light-duty fleet in 2020 and a 27 percent reduction in 2030.

CARB's economic analysis shows that as these expenditures occur, jobs and personal income increase. Jobs increase by 3,000 in 2010, by 53,000 in 2020, and by 77,000 in 2030, as compared to the baseline economy without the proposed regulation. Similarly, income grows by $170 million in 2010, by $4.7 billion in 2020, and by $7.3 billion 2030.

Greenhouse gas emissions from California light-duty vehicles are a small fraction of the global total. Thus the California regulation, viewed in isolation, will not wholly mitigate the potential consequences of climate change in California. Nevertheless, there are several compelling reasons to move forward, even while recognizing that by itself a California regulation will not solve the global climate change problem. First of all, the regulation is a "no regrets" policy that reduces climate change emissions from vehicles, while providing economic benefits to the state. Second, California is not acting in isolation. Other states in the United States, and other countries, have already taken or are contemplating steps to reduce greenhouse gas emissions from a variety of sectors and sources.

Finally, the longstanding technology-forcing role of California regulation should not be understated. There have been many instances where other jurisdictions have adopted motor vehicle controls that were pioneered in California. Thus there is potential for the regulation to spread to other jurisdictions and thereby add momentum to the existing initiatives that are under way around the globe.

ENERGY RESOURCES AND USE

California consumes nearly 9 EJ (slightly more than Italy's total consumption) of energy annually, including over 690 million barrels of oil (4.2 EJ). However, on a per capita energy consumption basis, it ranks 48th out of 50 U.S. states and has become a world leader in electricity created by renewable energy resources, and a recognized national leader in energy efficiency (CEC, 2006). In fact, the renewable energy industry can be credited with providing 30,000 jobs and generating $2 billion in tax revenues for the state.

Los Angeles and Southern California in general are highly reliant on petroleum and natural gas for their energy needs. Petroleum accounts for approximately 47 percent of energy consumption (mostly for transportation fuels), followed by natural gas at 28 percent (SCAG, 2006). Renewable (non-hydro), nuclear, and hydroelectric power account for 6.5 percent, 4.5 percent, and 4 percent, respectively, while coal accounts for less than 1 percent of total consumption.[1] Figure 10-7 provides a sectoral breakdown of energy consumption and illustrates that transportation is the dominant consumer. California is also highly reliant on energy imports, and in addition to electricity imports from the southwest United States (for southern California) and the Pacific Northwest, the state imports 63 percent of its petroleum and 84 percent of its natural gas from outside its borders (CEC, 2005). Moreover, nearly 42 percent of California's petroleum comes from foreign sources, and 23 percent of its natural gas comes from Canada.

Although California does possess petroleum reserves, consumption outpaced supplies long ago; and 2005 marked the first year that foreign petroleum supplies (excluding Alaska) made up a larger share than domestic supplies. Demand for transportation fuels increased 48 percent from 1985 to 2005 (CEC, 2005). In 2005, Southern California consumed approximately 8.8 billion gallons of fuel, and after an increase between 1995 and 2000, per capita vehicle fuel consumption has remained relatively constant, around 485 gallons annually (SCAG, 2006).

Figure 10-8 illustrates the region's electricity generation by source and reveals that natural gas is the primary fuel for electrical generation, followed by coal and nuclear. The two major electricity providers in the Southern California region are Southern California Edison (SCE) and LADWP. SCE is privately owned and provides nearly 70 percent of the region's electricity, while LADWP is

---

[1]These figures do not total 100 percent because the region is a net importer of electricity. Therefore, electrical imports (produced primarily by coal or natural gas) totaled 9 percent of energy use.

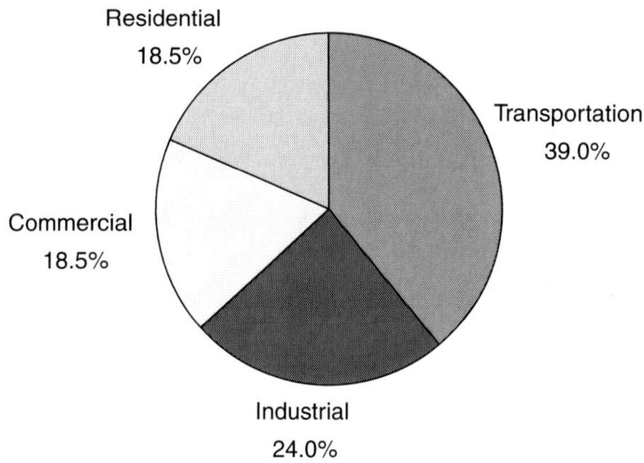

FIGURE 10-7 Energy consumption in Southern California by economic sector, 2005.
SOURCE: SCAG, 2006.

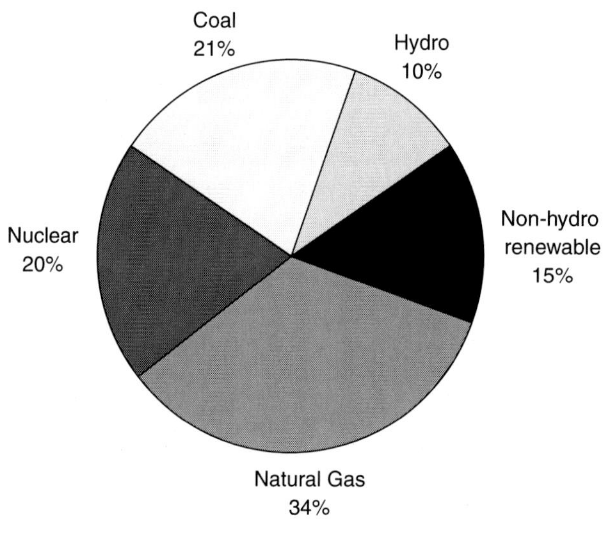

FIGURE 10-8 Electricity production in Southern California by source, 2005.
SOURCE: SCAG, 2006.

a considerably smaller and publicly owned utility, providing around 20 percent of the region's electricity including all of Los Angeles. Both utilities have set goals of reaching renewable energy generation targets of 20 percent by 2010; SCE's targets are guided by California's Renewable Portfolio Standard (RPS), which mandates that investor-owned utilities achieve specific renewable energy targets, while LADWP's target was established by the utility itself (CEC, 2007). LADWP and other publicly owned utilities are not currently bound by California's RPS.

California's aggressive policies mandating the use of renewable energy have led to other states adopting or considering similar standards. However, the development of these new resources has been slower than anticipated (CEC, 2005). California's similarly aggressive policies on energy efficiency have been more successful and are a major reason that Californians have been able to stabilize per capita electricity consumption since the 1970s, while the rest of the United States experienced an increase. Efficiency programs and standards have allowed California to reduce peak demand by approximately 22 percent over a business-as-usual scenario, and at a fraction of the cost of what additional generation might have required (Rosenfeld, 2007). Another important alternative which California is currently developing is distributed generation (see Chapter 5), and the most efficient and cost-effective form is co-generation through CHP (CEC, 2005). California has more than 9,000 MW of CHP capacity, though a major challenge to further development seems to be the lack of direct access to retail markets.

This and nearly all other energy issues fall under the purview of the California Energy Commission. The Energy Resources Conservation and Development Commission (as the Commission is formally called) was established by the Legislature in 1974 to address the energy challenges facing the state. Created by the Warren-Alquist Act which was signed into law by then-Governor Ronald Reagan, the Energy Commission is the state's principal energy policy and planning organization. Since 1974, successive administrations with bipartisan legislative support have enacted more than 100 separate laws to assist the Commission in implementing state energy policy.

The Governor appoints the five members of the Commission to staggered 5-year terms and selects a chair and vice chair from among the members every 2 years. The appointments require Senate approval. By law, one commission member must be selected from the public at large. The remaining commissioners represent the fields of engineering/physical science, economics, environmental protection, and law.

The Commission nominates and the Governor also appoints a Public Adviser, for a term of 3 years, who is responsible for ensuring that the public and other interested parties are adequately represented at all Commission proceedings. The Public Adviser's accessibility affords the citizens of California a unique opportunity to be a part of energy decision making which may affect their lives.

The Commission receives its funding from an electricity consumption surcharge collected by the electric utilities through customers' bills and is then transferred to the state treasury. CEC's budget for FY 2007/8 is $417.3 million. If an average residential customer uses 600 kWh of electricity per month, they would pay $0.132 per month toward the Energy Commission's operation. Federal money for specific energy-efficiency programs supplements the Commission's budget. The Energy Commission also manages the Public Good surcharge funds used for energy research and development and renewable energy programs. These funds are collected from customers of the investor-owned utilities. Half of the energy savings in California were made by separating utility profits from selling more energy (Chu, 2007).

POLLUTION AND ENERGY POLICIES AND THE APPROACH TO AIR QUALITY MANAGEMENT

California has adopted many emission standards which are more stringent than the U.S. standards. These include those for light- and medium-duty vehicles—exhaust and evaporative standards, for handheld and non-handheld small off-road equipment, personal watercraft, in-board motors for boats, and portable engines. Although Los Angeles has made significant progress by attaining health-based air quality standards for lead, SO_2, sulfates, NO_2, and CO, and for reducing peak ozone levels and PM, there are still many days of unacceptable ozone and particle levels.

California sets its own health-based ambient air quality standards, which are generally more stringent than those set by the U.S. Environmental Protection Agency (EPA). Even though North American agencies rely on many of the same human exposure and epidemiological studies, the standards of these agencies have striking differences (see Table 10-1). In addition, the use of allowable exceedances, spatial averaging of monitoring data, and natural (e.g., dust storms) and exceptional event (e.g., prescribed burn) exceptions can greatly reduce the stringency of these standards.

As noted above, California's air pollution control program began in 1959, when the California Motor Vehicle Pollution Control Board was created. Subsequently, under the Federal Air Quality Act of 1967, California was granted a waiver to adopt and enforce its own emission standards for new vehicles, in recognition of California's unique air quality need to set more stringent emission control requirements compared to the rest of the nation. The CARB formed in 1967 has the ability to set mobile source emission standards which are more stringent than the EPA, except for sources involved in interstate commerce: trains, planes, ships, and interstate trucking. Other states, like many in the northeastern United States, have taken advantage of their option to adopt California's mobile source emission standards.

CARB also sets regulations for consumer products, paints, and solvents, and identifies and controls toxic air contaminants. It coordinates the efforts of federal, state, and local authorities to meet ambient air quality standards, while minimizing the impacts on the economy. While local air quality management districts have the primary authority to control emissions from stationary and areas sources, CARB can assume this authority if local agencies do not develop or implement their air quality plans. Californians support and want air pollution control—65 percent of Californians put environmental protection over economic growth (although

TABLE 10-1 Ambient Air Quality Standards for North America

Pollutant	Averaging Period	United States	California	Mexico	Canada
SO_2	1 hour	—	655 μg/m³	350 μg/m³	160 μg/m³
	1 day	365 μg/m³	105 μg/m³	80 μg/m³	30 μg/m³
NO_2	1 hour	—	470 μg/m³	400 μg/m³	—
	1 year	100 μg/m³	—	—	60 μg/m³
PM_{10}	1 day	150 μg/m³	50 μg/m³	150 μg/m³	50 μg/m³
	1 year	—	20 μg/m³	50 μg/m³	—
$PM_{2.5}$	1 day	35 μg/m³	—	—	30 μg/m³
	1 year	15 μg/m³	12 μg/m³	—	—
Ozone	1 hour	235 μg/m³	180 μg/m³	216 μg/m³	100 μg/m³
	8-hour	160 μg/m³	150 μg/m³	—	—
CO	1 hour	40 mg/m³	23 mg/m³	—	34 mg/m³
	8-hour	10 mg/m³	10 mg/m³	13 mg/m³	—

California has accomplished both), and this has created a supportive legislature. For example, the California Legislature recently passed a bill to give CARB the authority to regulate greenhouse gas emissions to 1990 levels by 2020, a 30 percent reduction from business as usual.

The governor of California, with the consent of the State Senate, appoints the 11 members of CARB. It is an independent board when making regulatory decisions. Six of the members are experts in fields such as medicine, chemistry, physics, meteorology, engineering, business, and law. Five others are elected officials who represent regional air pollution control agencies—one each from the Los Angeles region, the San Francisco Bay area, San Diego, the San Joaquin Valley and another to represent other, more rural areas of the state. The first chairman was Professor Arie Haagen-Smit, who discovered how urban smog was created (Box 10-1), and the latest chairman, Ms. Mary Nichols, is an environmental lawyer with extensive experience in leadership positions at non-governmental, state, and federal organizations.

Except for the Chairman, the Board only works once per month and relies on its staff for technical input. The Board oversees a $150 million budget and a staff of over 1,100 employees located in Northern and Southern California. CARB oversees the activities of 35 local and regional air pollution control districts. These districts regulate industrial pollution sources. They also issue permits, develop local plans to attain healthy air quality, and ensure that the industries in their area adhere to air quality mandates. CARB provides financial and technical

BOX 10-1
Dr. Arie Haagen-Smit

Dr. Arie Haagen-Smit, known by many as the "father" of air pollution control, was a Dutch-born graduate of the University of Utrecht in the Netherlands and a professor of biochemistry at the California Institute of Technology in Pasadena (a suburb of Los Angeles) for 16 years before beginning his air pollution research in 1948. An avid gardener in the Los Angeles region, Dr. Haagen-Smit first became concerned about damage to his plants, such as discolored leaves and undersized flowers. His curiosity led to a series of experiments that uncovered the chemical interactions to form smog. He found that most of California's smog is a result of photochemistry: exhaust from motor vehicles and industrial facilities react with sunlight to create ozone. This break-through is the foundation on which today's nationwide air pollution standards are based. After serving as an original board member of the Motor Vehicle Pollution Control Board, formed in 1960, Dr. Haagen-Smit became the ARB's first chairman in 1968. Dr. Haagen-Smit died of lung cancer (ironically, he was a heavy smoker) two months after the CARB laboratory in El Monte was dedicated in his name in March 1977.

support to the 35 local districts. It is funded by vehicle registration fees and fees on stationary sources and consumer products. It also receives up to $166 million per year in incentive funds from fees on vehicle registration and new tire sales. This goes to diesel engine retrofits, car scrappage, and agricultural, port, and locomotive projects.

The SCAQMD, the district regulatory agency in charge of the SoCAB, is authorized to develop stationary source regulations and to set fines for violators. Thus, the biggest polluters pay the most toward funding the air pollution control effort. Also, businesses must pay annual fees for their operating permits. However, since motor vehicles account for more than half of this region's pollution, a surcharge was added in 1991 to the vehicle registration fee. Part of the surcharge goes to the SCAQMD to be used for air quality improvements involving mobile sources such as those promoting ridesharing, developing clean fuels, and as grants for programs intended to reduce vehicle emissions.

California has 4,000 air quality professionals at the state and local levels. Most of CARB's workforce are engineers and scientists, and about 20 percent have Ph.D.s and Master's degrees. CARB conducts its own vehicle testing programs and funds extramural research at a level of $5 million per year, taking advantage of the strong academic community in California and in other states. It also funds a technology demonstration and commercialization program, and the development of state-of-the-art emission, air quality, and macroeconomic models. The technology research demonstrates how reduced emissions are feasible, but the use of performance-based standards allows industry to come up with more cost-effective approaches. Enforcement and monitoring programs ensure that the emission standards are met. CARB has a requirement that the scientific underpinnings of all its regulations undergo scientific peer review. This is normally done by the University of California. Underlying this science-based approach is the willingness to move ahead in the face of some uncertainties.

Industrial sources must use the best available control technology to achieve the greatest feasible emission reductions. Box 10-2 provides a closer look at one such source. In addition to using advanced control technology in new factories, many older facilities have reduced their emissions by using retrofit equipment and switching to cleaner burning fuels.

Smaller, more personal air pollution sources, known as consumer products, also affect air quality. Products such as deodorants, hair spray, and cleaning products contain ozone-forming chemicals knows as volatile organic compounds (VOCs). In 1990, consumer products emitted about 264 tons of smog-forming

BOX 10-2
Southeast Resource Recovery Facility, SERRF

Due to an increasing population and growing amount of solid waste, the City of Long Beach, California, created a solid waste management strategy to reduce the amount of solid waste burial in landfills. Under the guidelines of this program, any solid waste that in not recycled through the City of Long Beach's sponsored curbside-recycling program is transported to the Southeast Resource Recovery Facility (SERRF). At this facility, metal is separated from waste after incineration, and as a result, SERRF collects an average of 825 tons of metal monthly and produces electricity as a by-product.

During the incineration process waste is pushed through the boiler, and the resulting ash is deposited in a "quench tank," where the ash is cooled as the tank is filled with water. After leaving the boiler, combustion gases travel through a pollution-control system where dry scrubbers are used to counterbalance acidic gases. This pollution-control system removes 95 percent of sulfur dioxide and hydrochloric acid from the gases. The gases are cleaned during another step of this process, as steady air currents flow through fiberglass bags where the gases are trapped.

The steam produced during the incineration stage is then used to power a turbine-generator, thus producing electricity. The electricity produced from waste burning is used to operate the facility, and the remaining electricity is sold for distribution. The same steam that is used to drive the generator is transferred to a condenser and converted into water to be used in the boiler stage of the waste incineration. SERRF provides over 35,000 homes with electricity, while controlling the amount of solid waste in this community. At the same time, SERRF uses a modern pollution-control system to reduce negative emissions' impacts on surrounding environments.

pollutants each day. This is more than all the refineries and gas stations in the state combined. California's clean air plan commits to an 85 percent reduction in ozone-forming pollution from consumer products.

In 1977, CARB appointed an independent panel of seven experts to review what was known about carcinogenic air pollutants in California. The panel recommended that follow-up research be done to explore further the relationship of cancer to air pollution and to determine the extent of the problem in California. California's air toxics program began in 1983 with the adoption of the Toxic Air Contaminant Identification and Control Act (AB 1807, Tanner). The act set up a process to identify a substance as a toxic air contaminant and, if necessary, develop one or more control measures to reduce emissions of that substance. In 1992 the Toxic Air Contaminant Identification and Control Act was further amended to integrate rules from the federal Clean Air Act. In some ways, this has been the most effective of Los Angeles' air pollution control programs, with at least 50 percent reductions in diesel PM, benzene, 1,3-butadiene, hexavalent chromium, perchloroethylene, and others during the 1990s.

As a result of CARB's and local air districts' work to limit air pollution, Californians today breathe the cleanest air since measurements have been recorded. The number of first-stage smog alerts (ozone > 0.20 ppm) in the Los Angeles area has been cut from over 200 per year in the 1970s to none today. This has occurred despite massive increases in population, the number of motor vehicles, and the distances they are driven.

California regulations led the way for the EPA and European Union motor vehicle emission standards that are now being adopted by many developing countries, particularly in Asia. Most of the world's population benefits from the fact that over 70 percent of the vehicles worldwide must comply with cleaner emissions standards (Michael Walsh, personal communication).

LESSONS LEARNED

CARB's technology-forcing emission standards have resulted in major advancements in emission-control technologies. Today's cleanest passenger car emits less than 1 percent of ozone precursor emissions compared to the emissions from a car produced in 1960. California's successful introduction of many emission-control programs has served as the basis for many similar U.S. programs. Through decades of emission-control success, these programs have significantly improved California's air quality, despite more than doubling the number of people and tripling the number of vehicles over the last four decades.

Two of the keys to CARB's success are the technical evaluations that go into its regulation development and the very open public process. CARB develops new emission test methods, and in some cases, proves that more stringent emission standards are achievable by funding or conducting technology demonstrations. It encourages participation by all stakeholders, including the public, industry, and communities that may be impacted by air pollution disproportionately from others. CARB meets with many stakeholders to hear concerns and to provide a mechanism for addressing their issues. It holds workshops that solicit suggestions and comments on initial issues. The technical data and assumptions are published in advance of the workshops. Regulations are first proposed in an initial report and additional workshops are held for public comment. CARB changes its proposal once significant issues are raised that warrant a revision. Once the regulation is adopted, it issues a formal response to all issues raised. The public has a chance to air their concerns directly to Board members. The Board reviews the technology and enforceability of regulations when necessary, to make sure that the regulations meet the expectation held at the time of adoption. CARB considers economic impacts of its regulations on California businesses and individuals, and regulations do not advantage or disadvantage California's manufactured products over products manufactured elsewhere in the United States or in the world.

Despite the increasing difficulty of pollutant emission reduction, technological advancements have kept control costs fairly steady. Figure 10-9 shows a 29-year timeline of the cost-effectiveness of various vehicle and fuel regulations, in dollars per pound of ozone precursor (VOC + NO_x). Most measures have cost less than \$2 per pound, which is considered to be quite reasonable in comparison to a benchmark of \$5 per pound for stationary and area source control measures. Air pollution control also has positive economic aspects. In 2001, the air pollution control industry in California generated \$6.2 billion in revenues and employed 32,000 people (EBI, 2004). The U.S. figures are \$27 billion in revenues and the employment of 178,000 people (EBI, 2004).

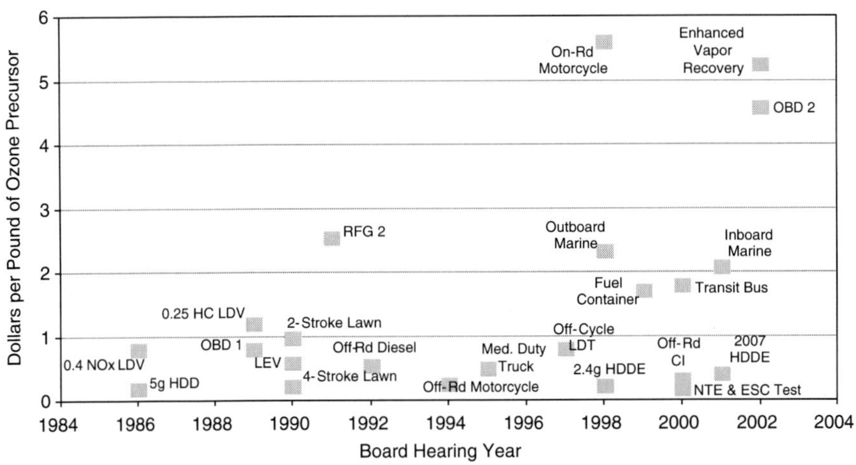

FIGURE 10-9 Timeline of CARB ozone precursor control costs.

Another groundbreaking aspect of California's air pollution control program were the efforts to independently verify the emission inventory. Past analyses of ambient air quality data reveal that, in the South Coast Air Basin, in 1987, (1) actual CO emissions were about 1.5 times higher than inventory estimates; (2) VOC emissions were about two times higher; (3) NO_X emissions were in agreement; (4) there was a large uninventoried source of unburned gasoline; (5) the agreement for VOC emissions was improving; and (6) these results were likely to apply nationwide. The good news is that VOCs, NO_X, and CO are all declining, and that the most recent CO emission inventory is in agreement with ambient measurements.

FUTURE DIRECTIONS

The region's rapid growth is expected to continue, with a further doubling of population projected for the next half-century (McConnell, 1992). Despite significant progress, Los Angeles' ozone and $PM_{2.5}$ levels are still among the highest in the United States, and per capita energy usage and greenhouse gas emissions are factors of 2-5 above those in European countries. In response, among other things, California is requiring zero-emitting cars, replacement (or retrofit with a particle filter) of every diesel engine, renewable fuels, and greenhouse gas emission reductions to 1990 levels by 2020, and 80 percent below 1990 by 2050.

Planned regulations for light-duty vehicles include a parts replacement program and improvements to the Smog Check program (i.e., more vehicles to test, loaded mode testing for gasoline trucks, and evaporative emission control testing to detect liquid leakage). For forklifts and other large spark-ignited equipment, CARB is working on lower emission standards for new equipment, as well as in-use reductions through catalyst retrofits. For heavy-duty vehicles, CARB has a broad range of controls to reduce emissions from both new and in-use vehicles (i.e., OBD, reduced idling, chip reflash,[2] gasoline tanker vapor recovery, and in-use inspections in environmental justice areas), but must go beyond these strategies to get the additional reductions needed.

For off-road compression ignition (CI) equipment, although California is preempted by federal law from controlling a significant fraction (~80 percent) of this equipment, it is a huge source of emissions and large reductions are needed. California will work with the EPA to establish more stringent nationwide standards for HC, NO_x, and PM from off-road CI engines, and to implement in-use strategies to get additional reductions. For marine engines, California plans to get reductions from existing harbor craft through cleaner engines and fuels. For the ports, reductions from land-based port emissions are planned, including from cargo handling equipment and locomotives, heavy trucks, and dredges. The ports

[2]Chip reflash refers to a program between CARB and diesel engine manufacturers to install engine control software upgrades on engines with software which normally allows the engine to switch to a more fuel-efficient but higher NO_x calibration during highway driving.

of Los Angeles and Long Beach estimate that port-related vessels and vehicles account for 12 percent of the region's PM, 9 percent of NO_x, and 45 percent of SO_x. California has set a target reduction of 85 percent diesel PM exposure (from 2000 to 2020), in spite of a projected tripling of container traffic, and has set aside $1 billion for a mitigation fund (San Pedro Bay Ports, 2007). According to the Clean Air Action Plan, the ports are obligated to include projects related to truck engine replacement and retrofit, ship cold ironing, and other emission-control enhancements on their shipyard equipment. CARB will set standards for additives to control engine deposits.

California has a goal of reducing diesel PM by 75 percent during this decade and 85 percent by 2020. This is being achieved with new emission standards, cleaner fuels, retrofits of existing engines, and enforcement programs. CARB and the EPA have adopted new vehicle standards that reduce emissions by 90 percent beginning in 2007. CARB will require aftertreatment on every diesel source where it is technically feasible. Low-sulfur fuel is required, as well as cleaner fuels like CNG and measures to reduce or to eliminate idling. Enforcement programs are used to minimize the effects of tampering and wear, especially in environmental justice communities.

The policy that people of all races and incomes need equal protection from the detrimental effects of pollution is known as environmental justice, and has emerged as an important issue in California over the past 5 years. The debate focuses on the need for community controls in addition to statewide measures. In California, people who live near busy roads are disproportionately Hispanic, Asian, and black, and from low-income families. Several Dutch studies found reduced lung function and higher asthma, hayfever, and wheezing rates for children living near heavy truck traffic.

California is also concerned about indoor sources of air pollution. Californians spend, on average, about 87 percent of their day indoors. During that time they are often exposed to air pollution levels higher than those outdoors. While the sources and risk reduction measures are known, CARB and other agencies have very little authority in this area.

REFERENCES

Blumenthal, D.L., W.H. White, and B. Smith. 1978. Anatomy of a Los Angeles smog episode: Pollutant transport in the daytime sea breeze regime. Atmospheric Environment 12:893-907.

Boucouvala, D. and R. Bornstein. 2003. Analysis of transport patterns during a SCOS97-NARSTO episode. Atmospheric Environment 37(Suppl 2):S73-S94.

Cal/EPA (California Environmental Protection Agency). 2002. Environmental Protection Indicators for California (EPIC).

CARB (California Air Resources Board). 2002. The 2002 California Almanac of Emissions & Air Quality. Sacramento, CA: California Air Resources Board, Planning and Technical Support Division.

CEC (California Energy Commission). 2005. 2005 Integrated Energy Policy Report, November.

CEC. 2006. California's Major Sources of Energy. http://www.energy.ca.gov/html/energsources.html.

CEC. 2007. 2006 Integrated Energy Policy Report Update, January.

Chu, S. 2007. The Energy Problem and What We Can Do About It. Lawrence Berkeley National Laboratory.

Croes, B.E. and E.M. Fujita. 2003. Overview of the 1997 Southern California Ozone Study (SCOS97-NARSTO). Atmospheric Environment 37(Suppl 2):S3-S26.

Dolislager, L.J. and N. Motallebi. 1999. Characterization of particulate matter in California. Journal of the Air & Waste Management Association 49:PM-45-56.

EBI (Environmental Business International, Inc.). 2004. The Economic Contribution of the California Air Pollution Control Industry. Air Pollution Research. Sacramento, CA: California Air Resources Board.

Edinger, J.G. and R.A. Helvey. 1961. The San Fernando convergency zone. Bulletin of the American Meteorological Society 42:626.

Green, M.C., R.C. Flocchini, and L.O. Myrup. 1992a. The relationship of the extinction coefficient distribution to wind field patterns in Southern California. Atmospheric Environment 26A:827.

Green, M.C., L.O. Myrup, and R.C. Flocchini. 1992b. A method for classification of wind field patterns and its application to Southern California. International Journal of Climatology 12:111.

Hayes, T.P., J.J.R. Kinney, and N.J.M. Wheeler. 1984. California Surface Wind Climatology. Sacramento: California Air Resources Board.

Keith, R.W. and B. Selik. 1977. California South Coast Air Basin Hourly Wind Flow Patterns. El Monte, CA: South Coast Air Quality Management District.

Knowles, N. and D.R. Cayan. 2004. Elevational dependence of projected hydrologic changes in the San Francisco estuary and watershed. Climatic Change 62:319-336.

LAEDC (Los Angeles County Economic Development Corporation). 2006. LA Stats 2006. http://www.laedc.org.

Lu, R. and R.P. Turco. 1994. Air pollution transport in a coastal environment. 1. 2-Dimensional simulations of sea-breeze and mountain effects. Journal of the Atmospheric Sciences 51(15):2285-2308.

Mass, C.F. and M.D. Albright. 1987. Coastal southerlies and alongshore surges of the west coast of North America: Evidence of mesoscale topographically trapped response to synoptic forcing. Monthly Weather Review 115:1707-1738.

McConnell, R. 1992. Population growth and environmental quality in California: An American laboratory. Population and Environment 14(1).

McElroy, J.L., E.L. Richardson, W.H. Hankins, and M.J. Pearson. 1982. Airborne Downward-Looking LIDAR Measurements During the South Coast Air Basin/Southeast Desert Oxidant Transport Study. Data Dept. Env. Mon. Systems Lab., U.S. Environmental Protection Agency, TS-AMD 82133.

Molina, L.T., M.J. Molina, R.S. Slott, C.E. Kolb, P.K. Gbor, F. Meng, R.B. Singh, O. Galvez, J.J. Sloan, W.P. Anderson, X.Y. Tang, M. Hu, S. Xie, M. Shao, T. Zhu, Y.H. Zhang, B.R. Gurjar, P.E. Artaxo, P. Oyola, E. Gramsch, D. Hidalgo, and A.W. Gertler. 2004. Air quality in selected megacities. Critical Review Online Version, Journal of the Air & Waste Management Association 54.

Moore, G.E., S.G. Douglas, R.C. Kessler, and J.P. Killus. 1991. Identification and tracking of polluted air masses in the South-Central Coast Air Basin. Journal of Applied Meteorology 30:715-732.

NRC (National Research Council). 1991. Rethinking the Ozone Problem in Urban and Regional Air Pollution. Washington, D.C.: National Academy Press.

O'Connor, S. and R. Cross. 2006. California's Achievements in Mobile Source Emission Control. EM, July.

Rosenfeld, A. 2007. Energy Efficiency in the U.S. and California. Presentation to committee, February 9, 2007.

Rosenthal, J. 1972. Point Mugu Forecasters Handbook. Pacific Missile Range Technical Publishers.

San Pedro Bay Ports. 2007. San Pedro Bay Ports Clean Air Action Plan Fact Sheet.

SCAG (Southern California Association of Governments). 2006. The State of the Region: 2006.

SCAQMD (South Coast Air Quality Management District). 1997. The Southland's War on Smog: Fifty Years of Progress Toward Clean Air. http://www.aqmd.gov/news1/marchcov.html.

Smith, T.B. and J.G. Edinger. 1984. Utilization of Remote Sensing Data in the Evaluation of Air Pollution Characteristics in the South Coast/Southeast Desert Air Basin. Contract No. ARB-R-84-236. Sacramento: California Air Resources Board.

Smith, T.B., D.L. Blumenthal, J.R. Stinson, and V.A. Mirabella. 1972. Climatological Wind Survey for Aerosol Characterization Program. Document MRI 72 FR 1000. Altadena, CA: Meteorology Research, Inc.

Smith, T.B., W.D. Saunders, and D.M. Takeuchi. 1984. Application of Climatological Analyses to Minimize Air Pollution Impacts in California. Prepared under Contract #AZ-119-32 for California Air Resources Board, Sacramento, CA, by Meteorological Research, Inc., Altadena, CA.

11

The Dalian Experience

PHYSICAL, ECONOMIC, AND SOCIETAL SETTING

Dalian, an important seaport, industrial, trade, and tourism center, is located in the southernmost part of the Liaodong Peninsula in Liaoning Province, northeast China (Figure 11-1). It borders the Yellow Sea to the east (300 km from the Korean peninsula), the Bohai Sea and Tianjin (a port city that provides access to Beijing) to the west, and the Shandong Peninsula to the south. Land area in Dalian totals 12,574 km^2, including 1,906 km of coastline. Dalian also has access to the river networks including the Biliu River, which is the urban district's primary water source.

Due to its location in the Northeast Asian economic zone, Dalian is an open, active, and rapidly developing city. It provides a link among the neighboring countries of North Korea, South Korea, Japan, and eastern Russia, and goods are transported from Dalian to Asia and Europe.

The terrain is high and wide to the north, and low and narrow to the south. The surface relief inclines from the central axis of the peninsula towards the Yellow and Bohai Seas, with a long and slow slope along the Yellow Sea. Changbai Mountain (1,132 m), an extension of the Qian Mountain Range, is the highest point in the region. Karst and coastal erosion landforms mark the lower terrain. The urban district extends along the valley area and is slowly expanding into the surrounding hillsides.

Atmospheric circulation in the city is dominated by westerly winds and subtropical systems. Dalian has a maritime climate with northerly and southerly monsoons dominant in winter and in summer, respectively. Northerly and southerly winds dominate in autumn and spring, respectively.

FIGURE 11-1 Dalian's location within China.

The Dalian administration governs six districts, three county-level cities, and one county (Figure 11-2). Since 1980, Dalian has administered four national-level open zones: a development zone, a high-tech park, a free trade zone, and a resort area. Such areas encourage foreign investment and provide opportunities to adopt innovative practices. The Dalian Environmental Protection Bureau, established in 1985, divides Dalian into three divisions for the purpose of environmental management: the city center (123 km^2), the urban district (2,414 km^2), and the new urban district (2,292 km^2) (DEPB, 2007). The city center includes four of Dalian's six districts—Zhongshan, Shahekou, Xigang, and Ganjingzi. The urban district contains six subsections, including the development zone and the high-tech park. Between the urban district and the city center is the new urban district.

With a population of ~6 million, Dalian's GDP was 229 billion RMB ($28.6 billion) in 2005 (DFTECB, 2007). Secondary and tertiary industries[1] contribute 45 and 46 percent, respectively, to total GDP. In 1984, the State Council autho-

[1]Primary industry refers to production of raw materials, such as agriculture or mineral extraction. Secondary industry involves processing raw materials, alternately referred to as manufacturing. Tertiary industry is also commonly referred to as the service industry and includes such industries as finance, tourism, and food service.

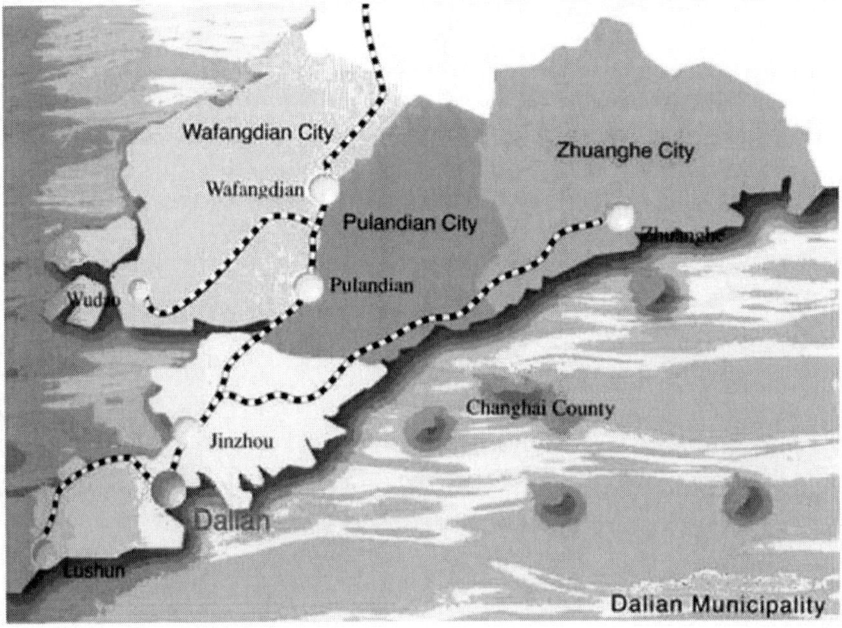

FIGURE 11-2 Dalian municipality.

rized Dalian to be among 14 "open coastal cities" (including Shanghai, Tianjin, and Guangzhou), effectively opening it up to outside investment, and in 1994 the State Council elevated it to a vice-provincial (or prefecture-level) city, granting local leaders more authority in economic decision making. In 1998, Dalian was designated as a National Model City for environmental protection and was also named China's Top Tourism City. In 1999, the United Nations (UN) confirmed Dalian as the Asia-Pacific Region's Leading Urban Governance Model City. In 2001, Dalian received the UN Environment Program's "Global 500 Award" and the UN's Human Habitat Award. In 2003, it won the China Habitat Environment Prize and the mayor at the time, Bo Xilai, was awarded the China Environment Prize.

Since 2002, Dalian has become a base for petrochemical production, electronics, informatics and computer software development, advanced machinery manufacturing, shipmaking, and international shipping. The Dalian port currently trades with over 300 ports in 160 regions and countries, accounting for 70 percent of cargo and 90 percent of containers in northeastern China. It is the largest petroleum and chemical port in China, and handles the most grain (mostly exports) in all of Asia. A planned railway in northeast China (the "Golden Passage") will connect to more than 10 ports in Russia and North Korea. A major international shipping center is expected to be completed by 2020.

Dalian is home to over 200 scientific research institutions, 22 universities, 250,000 scientific/technical personnel, and more than 200,000 students. In 2002, The Chinese Academy of Sciences' Institute of Chemical Physics and British Petroleum established a joint research center. Research areas include new methods for hydrogen storage, conversion of methane to hydrogen and aromatics, and fuel cell/hydrogen technology. It has developed fuel cell engines for minibuses and personal vehicles (DICP, 2006). In 2002, 50 kW fuel-cell bus engines were introduced with the goal of having these buses running in Beijing and other cities by 2008. It obtained a U.S. patent for a catalyst that could boost the yield of ultra-clean diesel from syngas.

SOURCES AND LEVELS OF AIR POLLUTION

Currently, the Dalian air quality monitoring network includes 10 stations that monitor inhalable particulate matter (PM_{10}), sulfur dioxide (SO_2), nitrogen dioxide (NO_2), and carbon monoxide (CO). Dalian's air quality ranked 16th nationally in 2005, based on the annual Air Pollution Index (API) rating. Major emissions sources are:

- **Coal burning—**a primary energy source in northern China.
- **Industrial processes—**primarily from steel plants, chemical manufacturers, and cement plants.
- **Motor vehicle exhaust—**as of 2005, Dalian's vehicle fleet (public and private) included 240,000 light-duty gasoline vehicles, 135,000 light- and heavy-duty diesel vehicles, 158,000 motorcycles, and 6,000 non-road vehicles.
- **Resuspended dust—**the 17 days with recorded API reaching 500 between 2001 and 2005 resulted from invasive dust storms. Local dust is estimated at 110,000 tons in 2005, ~25 percent of which originated from the city center.
- **Transport of pollutants**—from other parts of China and neighboring countries.

In 2005, there were 347 Class II days in Dalian, including 61 Class I (Excellent) days. Five dust storms (123 hours) were recorded in 2005. Annual dustfall in downtown Dalian averaged 15.7T/ km^2*M, which is double Liaoning Province's standard[2] (DEPB, 2005). Annual average concentrations are 85 $\mu g/m^3$ for PM_{10}, 44 $\mu g/m^3$ for SO_2, 32 $\mu g/m^3$ for NO_2, and 480 $\mu g/m^3$ for CO, in compliance with Class II standards. The pH value of the urban districts' rainfall ranged from 3.6 to 7.9 with ~25 percent of acid precipitation in the southern part of the

[2]Liaoning Province sets its own criterion for dustfall, though the province does not set standards for other air pollutants.

city. Acid rain was observed in the city for the first time in 2004, and it occurred frequently in 2005.

Daily air quality reports (see Table 11-1) and an annual environmental quality report are available from the *Dalian Daily*, from the Dalian EPB website (http://www.dlepb.gov.cn/), and from the State Environmental Protection Agency (http://www.sepa.gov.cn).

Table 11-2 shows that daily exceedances were highest for PM_{10} (7.2 percent), followed by SO_2 (2 percent), indicative of impacts from coal combustion. Although primary energy consumption from coal was reduced from 69 percent to 60 percent during 2001-2005, exceedance of SO_2 increased by ~4 percent. The highest levels occur in winter due to increased heating (November to March) coupled with a shallow surface layer. Table 11-3 (next page) shows that the major pollutant is PM_{10} in all seasons, with elevated SO_2 and CO concentrations during winter.

TABLE 11-1 Sample Daily Air Quality Report from Dalian Environmental Protection Bureau

Date: 06-30-2006

District	API	Main Pollutant	Air Quality
Xinghaisanzhan	60~70	PM_{10}	Very Good
Ganjingzi	70~80	PM_{10}	Very Good
QingniwaBridge	60~70	PM_{10}	Very Good
Fujiazhuang	40~50	None	Excellent
Qixianning	50~60	PM_{10}	Very Good
Lushun	50~60	PM_{10}	Very Good
Jinzhou	50~60	PM_{10}	Very Good
Development zone	60~70	PM_{10}	Very Good
Zhoushuizi	60~70	PM_{10}	Very Good
Jinshitan	40~50	None	Excellent

TABLE 11-2 Summary of Annual Exceedances of Air Pollutants from 2001 to 2005 (percent)

Year	SO_2	NO_2	PM_{10}	CO	≥ Grade II
2001	0.8	0.0	6.6	0.1	81.4
2002	1.3	0.0	6.7	0.0	89.7
2003	1.9	0.0	6.0	0.0	71.2
2004	1.3	0.0	7.9	0.1	76.3
2005	4.7	0.1	8.6	0.0	80.0
Average	2.0	0.0	7.2	0.0	79.7

TABLE 11-3 Average Seasonal Concentrations of Criteria Pollutants (2001-2005) in μg/m^3

Seasons	SO_2	NO_2	PM_{10}	CO
Spring	35	31	108	510
Summer	16	25	69	440
Autumn	29	29	78	580
Winter	69	31	86	720
Mean	37	29	85	560
Class I standard	20	40	40	4000
Class II standard	60	80	100	4000

Figure 11-3 illustrates elevated concentrations at the Ganjingzi district, followed by the Qinghiwa Bridge area.

Figure 11-4 shows a decreasing trend for total suspended particulate (TSP) with compliance to the Class II standard achieved after 1996. PM_{10} remained below Class II standard, and shows little variation from 2001 to 2005. SO_2, in Figure 11-5, has been in compliance with the Class II standard since 1995. Although a downward trend is found between 1993 and 2000, SO_2 increased from 1990 to

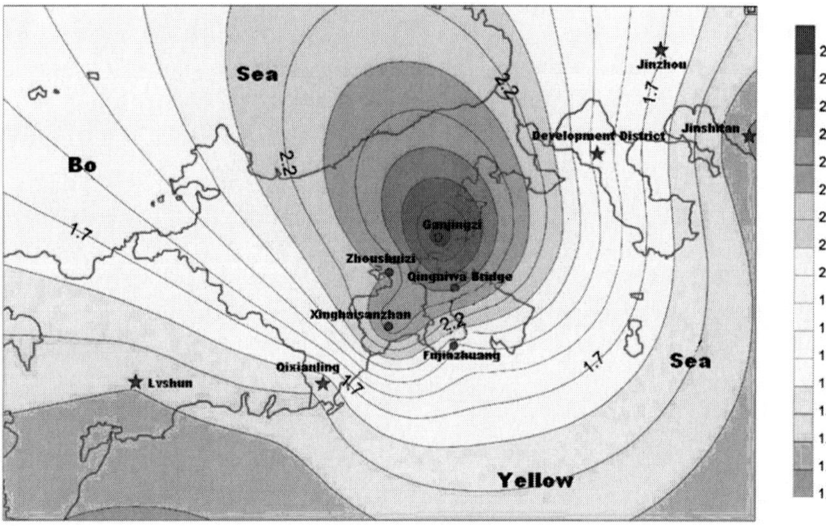

FIGURE 11-3 Spatial variations of SO_2, NO_2, and PM_{10}, interpolated from measurements averaged from 2001 to 2005.

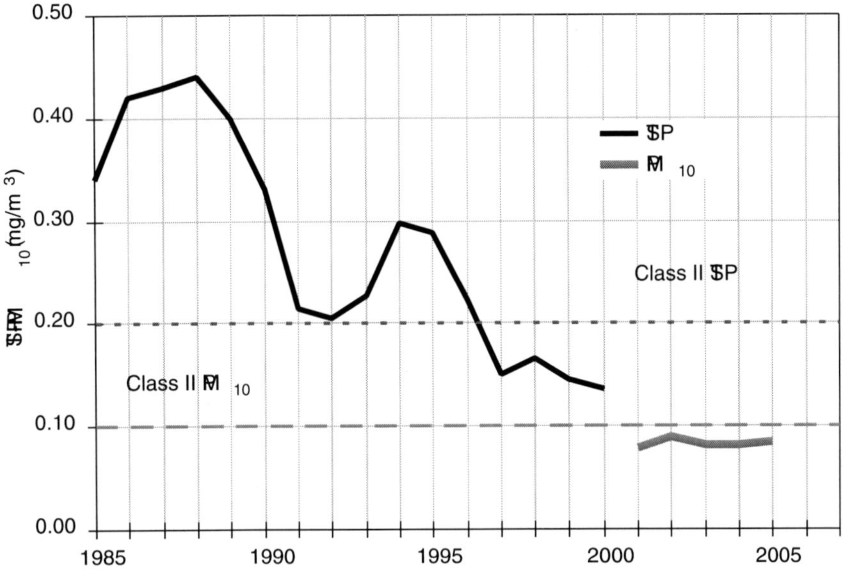

FIGURE 11-4 Trends of TSP and PM_{10} concentrations in Dalian since 1985. Also shown are Class II standards.

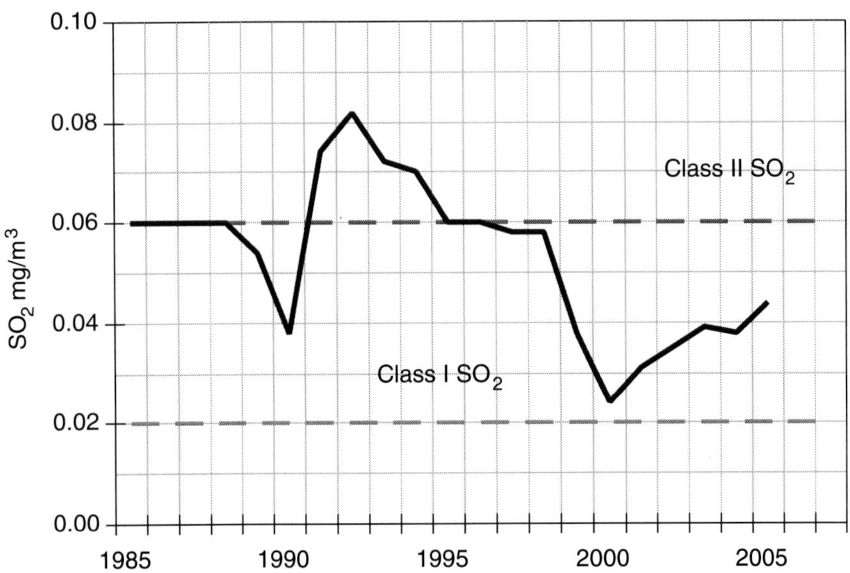

FIGURE 11-5 Trends of SO_2 concentrations in Dalian since 1988. Also shown are Class I and Class II standards.

1993, and from 2000 to 2005. NO_2 (not shown) increased along with TSP between 1991 and 1995; then it gradually descended toward the Class I standard. CO (not shown) remained stable and generally achieved the Class I standard.

Emissions estimates in Figure 11-6 show that 70.5 percent of SO_2 and 57.4 percent of soot dust originated from industries. Fugitive dust sources are not included in the inventory. In addition to local sources, Dalian is also affected by transported dust from Mongolia to the north and by steel production and coal combustion from the northern cities (e.g., Anshan, Liaoyang, Shenyang, Fuxin, Benxi, and Fushun).

By 2005, there were 441 registered industries reporting environmental statistics in Dalian. Table 11-4 shows that the top 29 sources accounted for 75.3 percent of total emissions. The thermoelectric power industry accounted for 49.5 percent and the petrochemical industry accounted for 25.8 percent of industrial SO_2 emissions, whereas the cement industry contributed to 23.4 percent of industrial TSP. These 29 enterprises have become the focal point of Dalian's air pollution

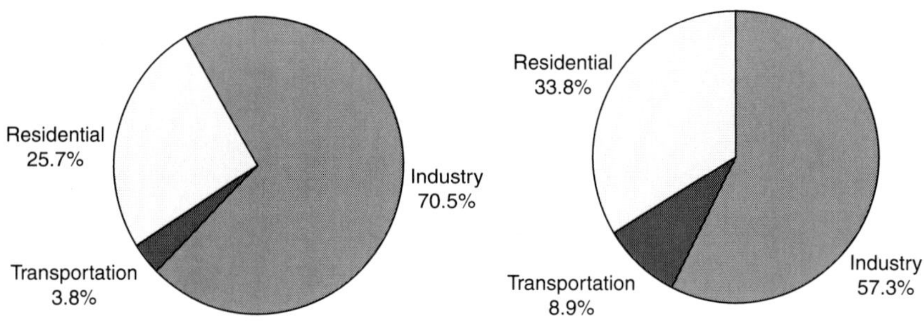

FIGURE 11-6 Distribution of 2005 emissions estimates for SO_2 and TSP.

TABLE 11-4 Dalian's Major Sources of SO_2, Soot, and Dust (2005)

	Standardized Pollution Load		SO_2		Fly Ash/ Suspended Dust	
Industry Unit	Number	Percent	Emission Amount/kt	Percent	Emission Amount/kt	Percent
Thermoelectric power	10	43.11	41.48	49.49	5.96	14.26
Petrochemical	3	22.75	21.65	25.83	3.94	9.43
Cement	12	7.48	1.27	1.51	9.79	23.42
Other major sources	4	1.94	—	—	0.77	1.84
Subtotal of the major sources	29	75.28	64.40	76.84	20.46	48.95
Total of the city	441	100.00	83.81	100.00	41.80	100.00

control strategy. Accordingly, the electric power, petrochemical, and building material industries have become the three key industries in the city targeted for pollution control.

Figure 11-7 shows that 62.2 percent of inventoried emissions are located in the city center, including 52.5 percent from the Ganjingzi District. Estimated 2005 emissions for SO_2 were 506.6 thousand tons (60.2 percent of the total).

Ganjingzi District, north to northwest of the city center, is the old industrial area. It is home to many large state-owned enterprises such as the Dalian Second Power Plant, the Dalian and Xiaoyetian cement plants, the Dalian Steel and Chemical Corporations, and the Dalian Oil Plant. Prevailing northerly winds during winter transport air pollution to the city center.

Dalian has attempted to limit coal consumption during rapid economic growth as shown in Figure 11-8. SO_2 emissions per 10,000 units of GDP in Dalian is 6.37 kg, lower than in Shanghai (7.2 kg), and in the nation (18.9 kg). An apparent downward trend is found for inventoried TSP emission (Figure 11-9), which follows a similar pattern to TSP concentrations as shown in Figure 11-4.

ENERGY RESOURCES AND USE

The primary fuels used in Dalian are coal, oil, and natural gas that are mostly imported from other areas. Dalian's local coal reserves are limited and have already been exploited. Renewable energy resources are also limited and are not widely utilized.

For 2003, consumption was 15 million tons of coal (crude fuel); 14.2 billion kWh for electricity; 208.5 billion m^3 for coal gas; 124,300 tons for liquefied petroleum gas (LPG); 1.1 million tons for fuel oil; and 180 × 105 GJ for heating. Total and per capita energy consumption in 2005 was 13 and 2.16 tce, respectively, and

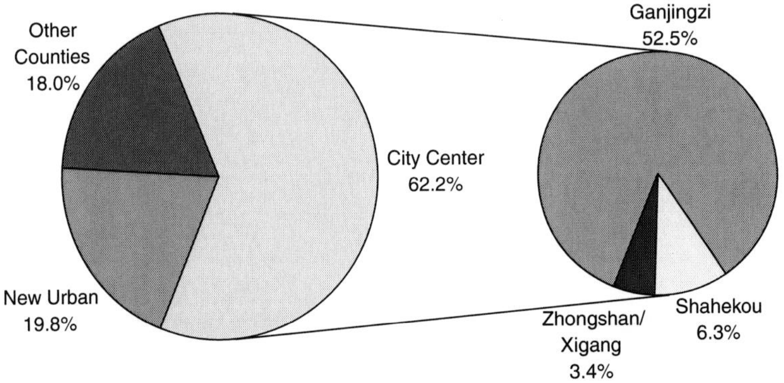

FIGURE 11-7 Standardized pollution load in different districts.

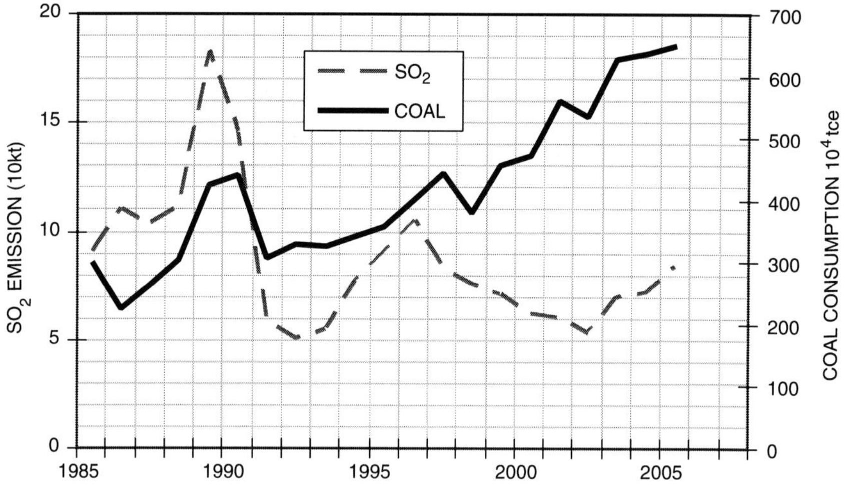

FIGURE 11-8 Relationship between industrial coal consumption and SO_2 emissions (1985-2005).

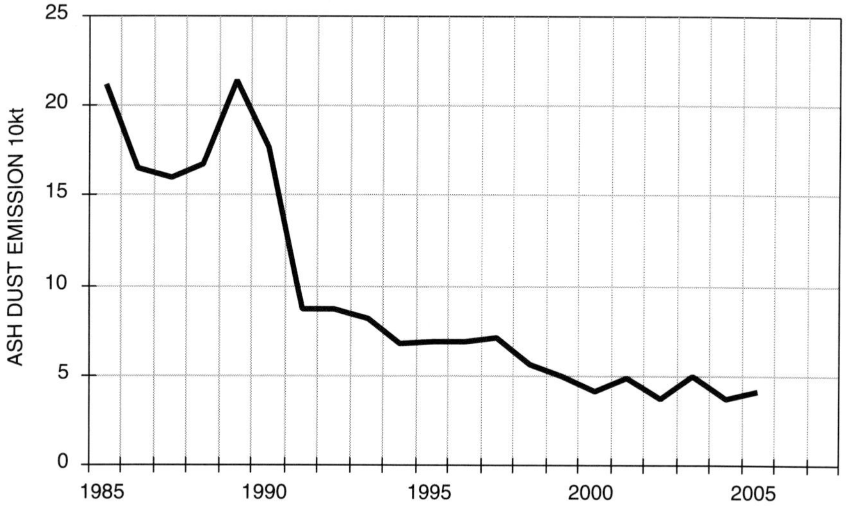

FIGURE 11-9 Trends of emission estimates for inventoried sources (1985-2005).

energy consumption per unit GDP was 0.6 tce/10,000 Yuan. As shown in Table 11-5, energy consumption per capita is 40.6 percent lower than in Beijing, but nearly double that of the nation.

Figure 11-10 shows that gasoline and diesel fuel account for 32.5 percent of terminal energy consumption, followed by electric power (19 percent) and coal (17.9 percent). Coal accounts for 59.3 percent of the primary energy fuel.

The contributions of primary, secondary, and tertiary industries to Dalian's GDP in 2003 are shown in Table 11-6. Figure 11-11 shows that secondary industry accounts for nearly half of terminal energy consumption, followed by transportation (~30 percent). It is apparent that secondary industries are highly energy intensive.

As of 2003, Dalian had 28 power plants, though this number encompasses 10 combined heat and power (CHP) plants, and 8 plants for private industrial supply, in addition to the more conventional coal-fired plants (3), wind farms (3), hydropower (3), and fuel oil (1) plants. The installed capacity was 2,416 MW.

TABLE 11-5 Energy Consumption Intensity

	Unit	Dalian	Liaoning	Beijing	National Average	World Average
Energy consumption per capita	Tce/capita 2003	2.16	—	3.64	1.14	2.05
GDP energy consumption	Tce/10000 Yuan GDP 2005	0.60	1.83	0.80	1.22	2.5

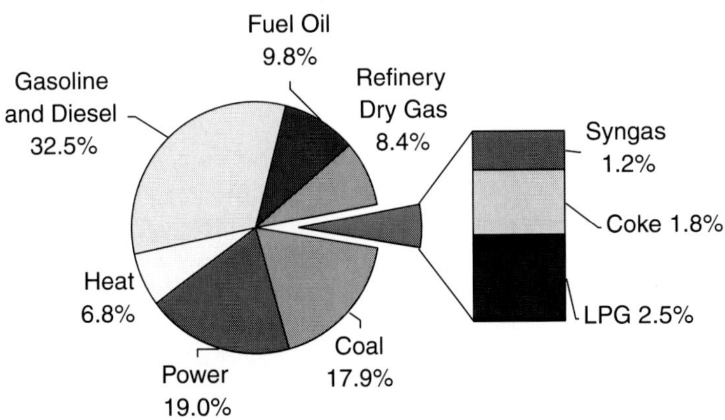

FIGURE 11-10 Main terminal energy consumption

TABLE 11-6 Comparison of Contribution Proportion of Each Industry to GDP of Dalian in 2003

Items	Primary[a] Industry	Secondary[b] Industry	Tertiary[c] Industry	Total
Domestic GDP (billion RMB)	14.56	78.21	70.49	163.26
Proportion/percent	8.9	47.9	43.2	100

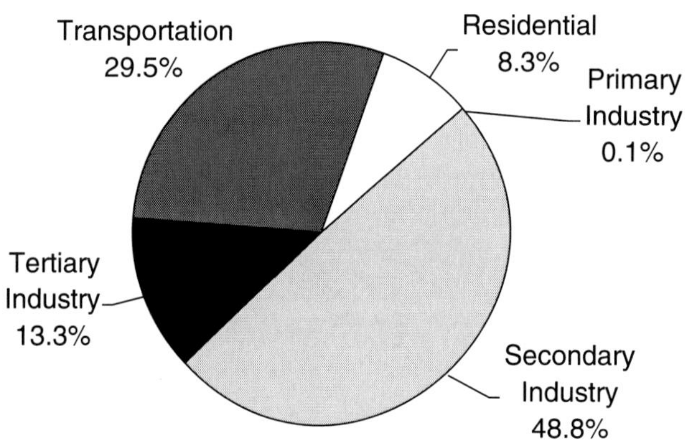

FIGURE 11-11 Structure of energy consumption in Dalian (2003).

For 2003 and 2004, the generated energy was 12.3 and 13.65 billion kWh; and annual power consumption was 14.2 and 15.5 billion kWh, respectively. Note that power production was slightly lower than power consumption, signaling a need to import electricity from other regions.

In 2003, daily production of synthetic coal gas was 730,000 m^3, with a total capacity of 208,500,000 m^3. Residential use consumed 137,000,000 m^3. The total capacity of LPG supply was 124,000 tons, of which residential use consumed 61,400 tons.

District heating is the primary means of heating in Dalian, and it is provided by steam from central heating plants. City heating, industrial purposes, and cooling utilize 80 percent, 18 percent, and 2 percent, respectively, of the total steam production from these plants. Since the early 1990s, Dalian has developed CHP co-generation and central heating systems. New energy-efficient equipment and building materials have been used. Renewable and clean energy sources are being explored, such as biogas, straw biomass conversion, wind power generation, solar energy water heaters, and seawater heat pumps. Energy conversion levels are low and energy efficiency needs to be improved, particularly since electric power consumption has been increasing rapidly. The exploitation of new and renewable energy sources is insufficient to meet rising energy needs.

POLLUTION AND ENERGY POLICIES AND THE APPROACH TO AIR QUALITY MANAGEMENT

Air pollution control measures have been implemented that include industrial source controls, new source prevention, centralized heating, gasification and co-generation, urban planning and restructuring, and vehicle emission controls. From 2001-2005, the city invested 12.9 billion RMB, or nearly 2.1 percent of GDP, into environmental protection (DEPB, 2005).

Industrial Source Control

Since 1985, 3,678 projects on emission source control were initiated at a cost of 1.12 billion RMB. More than 7K billion m^3 of sulfur-containing waste gas was treated with an estimated 8.73 million tons reduction in suspended dust (Table 11-7). Efforts for SO_2 reduction include flue gas desulfurization (wet and dry) and increased combustion efficiency through boiler improvements. Dust removal efforts included electrostatic precipitators, bag houses, and cement furnace improvement. In 2004, SO_2 emissions were 103,000 tons, ~20 percent below the target (125,000 tons) set by the State Council. As part of the joint anti-pollution/pollution prevention program between Japan and China (in Dalian, Chongqing, and Guiyang), Dalian received US$100 million to finance 13 projects on cleaner production and technological transformation. This resulted in annual

TABLE 11-7 Industrial Pollution Control Statistics, 1985-2005

Years	Investment (million Yuan)	Pollution Control Measures (Number)	Dust Reduction (thousand tons)
1986-1990	141.75	2491	960
1991-1995	94.76	616	1220
1996-2000	192.13	401	1390
2001-2005	693	170	5160
Total	1121.64	3678	8730

emissions reductions of 20,000 tons of SO_2, 20,000 tons of fly ash, and 1,000 tons of NO_x in Dalian.

Several small factories without pollution-control devices were closed to eliminate pollution. These included 4 lime factories, 11 charcoal factories and 150 small-scale charcoal ovens.

New Source Prevention

Since 1986, new sources are required to perform environmental impact assessments. New sources need to be in compliance with the *santongshi*[3] policy. In 1996, 14 projects were rejected due to potential pollution impacts and 206 factories were fined for failing to follow *santongshi*. In addition, *santongshi* encourages the substitution of new technologies for the old ones.

Centralized Heating, Gasification, and Co-generation

Since 1992, central heating and power co-generation has been encouraged. The Beihai, Xianghai, Taishan (Phase I), and Donghai (Phase II) thermoelectric plants have dismantled inefficient boilers and promoted central heating. In 2004, there were 37 districts receiving central heating, covering an area of 12 million m^2; 1,194 boilers and 887 chimneys were removed and the central to non-central heating ratio for downtown districts reached 74 percent (DEPB, 2005).

In the past, coal gas or LPG have been used for commercial and residential cooking. At present, the two gasification plants supply 100 percent of the city's cooking fuel, eliminating the use of coal briquettes (DEPB, 2007).

Urban Planning and Restructuring

Several projects (e.g., "May 24" [*wuersi*] soot treatment, "blue-sky jade-sea" [*lantianbihai*]) were initiated to remove 1,687 coal-fired low-capacity boilers (below 1 ton) and 860 chimneys in 1999. SO_2 and fly ash emissions were reduced by 2,550 and 31,000 tons, respectively. A coal distribution center was built in Dalian to discourage the use of high-sulfur and -ash coal. The center promotes and sells clean coal with sulfur content lower than 0.5 percent by weight and ash content lower than 10 percent. Similar coal standards are in place for major industries.

Since 1992, if corporations are in non-compliance for energy and water consumption by the specified deadline, they are required to implement mitigation plans, or to relocate their plants outside of urban areas (Bai, 2002). From 2001 to 2005, relocation from downtown to outside urban areas was completed for 85

[3]This is a Chinese policy to ensure that environmental regulations are addressed at the planning, construction, and operation stages of project development.

plants. Advanced processes and equipment were installed for city center coal-gas plants in order to maintain compliance.

Approximately half of Dalian's residents utilize public transportation for trips in the city, and though this share is beginning to give way to increased personal vehicle use, the government has invested $232 million to construct a light rail transit (LRT) network, in order to better serve urban areas and to guide future urban growth, with planned extensions nearing completion (Liu and Yang, 2006). In addition to the LRT network, Dalian has three tram lines and 87 bus routes to serve the nearly 3 million passenger trips taken daily.

Vehicle Emission Control

Dalian has implemented an annual motor vehicle emissions inspection program. It also conducts on-road emission detection to prohibit non-compliant vehicles from driving (DEPB, 2005). Unleaded gasoline and LNG gas have been in use since 1997 and 2000, respectively. Taxis with dual LNG and conventional fuel systems are now in use. Ethanol (10 percent in gasoline) use was required in Dalian beginning in 2000, and was further mandated as part of a province-wide regulation in 2004. Violators will be fined from US$600 to $3,600.

Dalian has increased the width, length, and quality of roads. Intelligent traffic management at the Dalian Traffic Direction and Control Center was implemented to improve traffic flow. From 2000 to 2005, highway road length increased from 549 to 640 km. In 2005 there were 425,000 motor vehicles registered in the city, with a projected annual increase of 15 percent per year, or approximately 700,000 motor vehicles in 2010 (DEPB, 2005).

Environmental Regulations and Civil Society

In 1991, the city promulgated its first environmental management statute, and in 2000 it called for "further control of air pollution." Presently, there are 20 local environmental protection rules, which play important roles in the environmental quality of the city.

Thirty-one corporations received ISO 14001 environmental management system certification, and 42 corporations were named Model Environmental Protection Corporations. The Dalian Environmental Protection Volunteer Association, established in 2003, encourages civilians to improve the environment by driving one day less per month, along with other activities.

FUTURE DIRECTIONS

Dalian's pollution prevention and control strategy, as well as its energy forecasts, are based on its 11th Five-Year Plan for 2006-2010. It is estimated that

Dalian's GDP will increase by 13 percent annually. Energy demand will also increase, as shown in Table 11-8.

There are three projected energy supply and demand issues over the next 10-20 years:

- Coal consumption for primary energy will decrease from 62.2 percent in 2003 to about 44 percent in 2020. However, coal will remain the main source of energy (Table 11-9).
- The energy consumption structure will change. Natural gas and nuclear energy consumption will increase, and the effects of energy-saving technologies will become apparent.
- Energy consumption will increase at a rate lower than that of GDP.

Air Quality Goals

Dalian's air quality goals for 2010 are shown in Table 11-10. The Chinese government is transitioning from "sacrificing" to "optimizing" the environment while growing the economy. By 2010, Dalian is expected to attain Class II air

TABLE 11-8 Forecasted Energy Demand in Dalian in Mtce

Category	2003	2005	2010(a)	2010(b)	2020
Terminal coal consumption	1.48	1.44	0.60	0.40	0.11
Coke	0.16	0.20	0.30	0.20	0.20
Oil-fired	0.87	0.72	0.50	0.30	0.23
Gasoline, diesel, and coal oil	2.90	3.64	2.73	2.35	3.15
LNG	0.21	0.39	4.91	5.46	11.38
Coal consumption for power generation	5.68	6.43	9.00	9.11	11.48
Other electricity	0.22	0.22	0.00	0.00	0.00
Gross	11.52	13.04	18.04	17.82	26.54
Energy growth rate per year (percent)		6.38	6.71	6.46	3.94
GDP grow rate (percent)		14.3	13	13	7

TABLE 11-9 Forecasted Coal Consumption in Dalian

Category	2003	2005	2010(a)	2010(b)	2020
Coal consumption (Mtce)	7.16	7.87	9.60	9.51	11.59
Proportion of coal in prime energy (percent)	62.15	60.35	53.22	53.37	43.67
Proportion of coal for power	79.33	81.70	93.75	95.79	99.05

TABLE 11-10 Current and Target Values for Air Quality and Emission Standards, and for Energy Consumption

No.	Items	Current value (2004)	Target value (2010)
1	Days of air quality reaching Class I standard	75	100
2	Days of air quality reaching or better than Class II standard	350	360
3	Annual average PM_{10} (mg/m^3)	0.086	≤0.075
4	Annual average SO_2 (μg/m^3)	0.038	≤0.030
5	Annual average NO_2 (μg/m^3)	0.036	≤0.040
6	Fly ash emissions (10,000 tons)	5.77	5
7	Industrial ash dust emissions (10,000 tons)	2	1.5
8	Total SO_2 emissions (10,000 tons)	10.3	9
9	SO_2 emission from medium or large sources (10,000 tons)	7.5	7
10	Opacity (percent)	87	93
11	City's central heating to coal consumption ratio/percent	73.9	85
12	Energy consumption per GDP (tons of coal/10,000 RMB)	0.6	0.45

quality standards for 360 days, which would include at least 100 Class I days. Total SO_2 emissions in Dalian should be reduced to <90,000 tons annually.

In order to achieve these energy and air quality targets, Dalian has set a number of specific goals, including:

- *Use domestic supercritical technologies for new coal-fired power plants.* Zhuanghe Power Plant will install two 600 MW generators which have been domestically developed based on foreign advanced technologies (MOST, 2006).
- *Construct Liaoning's first nuclear power plant.* The Hongyanhe nuclear power plant plans to begin operating by 2010. Starting and target capacities are 1,000 and 4,000 MW, respectively.
- *Build and expand coal-fired thermoelectric plants and expand central heating.* Ten CHP plants will increase installed capacity by 2,000 MW and will increase district heating areas by 10 million m^2. The central to non-central heating ratio of the city center will surpass 85 percent by 2010 and will reach 70 percent in the three northern cities and Changhai County.
- *Improve pollution-control equipment.* Desulfurization devices (with removal efficacy of ≥85 percent) will be installed for new and existing coal-fired power plants. Plants should adopt low-NO_x burners and leave space for installation of flue gas denitrification equipment. Desulfurization devices will be installed on existing power plants by 2010 (DEPB, 2005).

- *Phase out old wet dust removal devices.* A more efficient bag house or precipitator will be installed on new coal-fired plants. The average efficiency of dust removal equipment in the city should reach 98 percent in 2010.
- *Enforce industrial dust emissions.* By 2009, Dalian will close cement factories in four urban and suburban districts and will phase out cement corporations using polluting vertical kilns. 1,392 small coal-fired boilers and 925 chimneys will be removed. Small (<10 tons) heating coal-fired boilers are to be eliminated by 2010.
- *Ensure efficient dust control.* During demolition, water should be sprayed to minimize suspended dust. Earth and stone should be transported in bags and covered when in bulk. Windscreens and shelter belts should be installed around construction sites. Sand-gravel stacks and mortar mixers are forbidden; hillocks should be covered and roads should be hardened at construction sites.
- *Conduct air quality research.* This research should lead to specific dust control measures.
- *Establish a vehicle and traffic control management system.* By combining annual vehicle inspections, road surface monitoring, and parking management into a unit, Dalian will be able to manage vehicle emissions more efficiently. An automatic emission monitoring system will be established to inspect busy traffic sections.
- *Control vehicle fuel oil components and minimize vapor emissions at gas stations.*

Energy Sector Strategy

Dalian's energy and air quality management strategy builds on the principles of developing a circular economy (see Chapter 5, Box 5-1). The city plans to give priority to projects that save energy, improve efficiency, and encourage economic development. The government will facilitate this by exercising macroeconomic control to establish a market for energy efficiency. Specifically, it will:

- *Adjust the terminal energy structure.* Dalian seeks to decrease its reliance on coal while increasing electrification, gas-fired power generation, and central heating. The first planned LNG receiving station (expected to handle 3 million tons annually by 2009) will allow 3-4 thermal heating plants to operate on natural gas instead of coal (DEPB, 2005).
- *Develop a regional energy base.* As an international shipping center for petroleum, LNG, and coal, Dalian is expected to become a power generation base for northeast China. Related plans include developing petroleum refining, nuclear, and coal-fired power generating capacity.
- *Improve energy efficiency and the power supply.* The government will work with the residential, industrial, and commercial sectors to improve energy efficiency and guarantee supply, reliability, and safety.

• *Develop combined cooling, heating, and power (CCHP) and waste-derived power.* The city plans to establish distributed energy power stations (natural gas-fired) with the application of CCHP. Dalian has begun to convert 1,500 tons of municipal waste a day, the estimated daily household waste in urban areas. Incineration will generate 277 million kWh of electricity annually, and plans are in place to develop a facility with 20 MW capacity.

• *Demonstrate renewable and energy efficiency projects.* The city will build wind power plants with 200 MW capacity and will continue work on a tidal power demonstration project. It also plans to implement solar PV lighting systems to replace the current street lights. Dalian has applied solar water heaters in architecture projects since 2005, and energy-saving exterior insulation systems since 2006. It also uses heating and cooling systems with water heat pumps in seven districts.

• Power plants, chemical plants, and cement factories all release large quantities of waste heat. The city will work with these industries to develop energy-saving heating mechanisms and to inspire the application of new technologies for future waste heat utilization projects.

Analysis

Dalian provides an example of a Chinese city enjoying steady economic growth, while still maintaining high environmental standards. Measures taken to improve local environmental quality have likely benefited Dalian in attracting regional investment. The city also benefits from a strong scientific and technological capacity within the local universities and research institutes, aided by an active local bureau of science and technology. Dalian's proactive approach to guiding urban development has thus far allowed it to maintain a high percentage of public transportation use, though this will likely be a primary challenge in coming years as the development zones expand. From 1990 to 2000, Dalian's sprawl expanded some at the fringe of the existing urban district, but also in a linear fashion to the northeast, along Dalian Bay (Deng et al., 2005). Bicycle ridership (already low) and walking will likely decrease in share as automobile use increases in share, as an alternative to the LRT.

Energy security is an important concern in Dalian and will likely impact most decisions related to energy and air quality. Recent efforts to promote energy efficiency and renewable energy technologies should bring both short- and long-term benefits. In particular, by developing a market for energy efficiency, Dalian may be able to reduce the need for additional capacity, which in the U.S. experience has been extremely cost effective, in addition to its savings in potential air emissions. The city intends to develop into an energy base, and this will require attention to the unintended consequences, such as air pollution from petroleum refineries (directly emitted and fugitive leaks), added coal-fired capacity, and the shipping industry, which is a central component of the city's future development plans and is expected to increase dramatically. This latter source is still not well documented, and from Los Angeles and other port cities' experiences, the shipping industry is a major contributor to regional air pollution.

Dalian is focusing on air quality research to support dust control policies, but it will also benefit from research on $PM_{2.5}$, which is primarily from combustion sources. Having a better understanding of the ratio of $PM_{2.5}$ to total PM, as well as its sources, will help Dalian develop effective strategies to protect human health. NO_x and ozone monitoring and research are both in need of more attention, particularly given the large and rapidly increasing vehicle fleet. Considering Dalian's large service industry, and the amount of building construction taking place, there is a major opportunity to improve building energy efficiency. Building-integrated PV ought to be explored, and efficiency standards should be enforced for all new construction. This is one area in which Dalian could continue to be a national leader in environmental management and policy.

As Dalian expands and local air quality improves, it will need to increase its understanding of regional air quality. This will involve cooperation with other cities in the province, as well as potentially with Korea and Japan. Regional modeling and source apportionment (e.g., Wan et al., 2006) will aid Dalian in developing locally appropriate strategies which benefit regional air quality. Dalian's strong local science and technology capacity will be an asset in developing and implementing clean energy technologies, as well as research support for its air quality management.

REFERENCES

Bai, X. 2002. Industrial relocation in Asia: A sound environmental strategy? Environment 44(5):8-21.

Deng, X.Z., J.Y. Zhan, and R. Chen. 2005. The Patterns and Driving Forces of Urban Sprawl in China. Geoscience and Remote Sensing Symposium Proceedings, IEEE International, July 25-29.

DEPB (Dalian Environmental Protection Bureau). 2005. The Current Status and Development Plan of Dalian Environment Protection. Presentation to committee on October 28, 2005.

DEPB (Dalian Environmental Protection Bureau). 2007. Dalian Environmental Protection Bureau website, http://www.dlepb.gov.cn (accessed March 15, 2007).

DFTECB (Dalian Foreign Trade and Economic Cooperation Bureau). 2007. Dalian FTECB website, http://www.dalian-gov.net.

DICP (Dalian Institute of Chemical Physics). 2006. Dalian ICP website, http://www.dicp.ac.cn.

Liu, R. and F. Yang. 2006. Urban Development Directed by Light Rail Transit (LRT): A Case Study of Dalian, China. Proceedings of the 85th Annual Meeting of the Transportation Research Board, Paper No. 06-2870, January.

MOST (Ministry of Science and Technology). 2006. China Science and Technology Newsletter No. 437, April 30.

Wan, X.L., J.W. Chen, F.L. Tian, W.J. Sun, F.L. Yang, and K. Saiki. 2006. Source apportionment of PAHs in atmospheric particulates of Dalian: Factor analysis with nonnegative constraints and emission inventory analysis. Atmospheric Environment 40:6666-6675.

12

Findings and Recommendations

OVERARCHING FINDINGS AND RECOMMENDATIONS

1. Learn from experience.

Findings

For the United States, the GDP per capita and the energy use per capita are both several times higher than those for China. These comparisons reflect the cumulative effects of an enormous range of historical activities that contributed to building the U.S. economy. China has embarked on the same path in just the last 25 years and, as economic development has run ahead of pollution control, already is experiencing the adverse health, agricultural, environmental, and quality of life effects which have been largely ameliorated in the United States in the last 30 years.

Recommendations

a. *China should learn from the successes and failures of the United States and other developed countries in reducing the influence of energy use on air quality. Mistakes already made in the United States and elsewhere should be identified (as this report has attempted to do) and avoided in China.*
b. *Continued dialogue and information exchange among U.S. and Chinese scientists and policy makers should be promoted through professional organizations, government support programs, and the National Academies in both countries, to promote joint development of energy and pollution control strategies.*

2. *Recognize and respond to external costs of energy production and use.*

Findings

An important lesson learned is that air pollution damage imposes major economic costs, through premature mortality, increased sickness and lost productivity, as well as through decreased crop yields and ecosystem impacts. Most cost-benefit analyses in the United States show that emission reduction programs provide much greater benefits than their costs (Chapter 3). Emission controls are often less costly to implement than first envisioned. Appropriate programs can lead to economically efficient approaches for improving the environment, reducing costs further. Control costs are not purely costs, as they create opportunities (e.g., manufacturing and sales of pollution control and energy efficient equipment) that result in economic growth. As an example, air pollution control industries in the United States generated $27 billion in revenues and employed 178,000 people in 2001 (Chapter 10).

Recommendations

a. *Both countries need to improve permitting policies and economic mechanisms that reflect the external costs of pollution that are being paid by others (e.g., through adverse health effects and degraded quality of life). These might include the imposition of high enough taxes on emissions to make the addition of controls economically attractive, as well as rebates or subsidies to encourage use of higher efficiency and renewable technologies.*
b. *Subsidies must be carefully considered within a broader context, so as to avoid conflicting or divergent purposes. Subsidizing one energy source in the name of energy security can have an impact on other efforts to achieve air quality goals.*

3. *Establish and implement standards that protect human health.*

Findings

Excessive concentrations of SO_2, NO_2, and Pb have largely been reduced to levels that comply with health standards throughout the United States, but there are still many areas in China where these exceed ambient standards. Fine particulate matter ($PM_{2.5}$) and ozone (O_3) exceed healthful levels in many parts of the United States and China. In terms of premature death, there is roughly a 10 percent increase in adult mortality rates for every 10 $\mu g/m^3$ of annual-average $PM_{2.5}$, a 0.25-1 percent increase per 10 $\mu g/m^3$ 24-hour average PM_{10}, and 0.2-0.8

percent increase per 10 μg/m^3 increase in 1-hour peak ozone (Chapter 3). These require both local and regional emission reductions of directly emitted $PM_{2.5}$, sulfur dioxide (SO_2), oxides of nitrogen (NO_x), and volatile organic compounds (VOCs), which lead to secondary formation of ozone.

Recommendations

a. *Both the United States and China should adopt minimum standards based on healthful air quality, which may require revising currently accepted standards. Local governments should be able to enact more stringent local standards, but there should* not *be a sliding scale based on the level of economic development.*
b. *Greater efforts are needed to understand and reduce emissions from local sources in China where these standards are exceeded. These efforts would include source apportionment and dispersion modeling studies to determine source contributions, attainment studies to determine the needed emissions reductions, engineering design studies to evaluate control alternatives, enforcement to assure that the controls are implemented, and monitoring at the emissions source and receptor to assure that the reductions are successful.*
c. *Both countries need to evaluate and redesign compliance monitoring networks to better understand the precursor gases and local versus regional contributions to $PM_{2.5}$ and O_3.*
d. *Multiple complementary modeling and data analysis methods need to be applied in both countries to determine major contributors and to evaluate non-linearities in emission reduction efforts.*
e. *Regional emission caps currently in place in both the United States and China should be reexamined, as they may be insufficient to attain healthful ambient concentrations.*
f. *$PM_{2.5}$ control should be emphasized over, but not at the expense of, PM_{10} and O_3 reductions.*

4. Address pollution sources comprehensively.

Findings

Current pollution levels derive from a variety of energy uses and sectors on local and regional scales. All of these sectors must participate in solutions to pollution. Current inventories in both countries do not provide an accurate picture of emissions, especially for primary $PM_{2.5}$ and VOCs. Emission factors and chemical profiles for $PM_{2.5}$ and VOC derived from U.S. measurements probably do not represent the characteristics of Chinese sources, fuels, and operating conditions.

As demonstrated in Los Angeles, emission controls can be applied to many small and medium-size sources, including small engine exhaust, solvent use, refueling, ports and shipping, non-electrified locomotives, and vehicle fleets (e.g., buses, taxis, etc.), that collectively have a significant pollution impact (Chapter 10).

Available technical expertise, supply bottlenecks, financing, shortsighted economic decisions, and/or political opposition may limit the application of the best available control technology at a given time. Specific technologies, such as selective catalytic reduction for NO_x control and hybrid technologies for vehicles, are not yet broadly accepted or understood in China. The media can play a role in popularizing certain technologies (e.g., hybrid vehicles), thus increasing dissemination and potentially decreasing political opposition.

Recommendations

a. *Emission inventories must continue to be improved in both countries, with greater effort placed on developing real-world emission factors that can be related to available activity data for area and mobile sources.*
b. *Emission certification tests, while necessary to evaluate new engine and industrial designs, do not well represent real-world emission factors. Real-world testing methods are proven technology and should be more widely applied in the United States and China.*
c. *Fuel-based emission factors (i.e., mass of pollutant/mass of fuel consumed) provide a common basis for combustion emissions, are more easily related to available activity data (e.g., fuel sold in a certain area), and allow comparison among energy sectors. These can be more easily converted to other activity measures (e.g., g/km traveled, g/brake-horsepower-hour, g/BTU), and should be used in developing emissions inventories.*
d. *Emission inventories need to be evaluated and verified by independent means, such as receptor modeling. Special events (e.g., wildfires, dust storms, holiday celebrations) need to be included in inventories related to the time and location of occurrence.*
e. *There has to be participation in emissions reductions by all sectors, not just by the major industries. Enforcement and monitoring, as well as incentives, are needed to assure that emission reductions are implemented and maintained.*
f. *Incremental improvements should be made where possible, even if the best emission reduction technology is unaffordable at the current time.*
g. *Governments must improve policy incentives to adopt specific control technologies. Policies requiring the implementation of pollution controls are a positive first step, but these policies must be developed in tandem with appropriate incentives to overcome financial or other barriers.*

5. Strengthen SEPA's role in overseeing air quality planning and enforcement.

Findings

The United States has strong federal leadership and enforcement (EPA) for the attainment of ambient air quality and emission standards. This resulted from the realization that air pollution crossed political boundaries and that some states and localities were not sufficiently controlling their emissions. EPA maintains 10 regional offices to better interact with state and local agencies. There is a partnership between federal, state, and local agencies that addresses different types of emissions, with partial federal financing available to state and local pollution-control agencies. Federal highway funds can be withheld from areas that do not make good faith efforts to attain standards. In China, the central authority (SEPA) plays a minor role in air quality management in cities, with most activities carried out by local Bureaus of Environmental Protection (Chapter 4). The provinces have little motivation to reduce emissions that might affect neighboring regions. SEPA has, however, recently exercised more authority by halting construction on certain energy projects.

Federal and local energy and air pollution policies need to be better coordinated in both countries. Energy policies that appear to solve one set of problems may create other problems in terms of air pollution without this coordination. In the United States, DOE and EPA have primary responsibility for developing these policies, but interaction between the agencies has been limited. Their collaboration on the National Action Plan for Energy Efficiency is an example of improved coordination. In China, planning and policy making is even more diffuse, particularly with respect to energy policy. As has been stated elsewhere, pollution prevention through improved coordination is more cost-effective than pollution remediation.

Recommendations

a. *The Chinese government needs to expand SEPA's staff and influence over local air quality surveillance, management, and enforcement. Better coordination is needed between national and provincial authorities.*
b. *As in the United States, China needs formal emission reduction plans specific to cities and regions—plans that are independently evaluated and enforced at the national level. These plans should specify the activities that will bring areas into compliance with standards and that will keep areas already in compliance from becoming more polluted.*
c. *More incentives and enforcement for existing pollution reduction rules are needed in China. Penalties for violating air quality and emissions standards should not be a minor cost of doing business. Penalties should*

be sufficient to make it worthwhile to reduce emissions rather than violate the law.

d. *Environmental impact should be assessed on each major energy project in China prior to construction. Permits should include air quality emissions requirements. Enforcement should be by an independent agency. Compliance or non-compliance should be public. Companies should provide a composite analysis report to local governments.*

e. *Agencies should increase the amount of collaboration on issues of energy and air quality, to ensure that strategies for one sector do not conflict with another.*

6. Realize the potential of energy efficiency improvements.

Findings

Energy efficiency provides benefits for air quality and energy security while reducing costs. Energy efficiency can provide gains similar to or greater than specific pollution controls and can reduce the need for new power generators. Cost-effective technology is currently available to greatly improve energy efficiency across all energy use sectors. Efficiency measures adopted in the United States since 1973 now save $700 billion annually over business-as-usual growth, with little or no burden to the public (Chapter 5).

Recommendations

a. *The United States and China should consider evaluating the best energy efficiency standards for all energy sectors that have been formulated by each country, by their states and provinces, or by other countries. Efficiency standards, like air quality standards, will need to be properly enforced in order to be effective.*

b. *Consumer education and incentives should be increased in both countries to encourage the adoption of energy-saving technologies in all energy sectors.*

7. Promote efficient transportation systems and sustainable urban design.

Findings

The rapid growth of traffic in Dalian and in similar Chinese cities will repeat the air quality and energy consumption mistakes of Los Angeles and other U.S. cities if not better managed. Los Angeles actually dismantled its rail system to make way for highways. U.S. transportation and economic development policies have created the need to drive long distances and this is now occurring in China. In many cities the solution was to build more roads, and this pattern in being repeated throughout Chinese cities as well, but it has proven to be insufficient at alleviating congestion and, in some cases, only serves to increase personal vehicle use. Still, the personal vehicle is seen as a status symbol in both countries and is an important industry to both economies; therefore policies will likely need to focus on limiting vehicle miles traveled or improving efficiency, as opposed to limiting ownership through vehicle taxes or permitting fees. Modes such as light rail can greatly improve transportation efficiency and reduce emissions from the transportation sector, but require large up-front investments and are much more difficult to retrofit into an existing transportation infrastructure.

Recommendations

a. *Transit-oriented design and smart growth policies should be implemented to develop new urban areas or to redevelop existing areas, particularly in rapidly developing cities with high projected growth. Bus rapid transit (BRT) should be considered in a number of U.S. and Chinese cities, as it represents a low-cost (relative to subways and light-rail) transit system easily adapted to existing infrastructure, with proven success in other parts of the world.*
b. *Congestion pricing should be examined for possible implementation to discourage unnecessary driving in both countries. Parking fees and other disincentives should also be implemented to limit the use of personal vehicles in urban areas. Incentives such as use of high-occupancy vehicle (HOV) lanes should be considered to promote carpooling and hybrid vehicle use, as some states in the United States have done.*
c. *Traffic management systems, such as the system in place in Dalian, should be implemented in other Chinese cities, in order to manage the rapidly expanding vehicle fleets and to limit congestion.*

8. Accelerate improvements in fuel economy and reductions in mobile source emissions.

Findings

In a 2002 report, The U.S. National Academies examined the effectiveness and impact of the U.S. Corporate Average Fuel Economy (CAFE) standards legislated in 1975, and concluded that they had reduced oil consumption by about 2.8 million barrels per day (6.27 EJ per year), or about 14 percent, and that they had contributed to reduced emissions. China has more aggressive automotive fuel economy standards than the United States, which, if enforced, could do much to contain rising motor vehicle fuel demands and to reduce emissions. Hybrid vehicles represent an opportunity to significantly improve fuel efficiency in both countries. It is noted in the 2002 report, however, that other approaches, such as higher fuel taxes, tradable credits for fuel economy improvements, taxes on light-duty vehicles that fall below CAFE standards combined with rebates for vehicles exceeding the standards and/or standards based on vehicle attributes, such as weight, size, or payload—might be more successful at improving fuel economy (Chapter 5).

New vehicle emission standards remain too weak in China in spite of significant progress. Fuel quality remains poor, especially diesel fuel (Chapter 4). PM filters have been demonstrated in California and elsewhere to be very cost-effective for removing particulate, but require ultra-low-sulfur diesel fuel. Goods movement (shipping, trucking) is a large emitter that is not well documented or controlled in either country. The San Pedro Bay ports of Los Angeles and of Long Beach estimate that port-related vessels and vehicles account for 12 percent of the region's PM, 9 percent of NO_x, and 45 percent of SO_x. California has set a target reduction of 85 percent diesel PM exposure (from 2000 to 2020), in spite of a projected tripling of container traffic, and has set aside $1 billion for a mitigation fund (Chapter 10).

Recommendations

a. *The United States should examine the present CAFE standards or alternative incentives to improving fuel economy, to develop standards tailored to the U.S. market and vehicle stock.*
b. *China should enforce their fuel economy standards and consider other, possibly more effective alternatives as well.*
c. *Higher initial cost appears to be a major impediment to hybrid vehicles penetrating the market, and thus governments should consider preferential policies, such as tax deductions for individuals (as the United States has done) and government purchasing policies (such as Pittsburgh's), which may be most effective in China, given the high proportion of government-owned vehicles.*
d. *China should continue to increase its vehicle emission standards and to enforce those standards; China should also improve the quality of its refined fuels.*
e. *Additional measures should be implemented to reduce emissions from shipping and the movement of shipped goods to and from the ports in Dalian and in other Chinese and U.S. ports. Key stakeholders must be involved in designing, implementing, and evaluating these measures within a collaborative organizational structure for port pollution reduction.*

9. *Improve energy efficiency in buildings.*

Findings

Buildings are large energy users, especially for electricity and natural gas for heating. Therefore, they also represent a significant opportunity for energy savings (Chapter 5). "Green building" guidelines have been developed and shown to have paybacks on initial investments of 1-2 years. Lighting costs and power station emissions could be greatly reduced by using fluorescent lighting and by making better use of natural sunlight. In the United States, many energy-saving technologies and methods were developed in response to the 1972 oil embargo, but these were largely undone or undermined by utility deregulation in the 1990s. However, these experiences with demand-side management, integrated resource planning, and energy efficiency mandates are instructive and will be necessary components in future reductions of energy consumption. Examples include providing general information on energy efficiency opportunities, site-specific information involving facility inspections and rebates, or low-interest loans to implement specific efficiency measures.

Recommendations

a. *Building codes in both countries should be updated to require energy-saving technologies, for example, combined cooling, heating, and power (CCHP).*
b. *Subsidies, incentives, and low-cost financing should be enhanced in both countries to encourage up-front investments in energy-efficient technologies that will be paid back in future cost savings.*
c. *Both countries should allow or encourage utilities to decouple profits from energy sales. This is occurring to some degree in the United States, but needs to be accelerated and must be implemented in China.*

10. *Promote cleaner technologies for heat and power generation.*

Findings

Coal combustion will be a major component of energy production into the foreseeable future in the United States and China, owing to its abundance in both countries, which ameliorates energy security concern, its low relative cost and the longer lead time which would be required to develop a large scale alternate energy supply. Coal is primarily used to produce electricity, but it can also be used to create gaseous and liquid fuels as well as other feedstocks. Most trains are electrified in China, thus transportation is an important consumer of coal-generated power. Coal-fired generators operate for 50 years or more, and, therefore, decisions made today to install low-efficiency power sets will take many decades to correct in the future (Chapter 6).

Polygeneration plants provide an opportunity to efficiently provide power and coal-based liquid fuels. However, coal-based liquid fuels have not demonstrated reduced emissions vis-à-vis alternative fuels such as biofuels. Harnessing methane from coal mines has a number of co-benefits: it provides an additional source of energy for residential heat and power, removes an air pollutant and potent greenhouse gas from circulation before it reaches the atmosphere, and decreases a major safety risk associated with coalmine accidents.

Recommendations

a. *Incentives are needed in the United States and China to implement cleaner coal conversion technologies (e.g., IGCC), more efficient generation methods, and productive use of waste heat.*
b. *Coal washing and sieving rules should be implemented and enforced in all sectors of the coal industry in China, to reduce* SO_2 *and to increase combustion efficiency.*
c. *Residential/commercial coal burning should be further reduced in China by energy conservation measures and through replacement by natural gas or biogas.*
d. *Following the example of cities such as Huainan, coal-rich areas should implement systems to recover and make effective use of coalbed methane (CBM) and coke-oven gas.*
e. *Polygeneration plants must be considered within a framework accounting for possible carbon mitigation requirements in the future. Producing coal-based liquid fuels must be weighed against other potentially cleaner alternative fuel options, such as biofuels.*

11. Plan in advance for pollution control.

Findings

It is less costly to plan for and implement pollution controls up front than to install them later. Due to lack of knowledge of pollution effects and controls, the United States didn't act early enough to provide for emission controls on stationary and mobile sources. Retrofitting or closing down old industries, changing vehicle fleets, remodeling buildings, and changing attitudes has been costly and is not yet complete in the United States (Chapter 6).

Recommendations

a. *Better evaluation tools need to be developed and promulgated, specific to the United States and China, which assist project designers in evaluating the costs and benefits of different energy conservation/pollution control alternatives.*
b. *Projects need to be planned with the expectation that pollution controls and retrofits may be required, or deemed economical, in the future, even if benefits do not exceed the costs by today's standards.*
c. *Analyses are needed to evaluate total energy efficiency and pollution effects for the project's entire life cycle: material extraction/manufacturing, transportation, construction, use, and disposal. Environmental economics tools are also needed to perform these analyses on a recursive basis for the project's impacts on regional and national levels.*
d. *Since power plants have long life expectancies, adequate physical space and technological compatibility must be built in for future retrofits and conversions to cleaner and/or zero emissions technologies, advanced pollution abatement, and carbon capture technologies.*
e. *National and state/provincial governments should consider establishing capitalization funds to finance future improvements and retrofits for pollution controls.*

12. Accelerate development and use of renewable energy sources.

Findings

Renewable energy sources, including solar, wind, geothermal, waste-to-energy, and biofuels constitute important, but not large, fractions of energy portfolios in both countries. Several applications, such as solar water heating and wind-generated power, are economical in the long term, but require large up-front investments and have benefited in many cases from various financial incentives (Chapter 7). In terms of baseload generation, hydropower may continue to be the only reliable renewable energy source for decades. Wind-power capacity will continue to increase rapidly in both countries, but it will nonetheless continue to be an intermittent resource, as will solar energy. Renewable technologies can play an important role in distributed generation systems and thus represent a "no-regrets" choice as they replace a portion of energy supply which might otherwise have been provided by fossil fuel combustion. It is unclear whether some biofuels, including non-cellulose ethanol production, provide more renewable energy than they consume in non-renewable energy for their production.

Recommendations

a. *Both countries should continue to encourage the development, production and use of renewable energy wherever possible, through various policy instruments (i.e., renewable portfolio standards, tax rebates, preferential purchasing).*
b. *Both countries should support industrial-scale demonstrations to prove that cellulosic ethanol can be continuously economically viable, at a large scale. Early research appears promising, but is not yet conclusive.*

13. Expand public participation in Chinese air quality management efforts.

Findings

While much data and information about emissions, ambient concentrations, and energy use are publicly available in the United States (many of them over the Internet), such data are often sequestered in China (Chapter 4). Public and scientific scrutiny of these data have improved their quality and utility over time. The EPA has gone to great expense to convert older data management methods to modern web-based systems. Many of these modern concepts can be applied in China. Although China has made progress in reporting air quality indices to the public, the data needed for successful energy and air quality management are still difficult to obtain and analyze.

Reasonably accurate air quality forecast methods have been developed in the United States for public dissemination. Increasing public use is made of these forecasts to make personal decisions concerning exercise, travel, and health protection for susceptible populations. These forecasts are also used in the United States for intermittent pollution controls, such as domestic and agricultural burning restrictions and reduced driving periods. Forecasts are being implemented on a national basis using numerical simulations, and on a local basis for several cities using both empirical and prognostic models.

Non-governmental environmental advocates play a large role in promoting energy efficiency, renewable energy, and pollution reductions in the United States (Chapter 8), but not yet in China. Citizens' groups in the United States have also helped enforce pollution-control laws through the courts. Their activity is predicated on access to environmental information. China has growing numbers of NGOs working on environmental issues. In particular, Dalian has some volunteer groups dedicated to improving the environment.

Recommendations

a. *SEPA needs to convince public officials that the advantages of disseminating energy use, emissions, and air quality data outweigh the disadvantages. Such transparency will result in better data quality, by providing feedback on deficiencies to data generators.*
b. *Web-based data systems should be established in China to acquire, validate, and distribute air quality and meteorological data.*
c. *China should improve national and local air quality forecasting methods to better inform the public and to implement supplemental control measures during high-pollution periods.*
d. *SEPA and provincial agencies in China should continue to increase their efforts in outreach and education to engage the public in helping address air pollution problems, and to encourage public participation in environmental impact studies and decisions affecting the environment.*
e. *China should increase its efforts to enable citizens to use the country's legal system to address the harm they have suffered from unreasonable levels of pollution.*
f. *Local governments in China should encourage more volunteer groups focused on improving the environment.*

14. Improve capacity to address current and future issues through research and education.

Findings

Both countries have benefited from research, development, and technology transfer efforts in their universities, research institutes, and professional associations related to methods of energy production, pollution control, atmospheric processes, measurement systems, and simulation models (Chapters 4 and 8-11). These efforts also provide local expertise for states and provinces and train professionals needed for regulatory, industrial, and educational enterprises. Each country can benefit from the research and development results of the other.

Improved environmental and energy education and training is needed at all levels in both the United States and China. Outreach is needed to plant managers, engineers, technicians, workers, school children, parents, and regulatory agencies. University engineering programs need a quantitative component that addresses energy and pollution control options. Pittsburgh distributes public awareness brochures for cleaner biomass heating practices. Los Angeles publishes comic book-type instructional material on enforcement programs. The EPA has sponsored focused training courses, as have universities and professional associations.

Recommendations

a. *Both countries need to strengthen research and development in clean energy, energy efficiency, and air quality research. There is also a need for improved research across disciplines, in order to better understand the linkages between energy and air quality.*
b. *Research funding agencies in both countries should support formal bilateral programs that encourage joint efforts among U.S. and Chinese scientists and engineers.*
c. *As in the United States, expertise needs to be developed in provincial universities and research centers that do not yet have energy and environmental programs. The large research universities centered in the major eastern cities cannot accomplish all that is needed.*
d. *SEPA, provincial agencies, universities, and Chinese professional organizations need to create outreach materials and to conduct training sessions at all academic levels.*
e. *Chinese cities need to develop local and regional technical training centers and professional education centers, in order to build the capacity to operate and maintain pollution controls and advanced technologies.*

15. Expand cooperation on energy and air quality issues, including efforts to reduce greenhouse gas emissions.

Findings

In the fields of energy and air quality, there are a number of topics on which the United States and China can usefully collaborate. Numerous activities are already ongoing, between government agencies, universities, NGOs, research institutions, and within the private sector. These cooperative activities underscore the important strategic relationship between the United States and China, highlight our common interests in energy and air quality issues, and provide an opportunity to not only share lessons learned from the U.S. experience, but also to jointly address new and emerging challenges as both countries make a transition to sustainability. In addition to potential health and environmental benefits, there are also important economic benefits, as both countries represent large domestic markets for technologies and products.

Recommendations

a. *Given the existing interest in climate change, it is imperative that the United States and China begin substantial cooperation on issues to reduce greenhouse gas emissions. In addition to energy efficiency, there* *is great potential for collaborative research on improving CO_2 capture and sequestration technologies.*

b. *Energy efficiency cooperation should be taking place in all sectors, including national and local governments implementing efficiency codes and standards, industries instituting more efficient practices, and research on further improvements in technologies.*
c. *China will benefit from further cooperation on developing regional air quality management. Future activities should complement the ongoing work between Guangdong and Hong Kong, and efforts to develop SEPA's regional offices. Research universities and governments should also increase collaboration on measuring and monitoring $PM_{2.5}$ and O_3, as well as air quality forecasting.*
d. *As China begins mandating specific control technologies, it will be useful for the two countries to enhance programs focused on technology transfer and capacity building. The latter is critical, since installed technologies must be properly operated and maintained in order to be effective; this will almost certainly necessitate regional activities to engage local operators from the numerous cities throughout China.*
e. *Coal gasification appears to be in the interest of both countries, and therefore additional cooperation is needed to better apply technologies already in use for industrial purposes to commercial power generation.*
f. *The future will depend on reliable and affordable forms of renewable energy, and thus as the world's two largest energy consumers, the United States and China should enhance their collaboration on R&D for renewable energy technologies, both between NDRC and DOE, and within partnerships involving other research institutions and the private sector.*

Appendixes

Appendix A

Web-Based Resources on Energy and Air Quality

Data Sources for Policies, Laws, Regulations, and Guidelines

Title/Source	Description	Evaluation
SEPA index of Environmental Laws http://www.zhb.gov.cn/english/law.php3	Catalogues Chinese environmental laws through 2004, some in English	Useful index for laws prior to 2004, but is not being updated
EPA's Guide to Clean Air Act http://www.epa.gov/oar/oaqps/peg_caa/pegcaain.html	Comprehensive guide to 1990 Clean Air Act and subsequent changes	Organized well by category; plain language and figures make this a useful site for non-technical readers
EPA's Office of Air and Radiation http://www.epa.gov/oar/	The Office of Air and Radiation (OAR) develops technical policies and regulations for controlling air pollution	Links lead to information on Clean Air Act as well as other policies and regulations

Title/Source	Description	Evaluation
Global Village Beijing http://www.gvbchina.org/EnglishWeb/index.htm	One of the first NGOs in China; focused on environmental education with programs addressing energy efficiency, mercury emissions, and similar issues	Source for environmental news and campaigns in China, but little information on official policies
U.S. Department of Energy, Office of Environmental Management http://www.em.doe.gov/	Listing of laws and regulations affecting the U.S. DOE	Source for links for understanding how environmental laws and regulations impact a federal agency
SEPA Laws & Regulations http://www.sepa.gov.cn/english/chanel-3/chanel-3-end-2.php3?chanel=3&column=2 3	Chinese environmental regulations	Comprehensive list with links to text; many recent regulations
Cleaner Production in China-Environmental Legislation http://www.chinacp.com/eng/cnenvleg.html	NDRC maintained website (in English) with information pertaining to its programs	Provides extensive information related to cleaner production laws in China

Energy Information Sources

Title/Source	Description	Evaluation
Energy Information Administration http://www.eia.doe.gov/	Official energy statistics from the U.S. government	Starting point for information on energy supply, production, consumption, and forecasts in the U.S., as well as more limited international data
National Energy & Technology Laboratory http://www.netl.doe.gov/	U.S. DOE's fossil fuel research laboratory	Provides a portal to all information surrounding NETL, the only U.S. national laboratory devoted to fossil energy technology

Title/Source	Description	Evaluation
China Energy Group-Lawrence Berkeley National Laboratory http://china.lbl.gov/	Chinese Energy Databooks and other information particularly focused on energy efficiency	Easy to follow links for learning about the interaction of LBL with China and the research it is conducting.
National Renewable Energy Laboratory http://www.nrel.gov/	U.S. DOE's primary lab for renewable energy and energy efficiency R&D	Excellent portal for accessing information on U.S. national laboratory research on renewable energy technologies
National Bureau of Statistics (China) http://www.stats.gov.cn/english	The official source for energy production and consumption data in China	Official statistical data for China; Energy data are mixed in with other national data such as GDP
National Energy Leading Group http://www.chinaenergy.gov.cn/	Ministerial level office reporting directly to the State Council	Numerous links and scrolling news (in Chinese)
Energy Bureau, National Development and Reform Commission http://nyj.ndrc.gov.cn/	Supports the work of the Energy Leading Group (formerly was the primary energy office)	Regularly updated news and links (in Chinese)
Energy Research Institute http://eri.org.cn	National research organization studying comprehensive energy issues, reporting to the NDRC	Information in Chinese and English

Air Quality and Meteorological Information Sources

Title/Source	Description	Evaluation
U.S. EPA AirData http://www.epa.gov/air/data/geosel.html	Provides air quality maps, data, and reports for specific regions of the United States	Easy, step-by-step website for gathering air quality data and maps

Title/Source	Description	Evaluation
Visibility Information Exchange Web System (VIEWS) http://vista.cira.colostate.edu/views	Online exchange of air quality data, research, and ideas designed to support the Regional Haze Rule enacted by the U.S. EPA	Source for exchanging air quality data, research, and ideas. Guest book available for registering as an interested party in air quality
U.S. EPA Air and Climate Programs http://www.epa.gov/international/airandclimate/byregion/chinaair.html	Guide to U.S. EPA and SEPA joint programs on clean air and energy	Starting point for understanding how EPA (U.S.) and SEPA (China) are working together on clean air and energy projects
U.S. National Emissions Inventory http://www.epa.gov/ttn/chief/net/index.html	National database of air emissions information with input from numerous state and local air agencies as well as industry	Data and emissions factors contained on this site are used for purposes such as regulation setting and modeling. Year 2002 data was posted in February 2006
World Bank Air Quality Standards for Select Countries http://www.worldbank.org/html/fpd/em/power/standards/airqstd.stm	Standards for SO_2, NO_x, PM, and other pollutants in Asian countries, with comparisons to select OECD countries	Data is provided in easy to understand tables for quick reference and comparison
South Coast Air Quality Management District http://www.aqmd.gov	South Coast AQMD is the air pollution control agency for Southern California, including Los Angeles	Good resource for understanding how air quality management districts work and the future plans for SCAQMD

Title/Source	Description	Evaluation
California Air Resources Board http://www.arb.ca.gov/homepage.htm	ARB is the state agency responsible for protecting public health and the environment from air pollution effects	First stop for air quality information related to California. Website is well maintained with extensive information
Allegheny County Health Department, Office of Air Quality http://www.achd.net/airqual/index.php	ACHD issues pollution permits, monitors air quality, inspects sources, and investigates complaints in southwestern Pennsylvania	Useful portal with fact sheets, links to permit applications and general information on how the health department is involved with air quality in Pennsylvania
Pennsylvania Department of Environmental Quality, Bureau of Air Quality http://www.dep.state.pa.us/dep/deputate/airwaste/aq/default.htm	Comprehensive online resource for air quality information in the state of Pennsylvania	An example of the detailed level of air quality information that is available on the state level (Pennsylvania)
Dalian Environmental Protection Bureau http://www.dlepb.gov.cn/English/index.aspx	Dalian EPB's website	Basic Chinese site (English available) with daily air quality predictions
Huainan Environmental Protection Bureau http://www.hnhb.gov.cn/hnhb1/index.asp	Huainan EPB's website	Contains information collected at Environmental Monitoring Center, including air quality data
U.S. Regional Climate Centers http://www.wrcc.dri.edu/rcc.html	Directory of national and regional climate centers with links to each center	Easy map guides users to the appropriate regional climate center

Title/Source	Description	Evaluation
SEPA Daily Air Quality Report http://www.sepa.gov.cn/english/air-list.php3	Daily reports on air quality in 84 Chinese cities supplying SEPA with data (including Dalian)	Basic table with readings for 84 cities

Economic Analysis Models and Methods

Title/Source	**Description**	**Evaluation**
Economics and Cost Analysis Support http://www.epa.gov/ttn/ecas	Databases and models for cost, benefit, and economic impact analyses. Documents include analytical guidance and reports on conducting analysis of costs, benefits, and economic impacts of air quality management strategies, programs, and regulations	Free modeling programs available for download, including economic analysis models and emissions growth models
Center for Energy, Environmental, and Economic Systems Analysis http://www.dis.anl.gov/ceeesa/programs/environment_analysis.html	CEEESA is part of Argonne National Lab, responsible for GHG projections and analysis, emissions inventories, and environmental externalities	Free modeling programs available for download, ranging from power to environmental system analysis
World Bank, Environmental Management for Power Development http://www.worldbank.org/html/fpd/em/	A collaborative program coordinated by the World Bank to support the integration of environmental concerns into project and power system planning in developing countries	Useful information including downloadable software but in need of updating
Allowance Trading http://www.epa.gov/airmarkt/trading/index.html	Information on EPA Clean Air Market's trading concepts, program design considerations, and tracking of allowances	Starting point if considering a cap-and-trade scheme as part of a clean air market program. Site contains numerous relevant links

Industrial Source Emission Control Measures

Title/Source	Description	Evaluation
New and Emerging Environmental Technology http://neet.rti.org/	On-line repository for information on technologies that prevent, remove, sample, monitor, or model emissions	Excellent resource for finding technologies
RACT/BACT/LAER Clearinghouse http://cfpub.epa.gov/rblc/htm/bl02.cfm	Provides information on "best available" technologies for controlling air pollution from stationary sources	Good initial portal for guiding the user to additional resources on the best technologies for reducing air pollution from stationary sources
Environmental Protection Institute of Light Industry http://www.chinacp.com/eng/cporg/cporg_nlia.html	Research institute concentrating on technology development, pollution control, and environmental protection for light industry	Extension of the earlier Chinese cleaner production website focused on light industry

Mobile Source After-Engine Emission Control Measures and Mobile Source Fuels

Title/Source	Description	Evaluation
U.S. EPA Office of Transportation and Air Quality http://www.epa.gov/otaq	This program is charged with regulating air pollution from motor vehicles, engines, and the fuels used to operate them	Very rich site with links to key topic areas surrounding air quality and transportation
ARB Mobile Source Program http://www.arb.ca.gov/msprog/msprog.htm	California's statewide programs and strategies to reduce emissions from mobile sources	Start on this California Air Resources Board site if seeking information on California's mobile source emissions programs

Title/Source	Description	Evaluation
Clean Technologies Information Pool http://www.cleanairnet.org/infopool/1411/channel.html	Featured activity of the Clean Air Initiative; easily accessible and comprehensive information on cleaner buses, trucks, and fuels	Information is organized nicely by region including Asia, sub-Saharan Africa, and Latin America
U.S. EPA-Fuels and Fuel Additives http://www.epa.gov/OMS/fuels.htm	Information on fuel quality standards, programs and regulations	Start here for information related to conventional and alternative fuels for motor vehicles and engines
Clean Fuels Development Commission http://www.cleanfuelsdc.org/index.html	Non-profit organization supporting the development of cleaner technologies for mobile source fuels	Contains examples of how non-profits can partner with government agencies for energy-related projects

Appendix B

Alternative Energy Resources

OIL SHALE

Oil shale is sedimentary rock that is dark brown to black in color and high in organic matter. The organic matter is called *kerogen*, fossilized insoluble organic material that will yield liquid and gaseous hydrocarbons upon distillation. The kerogen can be converted into petroleum products by distillation. The shale must be heated, typically in a closed vessel (retort), to about 500°C to convert it into petroleum. Fisher assay is the most common method of ranking oil shale in terms of potential oil produced. Oil yields generally vary from 10 to 50 gallons per ton and some oil shale is as high as 65 gal/ton. Shales yielding more than 25 gal/ton are the most attractive and are considered to be potential resources.

Oil shale is generally shallower (<3,000 feet) than the deeper and warmer geologic zones required to form oil. The origins of oil shale can be categorized into three basic groups: terrestrial (organic origins similar to coal-forming swamps), lacustrine (organic origins from fresh or brackish water algae), and marine (organic origins from salt water algae, acri-tarchs, and dinoflagellates).[1]

Worldwide, the oil shale resource base is believed to contain about 2.6 trillion barrels, of which the vast majority of over 2.1 trillion barrels of oil equivalent, (including eastern and western shales), is located within the United States.[2] Due to differences in kerogen type (compared to western shale), eastern oil shale requires different processing. However, with processing technology advances, potential oil yields from eastern shales could someday approach yields from western shales.

[1]J.R. Dyni, "Oil Shale Deposits of the U.S.," *Oil Shale Journal*, 20:3, 2003.

[2]Ibid.

U.S. oil shales are concentrated in the western United States in the states of Colorado, Utah, and Wyoming, but sizable quantities also exist in the eastern United States. The most economically attractive deposits, containing an estimated 1.5 trillion barrels of oil equivalent, are found in the Green River Formation of Colorado in the Piceance Creek Basin, in Utah in the Uinta Basin, and in Wyoming in the Green River and Washakie Basins.

In particular:

- Colorado has 1.2 trillion barrels of oil shale resources, and five RD&D projects are currently being reviewed for Environmental Impact Statements (EISs) by the Bureau of Land Management (BLM) under their oil shale RD&D program. Shell has three projects, Chevron/Texaco has one, and EGL has one. The EIS is also under way for commercial leasing in 2007 or 2008.
- Utah has substantial oil shale resources and BLM has recently granted one company the right to proceed to a pilot project. The USGS has had an oil shale data compilation project in Utah for the last 2 years.
- In the eastern United States, oil shale underlies the Appalachian, Illinois, and Michigan Basins, predominantly in Devonian age deposits covering hundreds of thousands of acres from Illinois to New York to Alabama, and it is estimated that there are 189 billion barrels of oil equivalent in Eastern oil shale.[3] Kentucky has the largest outcrop of oil shale in the eastern United States and also has the largest amount of surface and near-surface oil shale. A two county area in eastern Kentucky was investigated in detail in the 1980s and was estimated to contain 4.4 billion barrels of oil equivalent with 1.3 billion barrels in a stripping ratio of 2.5:1.

The extent and characteristics of U.S. western oil shale resources, and particularly those in the Green River Formation, are well known and documented.[4] More than a quarter million assays have been conducted on core and outcrop samples for the Green River oil shale, and results have shown that the richest zone, known as the Mahogany zone, is located in the Parachute Creek member of the Green River Formation. This zone can be found throughout the formation.

A layer of volcanic ash several inches thick, known as the Mahogany marker, lies on top of the Mahogany zone and serves as a convenient stratigraphic event that allows oil shale beds to be correlated over extensive areas. Because of its relatively shallow nature and consistent bedding, the resource richness is well known, giving a high degree of certainty as to resource quality. By assay tech-

[3]Dyni, op. cit.

[4]Anton Dammer, Office of the Deputy Assistant Secretary for Petroleum Reserves, Office of Naval Petroleum and Oil Shale Reserves, U.S. Department of Energy Washington, D.C., *Strategic Significance of America's Oil Shale Resource, Volume II, Oil Shale Resources, Technology and Economics*, March 2004, pp. 2-5.

niques oil yields vary from about 10 gal/ton to 50 gal/ton and, for a few feet in the Mahogany zone, up to about 65 gal/ton.

When discussing oil shale resources, it is important to qualify the resource estimates by richness. Of the 1.5 trillion barrels of western oil shale resources, an estimated 418 billion barrels are in deposits that will yield at least 30 gal/ton in zones at least 100 feet thick,[5] and there are estimated resources of 750 billion barrels at 25 gal/ton in zones at least 10 feet thick.[6]

In general, room and pillar mining is likely to be used for resources that outcrop along steep erosions, and horizontal room and pillar mining has been used successfully by Unocal. Deeper and thicker ores will require vertical shaft mining, modified in situ, or true in situ recovery approaches. Because the pay zone is more than 1,500 feet thick in some places, it is conceivable that open pit mining could be applied even with 1,000 feet of overburden.

In recent years, Shell has experimented with a novel in situ process that shows promise for recovering oil from rich, thick resources lying beneath several hundred to 1,000 feet of overburden. There are locations that could yield in excess of 1 million barrels per acre and require, with minimum surface disturbance, fewer than 23 square miles to produce as much as 15 billion barrels of oil over a 40-year project lifetime. In addition, in the northern Piceance Creek basin, zones of high-grade oil shale also contain rich concentrations of nahcolite and dawsonite—high-value minerals that could be recovered through solution mining.

However, oil shale still faces a number of challenges to future development as a resource. Its recovery is still not economically practicable, and uncertain oil price forecasts and a lack of R&D into reducing production costs hamper its economic competitiveness. Both in situ and surface processes are energy intensive, thus while the recovered resources may satisfy one energy demand for liquid fuels, the net balance is much smaller and is an important consideration, particularly in light of efforts to reduce fossil fuel combustion. Current production methods release significantly more greenhouse gases than conventional crude oil production and refining. Retorting also requires large amounts of water and is a potential source of pollutants, and in the case of surface retorting, can also result in land and ecological disturbances.

ENHANCED OIL RECOVERY

Crude oil production occurs via a series of oil "crops" called primary (1st crop), secondary (2nd crop), and tertiary (3rd crop). Enhanced oil recovery (EOR) is a term often used to describe tertiary recovery, but should be reserved for the more advanced oil production technologies regardless of where the process occurs

[5]W.J. Culbertson and J.K. Pitman, "Oil Shale," in United States Mineral Resources, USGS Professional Paper 820, Probst and Pratt, eds., 1973, pp. 497-503.

[6]J.R. Donnell, "Geology and Oil-Shale Resources of the Green River Formation," *Proceedings, First Symposium on Oil Shale*, Colorado School of Mines, pp. 153-163, 1964.

in the sequence of oil crops. For example, thermally enhanced recovery of tars or heavy oils utilizes advanced technologies for the first or second crops of oil from a given resource.

Primary oil recovery is often the least efficient method in terms of the percent of original oil in place (OOIP) recovered, unless the reservoir has an active aquifer providing the driving force. Sometimes only 5 or 10 percent of OOIP is produced during primary recovery, especially in the case of low-pressure, shallow reservoirs with only small amounts of internal energy to force the oil out. After primary production has been completed, reservoirs require additional (secondary) energy sources to recover the oil left behind. Secondary oil recovery techniques historically have referred to the injection of gas or water to displace oil and drive it to a production well, and secondary recovery often yields as much as or more oil than primary recovery. Well-designed water floods may recover 20 to 40 percent of the OOIP, depending on oil and reservoir characteristics, leaving "residual oil" amounting to perhaps 50 percent of the OOIP.

Theoretically, EOR techniques offer prospects for producing up to 100 percent of the residual oil under nearly perfect reservoir conditions; however, practically speaking, the additional recovery is more likely to be similar to the amount of oil recovered during secondary recovery activities. Three major categories of EOR have been found to be commercially successful to varying degrees:

- Thermal recovery (e.g., steam injection) introduces heat into the reservoir to lower the oil's viscosity, thereby improving the oil's ability to flow from the reservoir. Thermal techniques account for over 50 percent of the U.S. EOR production.
- Gas injection uses gases such as natural gas, nitrogen, or carbon dioxide to displace additional oil from the reservoir or to dissolve in the oil causing it to expand while simultaneously lowering its viscosity, both of which improve the oil's ability to flow from the reservoir. Gas injection accounts for close to 50 percent of U.S. EOR production.
- Chemical injection may be used to enhance the characteristics of the water in a water flood, either to increase the water's viscosity, making it less likely to by-pass reservoir oil and leaving part of the oil behind, or to lower the interfacial tension between the water and the oil, "lubricating" the path for the oil to flow from the reservoir. Chemical techniques account for less than 1 percent of U.S. EOR production.
- Other processes, such as microbial EOR, are being researched, but do not currently contribute significantly to oil production.

Each of these techniques involves costs that are higher than typical conventional secondary recovery methods and involve additional risk because of the sensitivity of the processes to some of the reservoirs' unknown characteristics.

As shown in Table B-1, U.S. oil resources are very large. The problem is in recovering them. Discovered and documented resources amount to 582 billion bbls, 482 billion of light oil and about 100 billion of heavy oil. Approximately 208 billion bbls have been developed, leaving 374 billion bbls still in place. Of these 374 billion bbls of oil in place, at least 100 billion bbls are estimated to be producible via EOR. These numbers do not include projected reserves growth (RG), undiscovered resources (UR), residual oil zone resources (ROZ), or oil sands.

Beyond this point in the analysis, estimates of future oil recovery are based mostly on statistical analysis. While the statistical bases for the projections are sound and there is a statistically high probability that the projections will be borne out, there are no guarantees. That said, there could very well be another 430 billion bbls of oil produced in the future, including 179 billion bbls from undiscovered resources (UR), 111 billion from RG, and 10 billion from oil/tar sands, plus the 30 billion-bbl correction to the EOR potential from the four additional basins that were evaluated after Table B-1 was created—see note in the table.

TABLE B-1 U.S. Original, Developed and Undeveloped Domestic Oil Resources (billion barrels)[a]

	Original Oil in Place	Developed to Date	Remaining Oil in Place	Future Recovery[b]		
				Conventional Technology	EOR[c]	Total
I. Crude Oil						
1 Discovered	582	(194)	374	—	100	100
• Light oil	482	(189)	293	—	80	80
• Heavy Oil	100	(19)	81	—	20	20
2. Undiscovered	360	—	360	119	60	179
3. Reserve Growth	210	—	210	71	40	111
4. Residual Oil Zone	**100**	—	**100**	—	Unknown	Unknown
II. Tar Sands	80	—	80	—	10	10
TOTAL	1,332	(194)	1,124	190	210	400

NOTE: Above estimates do not include the additional resource potential outlined in 10 basin-oriented assessments or recoverable resources from residual oil zones, as discussed in related reports issued by DOE in February 2006. Accounting for these, the future recovery potential from domestic undeveloped oil resources by applying EOR technology is 240 billion barrels, boosting potentially recoverable resources to 430 billion barrels.

[a]Does not include oil shale.

[b]Technically recoverable resources rounded to the nearest 10 billion barrels.

[c]Based on six basin-oriented assessments released by DOE in April 2005.

SOURCE: Advanced Resources International. 2006. Undeveloped Domestic Oil Resources: The Foundation for Increasing Oil Production and a Viable Domestic Oil Industry. Prepared for the U.S. Department of Energy, Office of Fossil Energy.

The potential for enhanced oil recovery in the United States is increasing continuously with advances in technology. Reservoir modeling, especially for CO_2-EOR, has become extremely sophisticated with the increased capabilities of modern computers and with the development of advanced computer codes that are better capable of mimicking the physics and chemistry of enhanced oil recovery. Improved drilling and completion techniques are also contributing, providing better drilling efficiency and improved well control. New sensing devices and communication systems provide capability for real time analysis of field operations, including underground flow tracking and simulation, thus enhancing the ability to make intelligent decisions in a timely manner. The synergism of the advanced technologies allow a far better understanding and control of oil reservoirs, reservoir fluids, and the physics and chemistry of enhanced recovery.

CO_2-EOR is the "universal" enhanced recovery system, applicable to most reservoirs except the very shallow and the reservoirs with heavier oils, for which thermal technologies are more applicable. DOE recently sponsored a study to determine the CO_2-EOR potential for the reservoirs in 10 major U.S. basins (and essentially for the United States, since those basins hold the preponderance of U.S. oil resources). The results of the study are impressive, indicating that as much as 89 billion bbls of oil could be produced by applying modern and forthcoming advanced CO_2-EOR technologies. These estimates are based on assumptions that require the application of the very best technologies available, something that is not likely to happen in every case. Even so, the remaining resources offer a large target for CO_2-EOR, and even if only a portion of the 89 billion bbl estimate can be recovered, it is very much worth pursuing.

There are currently limited sources of low cost CO_2 and delivery infrastructure (pipelines) to supply CO_2 to the many oil fields in the United States with EOR potential. Coal to liquids and other alternative liquid transportation fuels production facilities are believed to be a key to unlocking the huge potential of U.S. EOR resources. These plants will be distributed across the United States, with many sited proximate to EOR-suited oil fields. CO_2 will be a residual product of alternative liquid fuel plants, and capturing the gas for sale will not only create economic value but will also demonstrate environmental stewardship. Thus, it is anticipated that these new liquid fuels manufacturing plants will be a source of low cost CO_2 for EOR operations.

The United States has limited existing CO_2 sources and pipelines currently delivering this strategic EOR gas, and even in these regions, low cost CO_2 is in short supply. Many of the basins showing large EOR potential have no existing supplies of CO_2. With more than three decades of experience with the process, companies are becoming more comfortable with using CO_2-EOR. If the price of oil remains high, there should be considerable incentive for companies to initiate new EOR projects, even though past experience has made investors leery of commitments to major projects.

Appendix C

Summary of PM Source-Apportionment Studies in China

Here we summarized the results from 11 typical studies. The following information was extracted from each publication: (1) sampling locations; (2) ambient sampling periods, frequencies, and durations; (3) source categories, source profiles, and methods of obtaining profiles; (4) chemical and physical properties quantified at source and receptor; (5) CMB solution and evaluation methods; (6) source contribution estimates. Since most of results don't reconcile with source modeling and emissions inventories, the description is omitted. This information is summarized in Table C-1.

We can draw several conclusions from the comparison of the studies from Table C-1.

1. *Geographic distribution*. Most of the studies were conducted in cities in north China including Beijing (Zhang Y.H. et al., 2004; Zheng et al., 2005; Song et al., 2006a, 2006b), Xi'an (Zhang X.Y. et al., 2001), Jinan (Feng et al., 2004), Yantai (Xu et al., 2001), Xining (Wang, 2006), Yinchuan (Sang et al., 2005). Studies in south China include Hong Kong (Lee et al., 1999; Ho et al., 2006), Nanjing (Hang et al., 2000), Chongqing (Tao et al., 2006).

2. *Study Objectives*. Most of these studies were undertaken to improve the source identification and support decision making. These studies were informational rather than regulatory; there was a desire by decision makers to understand the relative contributions from different source types. The result of source apportionment study like Xi'an has been adapted by Xi'an municipal government (Zhang X.Y. et al., 2001). Residential coal has been replaced by natural gas, gasoline in taxi cars has also been replaced by natural gas, and open burning has been prohibited. Air quality in Xi'an has been improved largely.

3. *Ambient Measurements*. All the studies used the chemical measurements of elements (17-36 elements from Na to U). Studies in Beijing (Zhang Y.H. et al., 2004; Zheng et al., 2005; Song et al., 2006a, 2006b), Jinan (Feng et al., 2004), Hong Kong $PM_{2.5}$ (Ho et al., 2006), and Chongqing (Tao et al., 2006) also used chemical measurements of water-soluble ions (chloride [Cl^-], nitrate [NO_3^-], sulfate [SO_4^{2-}], ammonium [NH_4^+], and sometimes sodium [Na^+], potassium [K^+], calcium [Ca^{2+}]), and carbon (organic [OC] and elemental carbon [EC]). Studies in Xi'an (Zhang X.Y. et al., 2001), and Hong Kong PM_{10} (Lee et al., 1999) used the measurements of elements and ions.

4. *Source Measurements and Profiles*. No area-specific source profile measurements were taken in studies of Beijing (Zhang Y.H. et al., 2004; Song et al., 2006a, 2006b), Yantai (Xu et al., 2001), Xining (Wang W., 2006), and Hong Kong (Lee et al., 1999; Ho et al., 2006). Only dust aerosol or dustfall samples from source-dominated microenvironments were taken in studies of Beijing (Zheng et al., 2005), Xi'an (Zhang X.Y. et al., 2001), Jinan (Feng et al., 2004), Yinchuan (Sang et al., 2005), Nanjing (Hang et al., 2000), and Chongqing (Tao et al., 2006). Other profiles like diesel engine exhaust, gasoline-powered vehicle exhaust were taken from earlier tests in the study area or similar areas (Feng et al., 2004; Zheng et al., 2005).

5. *Source Contribution Estimates*. The major sources including coal combustion dust, fugitive dust (soil dust), and construction dust accounted for 58 percent at Xi'an (Zhang X.Y. et al., 2001), 77 percent in Jinan (Feng et al., 2004), 67 percent in Yantai (Xu et al., 2001), 79.4 percent in Xining (Wang, 2006), 84.7 in Nanjing (Hang et al., 2000) for TSP fraction. They accounted for 72 percent in Jinan (Feng et al., 2004), 80 percent in Yinchuan (Sang et al., 2005), only 6.1 percent in Hong Kong (Lee et al., 1999) for PM_{10} fraction. Their percentage is 37.8 percent in Beijing (Zhang Y.H. et al., 2004), and 6-30 percent in Hong Kong (Ho et al., 2006) for $PM_{2.5}$ fraction. Coal is the dominant energy source and construction activities are serious in most of cities in north China. Strong wind and dry weather results in the large fugitive dust (soil dust) in TSP in these cities. These three sources are also dominant sources contributed to PM_{10} in cities in north China, but not in Hong Kong. Hong Kong is a developed city without intensive construction activities and coal utilization and coastal area with frequent precipitations, which lead to the small contribution from these three sources. In $PM_{2.5}$ fraction, their contribution decreased because the increasing contribution from secondary sources and vehicular exhaust in Beijing and Hong Kong.

TABLE C-1 Summary of PM Source Apportionment Studies Using CMB and Other Receptor Models in China

Study, Location, Period, and Measurements	Source Apportionment Method	Findings
Northern China		
Reference: **<u>Beijing $PM_{2.5}$</u>** study (Zhang Y.H. et al., 2004) When: 24-h samples were acquired during April 25-30, 2000, August 18-25, 2000, October 30-November 4 and January 9-14, 2001. Where: Three sites include Beijing Union University (BUU), Chinese Academy of Preventive Medicine (CAPM), and Chinese Research Academy of Environmental Sciences (CRAES). Ambient: Samples were acquired with a MOUDI-100 impactor, A-245 dichotomous sampler and a $PM_{2.5}$ sampler and a self-developed sampler. The samples were analyzed for mass, 19 elements (by ICP-AES), ions (NO_3^-, SO_4^{2-}, and NH_4^+ by IC), carbon (OC and EC by NIOSH), and organic compounds (including PAHs by Gas Chromatography/Mass Spectrometry). Source: No area-specific source profile measurements were taken.	Solution: CMB	Average CMB-calculated source contribution to $PM_{2.5}$ (in % mass): *Source Type* — *Annual* Coal combustion — 16.4 Vehicle exhaust — 5.6 Construction dust — 3.3 Fugitive dust — 18.1 Biomass burning — 4.5 Secondary sulfate and nitrate — 9.6 Organic matter — 15.0 Unexplained — 27.5 Average measured $PM_{2.5}$ mass (μg m^{-3}) — 122 Number in Average — Not reported

continued

TABLE C-1 Continued

Study, Location, Period, and Measurements	Source Apportionment Method	Findings				
Reference: **Beijing $PM_{2.5}$** study (Zheng et al., 2005; Song et al., 2006a, 2006b)	Solution: PCA/ APCA, UNMIX, PMF, and CMB	Average calculated source contribution to $PM_{2.5}$ (in % mass):				
When: 24-h samples were acquired once every 6 days in January, April, July, and October in 2000.		*Source Type*	*CMB*	*PMF*	*APCA*	*UNMIX*
Where: Five sites include Ming Tombs (OT), airport (NB), Beijing University (BJ), Dong Si EPB (XY), and Yong Le Dian (CH).		Secondary sulfates	16.7	16.0		
Ambient: Samples were acquired with Total Particle samplers and analyzed for mass, 19 elements (by XRF), ions (NO_3^-, SO_4^{2-}, and NH_4^+ by IC), carbon (OC and EC by NIOSH), and organic compounds (including PAHs by Gas Chromatography/Mass Spectrometry).		Secondary nitrates	10.7	15.0	23.1	28.0
Source: No area-specific source profile measurements were taken in PMF, APCA, and UNMIX studies (Song et al., 2006a, 2006b). Dust and coal emission profiles were composed and other profiles were taken from earlier tests in the study area or similar areas in CMB study (Zheng et al., 2005).		Secondary ammonium	6.4			
		Coal combustion	6.3	15.8	26.4	23.3
		Biomass aerosols	8.3	10.1		
		Motor vehicles	6.5	5.5	5.9	10.7
		Road dust [a]	12.3	7.0	7.1	8.3
		Industry		4.7	6.5	10.9
		Cigarette smoke	1.3			
		Vegetative detritus	1.0			
		Other organic matter	11.2			
		Unexplained	15.3	18.1	26.1	14.0
		Average measured mass (μg m-3) [b]	101	93	96	96
		Number in average	100	90	90	90

[a] Averaged in January, July, and October as a different dust signature used during April in CMB.

[b] CMB: an average of the measured $PM_{2.5}$ mass concentrations in 100 samples; PMF: the contributions of apportioned dust storms were subtracted from the CMB value (101 μg m^{-3}); PCA/APCS and UNMIX: averages of 90 samples (excluding 10 dust storm samples).

Reference: **Xi'an TSP** study (Zhang X.Y. et al., 2001)
When: 24-h samples were acquired from September 1996 to August 1997.
Where: Four sites include east, south, west and center sites. Ambient: Samples were acquired with bulk aerosol samplers and analyzed for mass, 20 elements (by PIXE), ions (by IC).
Source: Dust samples of resuspended road dust, construction dust and source-dominated samples from industrial, motor vehicle, night market and dumpling site were taken and measured.

Solution:
APCA/CEB

Average APCA-calculated source contribution to TSP (in % mass):

Source Type	*Annual*
Coal combustion	37
Fugitive dust	21
Motor vehicle	20
Agricultural & waste	12
Industrial	3
Unexplained	8
Average measured mass (μg m^{-3})	410
Number in Average	299

continued

TABLE C-1 Continued

Study, Location, Period, and Measurements	Source Apportionment Method	Findings		
Reference: **Jinan PM** study (Feng et al., 2004) When: 24-h samples were acquired from December 15-30 1999, April 30-May 6 2000, September 7-15, 2000. Where: Five sites includes Jinan Chemical Factory, Jinan Environmental Mornitoring Station, Shandan Seed Station, Jinan Machine Tool Factory and Official Resting Place. Ambient: TSP and PM_{10} samples were acquired with KB120 medium-vol sampler and analyzed for mass, 17 elements (by ICP-MS), ions (Cl^-, NO_3^-, SO_4^{2-}, and NH_4^+ by IC, Na^+ and K^+ by AAS), and carbon (OC and EC by TOR). Source: Dust samples from fugitive dust, soil dust, coal combustion, cement dust, and steel industry were taken and measured. Vehicular exhaust profile was used (Chow et al., 1994).	Solution: CMB	Average source contribution (in % mass):		
		Source Type	*TSP*	*PM_{10}*
		Fugitive dust	34	30
		Coal combustion	25	27
		Soil dust	18	15
		Motor vehicle exhausts	6	9
		Cement dust	2	3
		Unexplained	15	16
		Average measured mass (μg m^{-3})	304	175
		Number in average	no reported	no reported

Reference: **Yantai TSP** study (Xu et al., 2001)
When: 30-min samples were acquired.
Where: Three sites include east, west, and center stations.
Ambient: Samples were acquired with KB120 medium-vol samplers and analyzed for mass and 21 elements (by XRF).
Source: No area-specific source profile measurements were taken.

Solution: CMB

Average source contribution (in % mass):

Source Type	*Annual*
Construction dust	46
Residential coal combustion	21
Heavy vehicular exhaust	12
Coal burning boiler	10
Metal production plant	5
Marine aerosol	6
Mass	not reported
Number	101

Reference: **Xining TSP** study (Wang, 2006)
When: 30-min samples were acquired for 5 times during December 2001, May, August, and October 2002.
Where: Three sites include Environmental Mornitoring Station, Silu Hospital, and Medicine Storehouse.
Ambient: Samples were acquired with KB120 medium-vol samplers and analyzed for mass and 21 elements (by XRF).
Source: No area-specific source profile measurements were taken.

Solution: CMB

Average source contribution (in % mass):

Source Type	*Annual*
Coal combustion dust	37.0
Soil dust	27.0
Construction dust	15.4
Smelting dust	2.9
Mass	not reported
Number	45

continued

TABLE C-1 Continued

Study, Location, Period, and Measurements	Source Apportionment Method	Findings	
Reference: **Yinchuan PM_{10}** study (Sang et al., 2005)	Solution: CMB	Average source contribution (in % mass):	
When: 24-h samples were acquired for 5 times during January, April, July, and October 2002.		*Source Type*	*Annual*
Where: One site in Yinchuan Environmental Mornitoring Station.		Coal combustion dust	36.7
Ambient: Samples were acquired with Anderson PM_{10} samplers and analyzed for mass and 17 elements (by XRF).		Soil dust	33.9
Source: Dust samples from fugitive dust, soil dust, coal combustion, construction dust, and steel industry were taken and measured.		Construction dust	9.4
		Smelting dust	6.5
		Unexplained	13.5
		Mass	232
		Number	20
South China			

Reference: **Hong Kong PM_{10}** study (Lee et al., 1999)

When: 24-h samples were acquired once 6 days from 1992 to 1994.

Where: 11 sites include Central Western, Junk Bay, Taipo, Sham Shui Po, Shatin, Tsim Sha Tsui, Hong Kong South, Kwai Chung, Kwun Tong, Tsuen Wan, Mongkok.

Ambient: Samples were acquired with Anderson hi-vol samplers and analyzed for mass, 13 elements (by ICP-AES) and 6 ions (by IC)

Source: No area-specific source profile measurements were taken.

Solution: PMF

Average PMF-calculated source contribution to PM_{10} (in % mass):

Source Type	*Annual*
Secondary ammonium sulfate	37.8
Chloride depleted marine aerosols	14.3
Marine aerosols	6.9
Crustal/soil dust	6.1
Non-ferrous smelters	1.2
Vehicular emission	0.8
Particulater bromide	0.8
Particulater copper	0.6
Fuel oil burning	0.2
Unexplained	31.4
Mass	15.2
Number	1516

continued

TABLE C-1 Continued

Study, Location, Period, and Measurements	Source Apportionment Method	Findings		
Reference: **Hong Kong PM$_{2.5}$** study (Ho et al., 2006) When: 24-h samples were acquired once every 6 days from November 2000 to February 2001 and June 2001 to August 2001. Where: Two sites include PolyU and KT. Ambient: Samples were acquired with Anderson Instruments hi-vol samplers and analyzed for mass, 17 elements (by ICP-MS), ions (Cl^-, NO_3^-, SO_4^{2-}, and NH_4^+ by IC, Na^+ and K^+ by AAS), and carbon (OC and EC by TOR). Source: No area-specific source profile measurements were taken.	Solution: APCA	Average APCA-calculated source contribution to PM$_{2.5}$ (in % mass):		
		Source Type	*PolyU*	*KT*
		Diesel emission	47	4
		Secondary aerosol	18	
		Crustal matter	6	30
		Automobile emission + secondary aerosol	15	44
		Oil combustion	0	4
		unexplained	14	8
		Average measured mass (μg m^{-3})	41.7	43.9
		Number in average	0	29
Reference: **Nanjing TSP** study (Hang et al., 2000) When: Samples were acquired in October 1998, January, April, and July 1999. Where: Seven sites include Zhonghua Gate, Maigao Bridge, Ruijin Road, Xuanwu Lake, Zhongshan Tomb, Chaochang Gate, Shanxi Road. Ambient: 6-h samples were acquired with Kb-6A samplers and analyzed for mass and 17 elements by XRF Source: Dust samples from soil dust, coal combustion, construction dust, and steel industry were taken and measured.	Solution: CMB	Average source contribution (in % mass):		
		Source Type	*Annual*	
		Coal combustion dust	25.7	
		Soil dust	19.2	
		Construction dust	39.8	
		Smelting dust	1.8	
		Unexpained	13.5	
		Average measured mass (μg m^{-3})	no reported	
		Number	no reported	

Reference: **Chongqing TSP** study (Tao et al., 2006)
When: 11.5-h samples were acquired for two times once 6 days during July, October 2001, Janunary and April 2002.
Where: Seven sites include Beipei background site, Research academy of Environmental Science, No 2 Hospital, Shaping Meteorological Station, Nan'an Environmental Protection Office, Jiulong Environmental Protection Office and Yubei Environmental Protection Office.
Ambient: Samples were acquired with TH-150C medium-vol samplers and analyzed for mass, 36 elements (by XRF), ions (by IC) and carbon (OC, EC by MT-5 elemental analyzer).
Source: Dust samples from fugitive dust, coal combustion, construction dust, vehicular dust, and steel industry were taken and measured.

Solution: CMB

Average source contribution (in % mass):

Source Type	*Annual*
Coal combustion dust	18.0
Soil dust	30.0
Construction dust	25.0
Smelting dust	8.0
Vehicular dust	10.0
Unexpained	9.0
Average measured mass (μg m^{-3})	192
Number	336

Appendix D

Energy Conversion

ENERGY CONVERSION FACTORS

From one:	To:	EJ	Btce	Btoe	Tcm NG	Quad
Exajoule	*EJ*	1.000	0.033	0.022	0.025	0.948
Billion metric tons coal equivalent [2]	*Btce*	30.300	1.000	0.675	0.761	28.720
Billion metric tons oil equivalent [3]	*Btoe*	44.900	1.482	1.000	1.128	42.559
Trillion cubic meters natural gas [4]	*Tcm NG*	39.800	1.314	0.886	1.000	37.725
Quadrillion Btu	*Quad*	1.055	0.035	0.023	0.027	1.000

[1] These factors follow the U.S. convention of high-heat values.
[2] Chinese conversion factors for coal and other fuels are low-heat values. China typically converts all its energy statistics into "metric tons of standard coal equivalent" (tce); one tce equals 29.31 GJ (low heat), equivalent to 31.52 GJ/tce (high heat).
[3] China uses a conversion factor for its oil of 41.87 GJ/metric ton (low heat), equivalent to 44.07 GJ/t (high heat), assuming that low-heat values for oil are 95% of high-heat values.
[4] China uses a conversion factor for its natural gas of 38.98 GJ/thousand cubic meters (low heat), equivalent to 43.31 GJ/tcm (high heat), assuming that low-heat values for natural gas are 90% of high-heat values.

ABBREVIATIONS

Quad= quadrillion (10^{15}) British thermal units (Btu)
mtce = million ton of coal equivalent
mtoe = million ton of oil equivalent
bpd = barrels of oil per day

One barrel of oil = 0.136 tons of oil
One short ton (2000 lb.) = 0.907 metric tons
One cubic foot = 0.0283 cubic meters
One kilowatt hour (kWh) = 3.6×10^{6} J.
One million barrels of oil per day = 2.24 EJ per year

Adapted from NRC, Cooperation in the Energy Futures of China and the United States, *2000.*